Lecture Notes in Computer Science　　10424

Commenced Publication in 1973
Founding and Former Series Editors:
Gerhard Goos, Juris Hartmanis, and Jan van Leeuwen

Editorial Board

More information about this series at http://www.springer.com/series/7412

Michael Felsberg · Anders Heyden
Norbert Krüger (Eds.)

Computer Analysis of Images and Patterns

17th International Conference, CAIP 2017
Ystad, Sweden, August 22–24, 2017
Proceedings, Part I

 Springer

Editors
Michael Felsberg
Linköping University
Linköping
Sweden

Anders Heyden
Lund University
Lund
Sweden

Norbert Krüger
University of Southern Denmark
Odense
Denmark

ISSN 0302-9743   ISSN 1611-3349   (electronic)
Lecture Notes in Computer Science
ISBN 978-3-319-64688-6   ISBN 978-3-319-64689-3   (eBook)
DOI 10.1007/978-3-319-64689-3

Library of Congress Control Number: 2017947818

LNCS Sublibrary: SL6 – Image Processing, Computer Vision, Pattern Recognition, and Graphics

Printed on acid-free paper

This Springer imprint is published by Springer Nature
The registered company is Springer International Publishing AG
The registered company address is: Gewerbestrasse 11, 6330 Cham, Switzerland

Preface

We are very happy to present the contributions accepted for the 17th international Conference on Computer Analysis of Images and Patterns (CAIP 2017), which was held at Ystad Saltsjöbad, Ystad, Sweden, August 22–24.

CAIP 2017 was the 17th in the biennial series of conferences, which is devoted to all aspects of computer vision, image analysis and processing, pattern recognition, and related fields. Previous conferences were held, for instance, in Valletta, York, Seville, Münster, Vienna, Paris, etc. The contributions for CAIP 2017 were carefully selected based on a minimum of two but mostly three reviews. Among 144 submissions 72 were accepted, leading to an acceptance rate of 50%.

The conference included a tutorial on "Pose Estimation" by Anders G. Buch and a workshop on "Recognition and Action for Scene Understanding" (REACTS). Three keynote talks provided by world-renowned experts in the area of robotics (Markus Vincze), machine learning (Christian Igel), and image and video processing (Alan Bovik) were additional highlights.

The program covered high-quality scientific contributions in 2D-to-3D, 3D vision, biomedical image and pattern analysis, biometrics, brain-inspired methods, document analysis, face and gestures, feature extraction, graph-based methods, high-dimensional topology methods, human pose estimation, image/video indexing and retrieval, image restoration, keypoint detection, machine learning for image and pattern analysis, mobile multimedia, model-based vision, motion and tracking, object recognition, segmentation, shape representation and analysis, and vision for robotics.

CAIP has a reputation of providing a friendly and informal atmosphere, in addition to high-quality scientific contributions. We focused on maintaining this reputation, by designing a stimulating technical and social program that was hopefully inspiring for new research ideas and networking. We also hope that the venue at Ystad Saltsjöbad contributed to a fruitful conference.

We thank the authors for submitting their valuable work to CAIP. This is of course of prime importance for the success of the event. However, the organization of a conference also depends critically on a number of volunteers. We would like to thank the reviewers and the Program Committee members for their excellent work. We also thank the local Organizing Committee and all the other volunteers who helped us organize CAIP 2017.

We hope that all participants had a joyful and fruitful stay in Ystad.

August 2017

Michael Felsberg
Anders Heyden
Norbert Krüger

Organization

CAIP 2017 was organized by the Computer Vision Laboratory, Department of Electrical Engineering, Linköping University, Sweden.

Executive Committee

Conference Chair

Michael Felsberg   Linköping University, Sweden

Program Chairs

Anders Heyden   Lund University, Sweden
Norbert Krüger   University of Southern Denmark, Denmark

Program Committee

Muhammad Raza Ali, Pakistan
Furqan Aziz, Pakistan
Andrew Bagdanov, Spain
Donald Bailey, New Zealand
Antonio Bandera, Spain
Ardhendu Behera, UK
Michael Biehl, The Netherlands
Adrian Bors, UK
Kerstin Bunte, UK
Kenneth Camilleri, Malta
Kwok-Ping Chan, China
Rama Chellappa, USA
Dmitry Chetverikov, Hungary
Guillaume Damiand, France
Carl James Debono, Malta
Joachim Denzler, Germany
Mariella Dimiccoli, Spain
Francisco Escolano, Spain
Taner Eskil, Turkey
Giovanni Maria Farinella, Italy
Gernot Fink, Germany
Patrizio Frosini, Italy
Eduardo Garea, Cuba
Daniela Giorgi, Italy
Rocio Gonzalez-Diaz, Spain

Cosmin Grigorescu, The Netherlands
Miguel Gutierrez-Naranjo, Spain
Michal Haindl, Czech Republic
Anders Heyden, Sweden
Atsushi Imiya, Japan
Xiaoyi Jiang, Germany
Maria-Jose Jimenez, Spain
Dakai Jin, USA
Martin Kampel, Austria
Nahum Kiryati, Israel
Reinhard Klette, New Zealand
Gisela Klette, New Zealand
Andreas Koschan, USA
Ryszard Kozera, Australia
Walter Kropatsch, Austria
Guo-Shiang Lin, Taiwan
Agnieszka Lisowska, Poland
Josep Llados, Spain
Rebeca Marfil, Spain
Manuel Marin, Spain
Heydi Mendez, Cuba
Eckart Michaelsen, Germany
Majid Mirmehdi, UK
Matthew Montebello, Malta
Radu Nicolescu, New Zealand

Mark Nixon, UK
Darian Onchis, Austria
Mario Pattichis, Cyprus
Nicolai Petkov, The Netherlands
Gianni Poggi, Italy
Pedro Real, Spain
Paul Rosin, UK
Samuel Rota Bulo, Italy
Robert Sablatnig, Austria
Alessia Saggese, Italy
Hideo Saito, Japan
Albert Salah, Turkey
Lidia Sanchez, Spain
Angel Sanchez, Spain
Gabriella Sanniti di Baja, Italy

Sudeep Sarkar, USA
Klamer Schutte, The Netherlands
Giuseppe Serra, Italy
Francesc Serratosa, Spain
Antonio Jose Sanchez Salmeron, Spain
Akihiro Sugimoto, Japan
Bart ter Haar Romeny, The Netherlands
Bernie Tiddeman, UK
Klaus Toennies, Germany
Ernest Valveny, Spain
Thomas Villmann, Germany
Michael Wilkinson, The Netherlands
Richard Wilson, UK
Christian Wolf, France
Weiqi Yan, New Zealand

Invited Speakers

Markus Vincze   Technische Universität Wien, Austria
Christian Igel   University of Copenhagen, Denmark
Alan Bovik   The University of Texas at Austin, USA

Tutorial

Anders G. Buch   University of Southern Denmark, Denmark

Additional Reviewers

M. Al-Sarayreh
F. Castro
F. Diaz-del-Rio
J. Dong
A. Eldesokey
R. Grzeszick
Q. Gu

F. Hagelskjær
J. Hilty
C. Istin
T.B. Jørgensen
F. Kahn
S.F. Noor
D. Onchis

M.A. Pascali
P. Radeva
N. Saleem
S. Sudholt
J. Toro
M. Wallenberg
S. Zappala

Sponsoring Institutions

Computer Vision Laboratory, Department of Electrical Engineering, Linköping University
Swedish Society for Automated Image Analysis
The Swedish Research Council
Springer Lecture Notes in Computer Science
And all our industrial sponsors listed in the conference program booklet

Contents – Part I

Image/Video Indexing and Retrieval

Shape Representation and Analysis

Biomedical Image Analysis

Contents – Part II

Biometrics

Machine Learning

Image Restoration I

Poster Session III

Vision for Robotics

A Novel Approach for Mobile Robot Localization in Topological Maps Using Classification with Reject Option from Structural Co-occurrence Matrix

Suane Pires P. da Silva, Leandro B. Marinho, Jefferson S. Almeida, and Pedro Pedrosa Rebouças Filho$^{(\boxtimes)}$

Programa de Pós-Graduação em Ciência da Computação (PPGCC), Instituto Federal de Educação, Ciência e Tecnologia do Ceará (IFCE), Fortaleza, CE, Brazil
suanepires@lapisco.ifce.edu.br, pedrosarf@ifce.edu.br

Abstract. Location is an elemental problem for mobile robotics due the importance of determining a position of the robot in space. This knowledge along with the environment map are basic information for robot mobility. In this paper, a new approach for navigation and location of mobile robots on topological maps using classification with reject option in attributes obtained from a Structural Co-occurrence Matrix (SCM) is proposed. Furthermore, we compare our approach with others state-of-the-art extractors, such as Statistical Moments, Gray-Level Co-occurrence Matrix (GLCM) and Local Binary Patterns (LBP). Structural Co-Occurrence Matrix was evaluated with the Average, Gaussian, Laplacian and Sobel filters. Regarding to classifiers, Bayesian classifier, Multilayer Perceptron (MLP) and Support Vector Machines (SVM) were analyzed. The descriptors Scale Invariant Feature Transform (SIFT) and Speed Up Robust Features (SURF) were also used. According to results, SCM was the fastest feature extractor with 0.117 s and accuracy of 100% in navigation test, showing the relevance of our approach in the mobile robot localization.

Keywords: Robot localization · Topological maps · Classification with reject option · Structural co-occurrence matrix

1 Introduction

Mobile robotics is a growing and challenging area of research, with applications in different activities, and mobile robot location is one of major difficulties in this field. In [1], a image descriptor is proposed with approach in problem of real-time scene context classification on mobile devices as in robot navigation systems, for example. In [2], a scene categorisation engine in which the holistic representation of the scene is built exploiting features extracted on discrete cosine transform (DCT) domain is proposed and also represents a solution to the problem of location and navigation of mobile robots.

© Springer International Publishing AG 2017
M. Felsberg et al. (Eds.): CAIP 2017, Part I, LNCS 10424, pp. 3–15, 2017.
DOI: 10.1007/978-3-319-64689-3_1

There are two main types of methods to localization and mapping of mobile robots, geometric and topological. In geometric method, the entire navigation environment is depicted in a coordinate system, as in [3]. On the other hand, in topological method the total space is configured in a graph, not restricted to inflexible geometric information, as presented in [4]. There is also the hybrid approach, where both techniques described before are employed simultaneously, as in [5].

Localization of a mobile robot in the environment is paramount for its navigation. To achieve this, several technologies can be employed, according to the environment considered. In outdoor environments, the Global Positioning System (GPS) is a precise form of navigation and can be used for place recognition [6]. However, in indoors environments, this system is not suitable for the application [7] and, because of that difficulty, other alternatives were conceived. Some of the most commonly used processes for locating indoor mobile robots are ultrasonic, Radio Frequency Identification (RFID), Wireless Local Area Network (WLAN), inertial navigation and image recognition [8]. Image recognition has been increasingly exploited because they do not suffer sound interference, such as ultrasound, or coverage limit, similar to Bluetooth [9]. In addition, image-based systems do not require changes in the environment.

In this paper, a new approach for navigation and localization about the images analyzed of mobile robots on topological maps using classification with reject option in attributes obtained from a structural co-occurrence matrix (SCM) is proposed. SCM is a rotation-invariant feature extraction technique reasoned on a structural concept using co-occurrence statistics, presented as advantage to introduce a previous knowledge about the images analyzed, enhancing the details detection [10]. Furthermore, we perform a study among several feature extractors and classifiers consolidated in the literature, emphasizing the robustness and efficiency according to accuracy and processing time because these properties are fundamental in recognition systems aimed at applications in the real world. An high-resolution camera, GoPro®, was employed for robot navigation in an indoor environment. The results show that SCM obtained an average accuracy of 100% during navigation and extraction time of 0.117 s.

2   Review Feature Extraction Techniques

In this work, a region of interest (ROI) is composed of the entire image. Based on ROI, the attributes are extracted to be subsequently applied in the classification through machine learning techniques. Next, a brief presentation of the feature extraction techniques used is presented.

The Gray-Level Co-occurrence Matrix (GLCM) is based on a method created by Haralick [11], where his main focus is on texture analysis. The method consists of a second order statistical process, once the co-occurrences between pairs of pixels is analyzed, [11]. GLCM is a square matrix that stores references of the relative intensities of the pixels belonging to an image [11].

Local Binary Patterns (LBP) are intelligible and powerful texture descriptors. The elementary LBP operator, created by [12], binds a label to each of the measured pixels. This label corresponds to a binary number. The determination of the label mentioned is performed by comparing the value of each pixel next to the pixel under analysis, considering a fixed radius and the value of a threshold, which is stipulated based on the value of the central pixel, for which the label is defined [12].

Moments can be interpreted as scalar-type quantities, whose purpose is to describe an application, in addition to extracting its main characteristics [13]. Thus, they emphasize relevant parameters for the identification of the object of interest. Statistical Moments (SM) stands out as a very useful technique for the extraction of attributes in images, detailing the spatial ordering of the points belonging to the image or surface of interest [14].

Scale Invariant Feature Transform (SIFT) is a descriptor of characteristics developed by [15]. SIFT is invariant to scale and rotation and partially invariant to change in lighting. This method presents another great advantage, which consists in obtaining specific attributes, allowing its compaction with a large database of images.

Speed Up Robust Features (SURF) is a descriptor established by [16] to detail strategic points in images. This technique is considered an improvement of the SIFT because it consists of a more efficient implementation of this descriptor [16]. SURF is robust to noise and invariant to the rotation, these characteristics can be very relevant in scene recognition.

Structural Co-occurrence Matrix (SCM) is a rotation-invariant method based on co-occurrence statistics focused on the structural analysis of discrete signals by considering the connection between low-level structures of two discrete n-dimensional signals [10]. This method organizes, in a two-dimensional histogram, co-occurrences between structures of the input signals. In other words, SCM consists of a matrix that stores differences of the structures of two input signals [10]. In the image analysis, SCM easily perceives the structural differences, even in cases where the image histograms are similar [10]. SCM also provides prior knowledge about images considered, increasing its ability to detect details [10].

3 Review Machine Learning Techniques

The attributes obtained from the feature extractors are used to classification. In this section, the machine learning methods employed for this purpose are briefly described.

Bayesian Classifier is qualified as a statistical type and used for the classification of objects according to the probability that each one of them will fit a given class [17]. Consists of a supervised machine learning method and based on the Bayes Decision Theory [17]. Bayesian classifier labels the samples by calculating the a posteriori probability, based on conditional densities and a priori probabilities. Gaussian or normal density is one of the most used density functions because it provides a low computational cost, besides molding itself to several applications [17].

Multi-layer Perceptron (MLP) is a combination of perceptrons for solving nonlinearly separable problems. In the input layer, the vector formed by the information to be analyzed is shown to the network, giving impulses that will multiply later in the following layers [18]. The responses from each neuron of the hidden layers will represent the inputs of the consecutive layer and thus, through the links between the neurons, the impulses travel and are estimated by their corresponding weights [18]. Results on the neurons connected to the output are determined according to an activation function. By this process, therefore, a conclusive solution in the output layer is found, from the vector considered in the input layer [18].

Support Vector Machines (SVM) is a classifier based in the Theory of Statistical Learning idealized by [19]. The primary function of SVM is to define classes with surfaces that increase the distance between them. This technique delimits linearly the models in space and their inputs are transformed into a vector of large features. These models are called Support Vectors. In essence, the SVM was developed to solve binary issues, however, it also presents methods for solving multiclass issues [20]. However, because it becomes complex when applied to multiclass problems, approaches such as one-versus-one [21] and one-versus-all [21] are examples of SVM variations for this purpose.

4 Methodology

In this section, the methodology adopted to localize the robot in topological map using classification with reject option from Structural Co-occurrence Matrix is described. The navigation and location system begins with the capture of the image. After that, feature extraction is performed and the feature vector is created. In this paper, SCM, Statistics Moments, GLCM and LBP are evaluated. The feature vector generated in previous step is used to perform the recognition of the environment. In the classification step, MLP, SVM and Bayesian classifier are evaluated. Both SIFT and SURF were analyzed separately. Figure 1 shows an overview of the proposed approach.

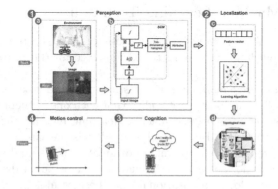

Fig. 1. Flowchart of the proposed approach to mobile robot navigation.

We used a robot with an high resolution camera, GoPro®, to locomotion in indoor environment, in this case an apartment, see Fig. 2(c). This environment was chosen because the features which contribute to the localization and navigation of the mobile robot using an approach based on computer vision.

Regarding to the topological map, the nodes were numbered 1 to 6. Pictures were taken of strategic points relatives to classes numbered 1 to 15. The paths of the robot are represented by the edges in the map. Figure 2(a) presents the topological map of the environment and Fig. 2(b) shows your perspective view.

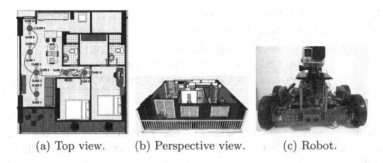

(a) Top view.   (b) Perspective view.   (c) Robot.

Fig. 2. Topological map of the environment: (a) top; (b) perspective view; and (c) the autonomous mobile robot.

We evaluated the Structural Co-occurrence Matrix with two low-pass filter, Average and Gaussian, and two high-pass filter, Laplacian and Sobel. Statistical Moments was assessed with order 0, 1, 2 and 3. The distance $D = 1$ and direction $\theta = 0$ was used for the calculation of GLCM. In LBP, uniform pattern is utilized. The amount of attributes generated by each feature extractor was 10, 14, 59 and 8, respectively, by SM, GLCM, LBP and SCM.

In training phase of the classifiers, hyper-parameters for SVM and MLP are selected using 10-fold cross-validation. SVM is learned using the range $[2^{-2}, 2^{11}]$ for the hyper-parameters and grid search, with a linear and radial basis kernel (RBF). MLP is optimized using Levenberg-Marquardt and validated on a range of hidden units from 1 to 50. Normal or Gaussian probability density function is used in the Bayesian classifier.

We adopted ten distinct routes in the apartment to the navigation evaluation. Table 1 shows the commands used to move the robot, as well, start and end locations. Figure 3(a) exemplifies the route 9 from Table 1. In this route, the robot begins at class 2 (node 3) and receives two commands to go straight ahead, after other command to turn right. Next, the robot turn right and receives more two commands to go straight ahead. Finally, the robot turn left and reaches class 3 (node 3).

The reliability of the classifier in some decision systems can be increased if an element that is not considered sufficiently reliable is automatically rejected as in [22]. In [23], distance information of the topological map are used to improve

Table 1. Commands to move the robot and the start and end points used in the navigation. Go straight ahead (GSA), turn left (TL), turn right (TR) and turn 180 degrees (T180).

Route	Start (Class)	Commands	End (Class)
1	8	GSA, GSA, GSA	4
2	1	GSA, GSA, TR	9
3	5	GSA and TR, GSA, GSA	12
4	5	GSA, GSA, GSA and TR, TL	10
5	8	GSA, GSA and TL, GSA, GSA	12
6	13	GSA, GSA and TL, GSA, T180, GSA	1
7	8	GSA, GSA, TL, GSA, GSA	12
8	9	TR, GSA, GSA and TL, GSA, T180, GSA and TL	3
9	2	GSA, GSA and TR, TR, GSA, GSA and TL	3
10	12	T180, GSA, GSA and TL, GSA, T180, GSA, GSA, GSA	10

automatically classification step, which is called localization with reject option. In this approach, the result from each classifier is a sorted sequence, wherein the first class is the most likely to be the sample and so on. In this paper, we adopted the localization with reject option along with a Structural Co-occurrence Matrix to extract the features of the environment during the mobile robot navigation. Figure 3(b) depicts a alleged situation to demonstrate the functioning of classification with reject option in the proposed approach. Robot is located in the place referring to class 13 (node 6) and commands are sent so that it goes to class 15 (node 5). After arriving in class 15 (node 5), the response given by the classifier indicates a high probability of the place being class 9 (node 1) followed by the probability of being of class 15 (node 5), which can lead to a wrong localization. Therefore, the first suggestion of classification would be ignored by the reject

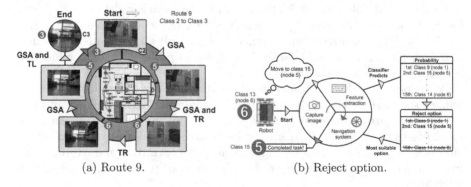

(a) Route 9.   (b) Reject option.

Fig. 3. (a) Demonstration of route 9. (b) Alleged situation with operation of reject option.

option and the next suggestion would be chosen because it is impracticable to reach desired class directly from initial position considered.

Another important contribution of this paper is production of an image database which can still be used in future works. The database is composed of 600 images, 40 images per class, with resolution of 4000 × 3000 pixels. The database is available at http://lapisco.ifce.edu.br/?page_id=252. In order to ascertain the performance of each classifier and confront them by means of a consolidated process, we used two evaluation metrics, Accuracy (Acc) and F1-Score (F1S).

5   Results

In this section, we present the results achieved by our approach for robot localization in topological maps using classification with reject option from Structural Co-occurrence Matrix and others feature extraction techniques. Firstly, the influence of feature extractors along with the classifiers in the images was discussed. Subsequently, the navigation tests were performed. The results of the navigation tests were obtained from 10 executions of each of the routes.

Table 2 shows the average and standard deviation values for F1-Score and Accuracy obtained by the machine learning and feature extraction techniques using the images of the environment. The experiments were computed on an iMac 2.5 GHz processor Core i5 with 4 GB RAM.

According to Table 2, the combination that obtained the highest Acc and F1S was the Statistical Moments with Bayesian classifier, with 99.94% in both. In general, the results have reached values greater than 95%. Regarding to feature extraction techniques, LBP was the best method with Acc and F1S above 99%, including the combination with SVM using linear kernel.

Another important analysis is about the computational cost, once the problem is a real application. Table 3 presents accuracy, training time, testing time and attribute extraction time, which are important indicators for embedded applications. Bayesian classifier obtained the shortest training times, with 0.003 s when combined with SCM-Sobel and Statistical Moments and 0.004 s when combined with SCM-Average. With regard to testing time, Bayesian classifier also obtained the shortest times, with 47.3 µs and 49.5 µs when associated with Statistical Moments and SCM-Gaussian, respectively. SIFT and SURF descriptors were the slowest, taking 38111630.3 µs and 3432554.7 µ respectively to finish test step.

In the last column of Table 3, we can note that SCM was the fastest among applied feature extractors, obtaining extraction time of 0.117 s and 0.118 s in their versions with Gaussian and Sobel filter and with Average and Lapacian filter, respectively. The values of the extraction times of the SCM are highlighted in bold in Table 3.

Finally, a navigation test using methods was performed. Figure 4 shows the results of the navigation test, considering average accuracy with and without reject option obtained by the feature extractors and classifiers. It is important to mention that a route is considered correct if the classification system finds

Table 2. F1-Score (F1S) and Accuracy (Acc) obtained by features extraction and classifiers.

Feature	Classifier	Setup	F1S (%)	Acc (%)
SCM-Avg	Bayes	Normal	99.00 ± 0.65	99.00 ± 0.66
	MLP		95.42 ± 3.72	95.08 ± 4.62
	SVM	Linear	84.00 ± 4.37	84.58 ± 4.03
		RBF	98.43 ± 0.99	98.42 ± 1.00
SCM-Gau	Bayes	Normal	99.02 ± 1.13	99.00 ± 1.17
	MLP		93.74 ± 4.40	92.33 ± 6.13
	SVM	Linear	88.56 ± 1.98	88.58 ± 2.08
		RBF	97.20 ± 2.34	97.17 ± 2.40
SCM-Lap	Bayes	Normal	99.25 ± 0.73	99.25 ± 0.73
	MLP		96.20 ± 2.34	95.58 ± 3.38
	SVM	Linear	92.88 ± 2.01	92.75 ± 2.19
		RBF	96.99 ± 0.89	97.00 ± 0.90
SCM-Sob	Bayes	Normal	99.26 ± 0.98	99.25 ± 1.00
	MLP		96.66 ± 5.56	95.42 ± 8.56
	SVM	Linear	95.74 ± 1.49	95.75 ± 1.49
		RBF	98.43 ± 1.13	98.42 ± 1.14
SM	**Bayes**	**Normal**	**99.94 ± 0.21**	**99.94 ± 0.21**
	MLP		97.71 ± 3.10	98.00 ± 2.32
	SVM	Linear	96.03 ± 2.08	96.12 ± 1.99
		RBF	98.71 ± 1.14	98.71 ± 1.16
GLCM	Bayes	Normal	96.82 ± 0.99	96.83 ± 1.00
	MLP		98.95 ± 1.04	98.96 ± 1.01
	SVM	Linear	98.05 ± 1.06	98.08 ± 1.02
		RBF	98.30 ± 0.94	98.29 ± 0.95
LBP	Bayes	Normal	99.58 ± 0.51	99.58 ± 0.51
	MLP		98.88 ± 3.04	99.17 ± 2.16
	SVM	Linear	99.42 ± 0.72	99.42 ± 0.72
		RBF	99.62 ± 0.51	99.63 ± 0.50
SIFT			99.36 ± 0.97	99.37 ± 0.94
SURF			96.86 ± 2.43	97.03 ± 2.27

all classes along the route. When we analyze navigation results in Fig. 4, we can observed that SCM-Average along with Bayesian classifier and SVM (RBF) obtained 100% accuracy and reject option confirmed the path as well as SCM-Laplacian and Statistical Moment combined with Bayesian classifier.

We can still observe that localization with reject option also increased accuracy of all combinations of methods. For example, average accuracy of the

Table 3. Accuracy (Acc), training time, testing time and feature extraction time for all features extraction in real environment obtained by the best classifiers.

Classifier	Acc (%)	Training time (s)	Testing time (μs)	Extraction time (s)
SCM - Average				
Bayes (Normal)	99.0 ± 0.7	0.004 ± 0.0	67.9 ± 19.7	**0.118 ± 0.01**
MLP	95.1 ± 4.6	16.089 ± 3.2	121.8 ± 4.9	
SVM (Linear)	84.6 ± 4.0	187.713 ± 94.5	51.0 ± 0.3	
SVM (RBF)	98.4 ± 1.0	1.564 ± 0.2	56.7 ± 0.6	
SCM - Gaussian				
Bayes (Normal)	99.0 ± 1.2	0.829 ± 2.6	49.5 ± 7.1	**0.117 ± 0.01**
MLP	92.3 ± 6.1	83.889 ± 30.3	125.0 ± 15.6	
SVM (Linear)	88.6 ± 2.1	841.101 ± 106.7	49.2 ± 0.5	
SVM (RBF)	97.2 ± 2.4	1.380 ± 0.1	55.3 ± 1.8	
SCM - Laplacian				
Bayes (Normal)	99.2 ± 0.7	0.059 ± 0.2	81.147 ± 14.45	**0.118 ± 0.01**
MLP	95.6 ± 3.4	21.213 ± 7.6	133.5 ± 8.1	
SVM (Linear)	92.8 ± 2.2	34.291 ± 5.0	50.6 ± 1.1	
SVM (RBF)	97.0 ± 0.9	0.482 ± 0.0	60.0 ± 2.5	
SCM - Sobel				
Bayes (Normal)	99.2 ± 1.0	0.003 ± 0.0	51.2 ± 4.5	**0.117 ± 0.01**
MLP	95.4 ± 8.6	46.905 ± 24.3	119.9 ± 2.3	
SVM (Linear)	95.7 ± 1.5	4.703 ± 1.6	50.5 ± 1.7	
SVM (RBF)	98.4 ± 1.1	0.446 ± 0.0	93.1 ± 4.3	
Statistical Moments				
Bayes (Normal)	99.9 ± 0.2	0.003 ± 0.0	47.3 ± 3.2	0.165 ± 0.01
MLP	98.0 ± 2.3	3.334 ± 0.6	305.4 ± 4.6	
SVM (Linear)	96.1 ± 2.0	162.444 ± 21.6	3917.9 ± 72.2	
SVM (RBF)	98.7 ± 1.2	1.039 ± 0.1	5196.6 ± 51.2	
GLCM				
Bayes (Normal)	96.8 ± 1.0	0.005 ± 0.0	65.6 ± 4.5	0.494 ± 0.03
MLP	99.0 ± 1.0	11.713 ± 3.9	351.3 ± 9.8	
SVM (Linear)	98.1 ± 1.0	6.509 ± 1.3	3973.7 ± 40.3	
SVM (RBF)	98.3 ± 1.0	0.486 ± 0.0	5393.9 ± 66.7	
LBP				
Bayes (Normal)	99.6 ± 0.5	0.007 ± 0.0	108.8 ± 70.5	0.311 ± 0.02
MLP	99.2 ± 2.2	5.626 ± 0.8	403.2 ± 186.5	
SVM (Linear)	99.4 ± 0.7	0.731 ± 0.1	4251.7 ± 49.3	
SVM (RBF)	99.6 ± 0.5	0.418 ± 0.0	5216.4 ± 53.7	
SIFT	99.4 ± 0.9	181.310 ± 1.5	38111630.3 ± 902798.9	15.654 ± 0.33
SURF	97.0 ± 2.3	40.732 ± 0.3	3432554.7 ± 219154.8	3.091 ± 0.01

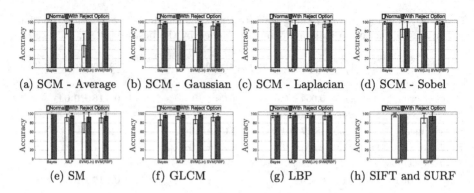

(a) SCM - Average (b) SCM - Gaussian (c) SCM - Laplacian (d) SCM - Sobel

(e) SM   (f) GLCM   (g) LBP   (h) SIFT and SURF

Fig. 4. Average accuracy of route testing obtained by the feature extraction techniques and classifiers with and without reject option.

SCM-Average and SCM-Sobel both with SVM (Linear) was less than 50% and 80%, respectively, and with reject option reached 100%. In other words, our approach of classification with reject option from Structural Co-occurrence Matrix proves to be effective and reliable for mobile robotic navigation. According to the detailed results by route without and with the reject option present in

Table 4. Accuracy by route obtained from the feature extraction techniques and classifiers **without reject option**.

Feature	Classifier	Setup	Route 1	Route 2	Route 3	Route 4	Route 5	Route 6	Route 7	Route 8	Route 9	Route 10
SCM-Avg	Bayes	Normal	100.0±0.0	100.0±0.0	100.0±0.0	100.0±0.0	100.0±0.0	100.0±0.0	100.0±0.0	100.0±0.0	100.0±0.0	100.0±0.0
	MLP		100.0±0.0	70.0±48.3	90.0±31.6	80.0±42.2	80.0±42.2	90.0±31.6	100.0±0.0	80.0±42.2	100.0±0.0	70.0±48.3
	SVM	Linear	50.0±52.7	80.0±42.2	90.0±31.6	70.0±48.3	30.0±48.3	60.0±51.6	40.0±51.6	10.0±31.6	30.0±48.3	30.0±48.3
		RBF	100.0±0.0	100.0±0.0	100.0±0.0	100.0±0.0	100.0±0.0	100.0±0.0	100.0±0.0	100.0±0.0	100.0±0.0	100.0±0.0
SCM-Gau	Bayes	Normal	100.0±0.0	100.0±0.0	100.0±0.0	100.0±0.0	90.0±31.6	100.0±0.0	100.0±0.0	80.0±42.2	80.0±42.2	100.0±0.0
	MLP		100.0±0.0	100.0±0.0	0.0±0.0	0.0±0.0	100.0±0.0	0.0±0.0	100.0±0.0	90.0±31.6	90.0±31.6	0.0±0.0
	SVM	Linear	100.0±0.0	90.0±31.6	50.0±52.7	70.0±48.3	20.0±42.2	100.0±0.0	60.0±51.6	30.0±48.3	50.0±52.7	50.0±52.7
		RBF	100.0±0.0	100.0±0.0	90.0±31.6	90.0±31.6	80.0±42.2	100.0±0.0	100.0±0.0	80.0±42.2	80.0±42.2	100.0±0.0
SCM-Lap	Bayes	Normal	100.0±0.0	100.0±0.0	100.0±0.0	100.0±0.0	100.0±0.0	100.0±0.0	100.0±0.0	100.0±0.0	100.0±0.0	100.0±0.0
	MLP		100.0±0.0	100.0±0.0	100.0±0.0	90.0±31.6	50.0±52.7	100.0±0.0	80.0±42.2	80.0±42.2	90.0±31.6	80.0±42.2
	SVM	Linear	60.0±51.6	90.0±31.6	80.0±42.2	90.0±31.6	30.0±48.3	50.0±52.7	60.0±51.6	80.0±42.2	80.0±42.2	20.0±42.2
		RBF	100.0±0.0	100.0±0.0	100.0±0.0	100.0±0.0	80.0±42.2	100.0±0.0	100.0±0.0	90.0±31.6	90.0±31.6	100.0±0.0
SCM-Sob	Bayes	Normal	100.0±0.0	100.0±0.0	100.0±0.0	100.0±0.0	100.0±0.0	100.0±0.0	90.0±31.6	100.0±0.0	100.0±0.0	100.0±0.0
	MLP		100.0±0.0	80.0±42.2	60.0±51.6	100.0±0.0	90.0±31.6	100.0±0.0	50.0±52.7	100.0±0.0	90.0±31.6	80.0±42.2
	SVM	Linear	90.0±31.6	50.0±52.7	60.0±51.6	90.0±31.6	70.0±48.3	90.0±31.6	50.0±52.7	100.0±0.0	80.0±42.2	60.0±51.6
		RBF	100.0±0.0	90.0±31.6	100.0±0.0	100.0±0.0	100.0±0.0	100.0±0.0	100.0±0.0	100.0±0.0	100.0±0.0	100.0±0.0
SM	Bayes	Normal	100.0±0.0	100.0±0.0	100.0±0.0	100.0±0.0	100.0±0.0	100.0±0.0	100.0±0.0	100.0±0.0	100.0±0.0	100.0±0.0
	MLP		100.0±0.0	100.0±0.0	90.0±31.6	100.0±0.0	80.0±42.2	90.0±31.6	100.0±0.0	90.0±31.6	80.0±42.2	90.0±31.6
	SVM	Linear	100.0±0.0	100.0±0.0	90.0±31.6	100.0±0.0	90.0±31.6	60.0±51.6	100.0±0.0	60.0±51.6	70.0±48.3	40.0±51.6
		RBF	100.0±0.0	100.0±0.0	100.0±0.0	100.0±0.0	70.0±48.3	90.0±31.6	90.0±31.6	80.0±42.2	100.0±0.0	80.0±42.2
GLCM	Bayes	Normal	100.0±0.0	70.0±48.3	80.0±42.2	80.0±42.2	80.0±42.2	100.0±0.0	100.0±0.0	100.0±0.0	90.0±31.6	70.0±48.3
	MLP		100.0±0.0	90.0±31.6	100.0±0.0	100.0±0.0	80.0±42.2	100.0±0.0	90.0±31.6	100.0±0.0	100.0±0.0	90.0±31.6
	SVM	Linear	100.0±0.0	90.0±31.6	80.0±42.2	100.0±0.0	90.0±31.6	100.0±0.0	80.0±42.2	80.0±42.2	80.0±42.2	80.0±42.2
		RBF	100.0±0.0	80.0±42.2	80.0±42.2	100.0±0.0	90.0±31.6	100.0±0.0	100.0±0.0	100.0±0.0	90.0±31.6	90.0±31.6
LBP	Bayes	Normal	100.0±0.0	100.0±0.0	100.0±0.0	90.0±31.6	100.0±0.0	100.0±0.0	90.0±31.6	90.0±31.6	100.0±0.0	100.0±0.0
	MLP		100.0±0.0	100.0±0.0	90.0±31.6	100.0±0.0	100.0±0.0	100.0±0.0	90.0±31.6	90.0±31.6	100.0±0.0	100.0±0.0
	SVM	Linear	100.0±0.0	100.0±0.0	100.0±0.0	100.0±0.0	90.0±31.6	100.0±0.0	90.0±31.6	100.0±0.0	100.0±0.0	100.0±0.0
		RBF	100.0±0.0	100.0±0.0	100.0±0.0	100.0±0.0	100.0±0.0	100.0±0.0	80.0±42.2	80.0±42.2	100.0±0.0	100.0±0.0
	SIFT		100.0±0.0	100.0±0.0	100.0±0.0	100.0±0.0	90.0±31.6	100.0±0.0	100.0±0.0	100.0±0.0	90.0±31.6	100.0±0.0
	SURF		100.0±0.0	100.0±0.0	100.0±0.0	100.0±0.0	70.0±48.3	90.0±31.6	90.0±31.6	80.0±42.2	100.0±0.0	80.0±42.2

Table 5. Accuracy by route obtained from the feature extraction techniques and classifiers **with reject option**.

Feature	Classifier	Setup	Route 1	Route 2	Route 3	Route 4	Route 5	Route 6	Route 7	Route 8	Route 9	Route 10
SCM-Avg	Bayes	Normal	100.0±0.0	100.0±0.0	100.0±0.0	100.0±0.0	100.0±0.0	100.0±0.0	100.0±0.0	100.0±0.0	100.0±0.0	100.0±0.0
	MLP		100.0±0.0	80.0±42.2	100.0±0.0	90.0±31.6	100.0±0.0	90.0±31.6	100.0±0.0	100.0±0.0	100.0±0.0	100.0±0.0
	SVM	Linear	100.0±0.0	100.0±0.0	100.0±0.0	100.0±0.0	100.0±0.0	100.0±0.0	100.0±0.0	100.0±0.0	100.0±0.0	100.0±0.0
		RBF	100.0±0.0	100.0±0.0	100.0±0.0	100.0±0.0	100.0±0.0	100.0±0.0	100.0±0.0	100.0±0.0	100.0±0.0	100.0±0.0
SCM-Gau	Bayes	Normal	100.0±0.0	100.0±0.0	100.0±0.0	100.0±0.0	100.0±0.0	100.0±0.0	100.0±0.0	100.0±0.0	90.0±31.6	100.0±0.0
	MLP		100.0±0.0	100.0±0.0	0.0±0.0	0.0±0.0	100.0±0.0	0.0±0.0	100.0±0.0	90.0±31.6	90.0±31.6	0.0±0.0
	SVM	Linear	100.0±0.0	100.0±0.0	100.0±0.0	90.0±31.6	100.0±0.0	100.0±0.0	100.0±0.0	100.0±0.0	90.0±31.6	100.0±0.0
		RBF	100.0±0.0	100.0±0.0	90.0±31.6	90.0±31.6	100.0±0.0	100.0±0.0	100.0±0.0	100.0±0.0	90.0±31.6	100.0±0.0
SCM-Lap	Bayes	Normal	100.0±0.0	100.0±0.0	100.0±0.0	100.0±0.0	100.0±0.0	100.0±0.0	100.0±0.0	100.0±0.0	100.0±0.0	100.0±0.0
	MLP		100.0±0.0	100.0±0.0	100.0±0.0	100.0±0.0	70.0±48.3	100.0±0.0	90.0±31.6	100.0±0.0	100.0±0.0	80.0±42.2
	SVM	Linear	90.0±31.6	90.0±31.6	100.0±0.0	100.0±0.0	100.0±0.0	90.0±31.6	100.0±0.0	100.0±0.0	100.0±0.0	90.0±31.6
		RBF	100.0±0.0	100.0±0.0	100.0±0.0	100.0±0.0	80.0±42.2	100.0±0.0	100.0±0.0	90.0±31.6	100.0±0.0	100.0±0.0
SCM-Sob	Bayes	Normal	100.0±0.0	100.0±0.0	100.0±0.0	100.0±0.0	100.0±0.0	100.0±0.0	100.0±0.0	100.0±0.0	100.0±0.0	100.0±0.0
	MLP		100.0±0.0	80.0±42.2	60.0±51.6	100.0±0.0	90.0±31.6	100.0±0.0	60.0±51.6	100.0±0.0	90.0±31.6	80.0±42.2
	SVM	Linear	100.0±0.0	100.0±0.0	100.0±0.0	100.0±0.0	100.0±0.0	100.0±0.0	100.0±0.0	100.0±0.0	100.0±0.0	100.0±0.0
		RBF	100.0±0.0	90.0±31.6	100.0±0.0	100.0±0.0	100.0±0.0	100.0±0.0	100.0±0.0	100.0±0.0	100.0±0.0	100.0±0.0
SM	Bayes	Normal	100.0±0.0	100.0±0.0	100.0±0.0	100.0±0.0	100.0±0.0	100.0±0.0	100.0±0.0	100.0±0.0	100.0±0.0	100.0±0.0
	MLP		100.0±0.0	100.0±0.0	90.0±31.6	100.0±0.0	100.0±0.0	90.0±31.6	100.0±0.0	100.0±0.0	90.0±31.6	90.0±31.6
	SVM	Linear	100.0±0.0	100.0±0.0	90.0±31.6	100.0±0.0	90.0±31.6	100.0±0.0	100.0±0.0	100.0±0.0	80.0±42.1	70.0±48.3
		RBF	100.0±0.0	100.0±0.0	100.0±0.0	100.0±0.0	70.0±48.3	100.0±0.0	90.0±31.6	100.0±0.0	100.0±0.0	90.0±31.6
GLCM	Bayes	Normal	100.0±0.0	100.0±0.0	90.0±31.6	100.0±0.0	90.0±31.6	100.0±0.0	100.0±0.0	100.0±0.0	100.0±0.0	90.0±31.6
	MLP		100.0±0.0	100.0±0.0	100.0±0.0	100.0±0.0	90.0±31.6	100.0±0.0	90.0±31.6	100.0±0.0	100.0±0.0	90.0±31.6
	SVM	Linear	100.0±0.0	90.0±31.6	100.0±0.0	100.0±0.0	100.0±0.0	100.0±0.0	100.0±0.0	100.0±0.0	100.0±0.0	100.0±0.0
		RBF	100.0±0.0	80.0±42.1	90.0±31.6	100.0±0.0	90.0±31.6	100.0±0.0	100.0±0.0	100.0±0.0	90.0±31.6	90.0±31.6
LBP	Bayes	Normal	100.0±0.0	100.0±0.0	100.0±0.0	100.0±0.0	100.0±0.0	100.0±0.0	90.0±31.6	90.0±31.6	100.0±0.0	100.0±0.0
	MLP		100.0±0.0	100.0±0.0	90.0±31.6	100.0±0.0	100.0±0.0	100.0±0.0	100.0±0.0	90.0±31.6	100.0±0.0	100.0±0.0
	SVM	Linear	100.0±0.0	100.0±0.0	100.0±0.0	100.0±0.0	100.0±0.0	100.0±0.0	90.0±31.6	90.0±31.6	100.0±0.0	100.0±0.0
		RBF	100.0±0.0	100.0±0.0	100.0±0.0	100.0±0.0	100.0±0.0	100.0±0.0	90.0±31.6	90.0±31.6	100.0±0.0	100.0±0.0
SIFT			100.0±0.0	100.0±0.0	100.0±0.0	100.0±0.0	100.0±0.0	100.0±0.0	100.0±0.0	100.0±0.0	100.0±0.0	100.0±0.0
SURF			100.0±0.0	100.0±0.0	100.0±0.0	100.0±0.0	70.0±48.3	100.0±0.0	90.0±31.6	100.0±0.0	100.0±0.0	90.0±31.6

Tables 4 and 5, respectively, we can verify that average accuracy was from 87.8% to 97.5% with the localization with reject option.

6   Conclusion

In this work, we proposed an novel approach for localization and navigation of mobile robots in topological maps using classification with reject option from Structural Co-occurrence Matrix (SCM). SCM is a method of feature extraction that consists on a structural perspective using co-occurrence statistics [10]. This feature extractor is rotation-invariant and has the advantage of inserting a prior knowledge about the images analyzed, optimizing the details detection [10]. Furthermore, we performed an evaluation among others feature extractors and machine learning techniques consolidated in the interest task.

Adopting the same requirements used in [23] and considering the results of the our approach, we can conclude that SCM is a suitable feature extractor to robot localization in topological maps using classification with reject option. SCM obtained the shortest extraction time, finalizing its task in 0.117 s, and accuracy of 100% in the navigation tests when combined with Bayesian classifier, SVM (linear) and SVM (RBF). Furthermore, reject option confirmed or increased accuracy of all SCM combinations, demonstrating to be an effective and reliable solution for navigation of mobile robots. For example, SCM-Average and

SCM-Sobel both with SVM (Linear) were less than 50% and 80%, respectively, to 100% of accuracy when considered reject option.

Another important contribution of this paper is the creation of a image database, which can be used in future works to evaluate others machine learning and feature extraction techniques for localization and navigation task. About future work, other machine learning technique that can be applied is the Optimum Path Forest (OPF) presented in [24], in addition to incorporating other types of filters to SCM.

References

1. Farinella, G.M., Ravì, D., Tomaselli, V., Guarnera, M., Battiato, S.: Representing scenes for real-time context classification on mobile devices. Pattern Recogn. **48**(4), 1086–1100 (2015)
2. Farinella, G.M., Battiato, S.: Scene classification in compressed and constrained domain. IET Comput. Vis. **5**(5), 320–334 (2011)
3. Jiang, R., Yang, S., Ge, S.S.: Geometric map-assisted localization for mobile robots based on uniform-Gaussian distribution. IEEE Robot. Autom. Lett. **2**(2), 789–795 (2017)
4. Jiang, J.-H., Le, H.-L., Shie, S.-C.: Lightweight topological-based map matching for indoor navigation. In: 30th International Conference on Advanced Information Networking and Applications Workshops AINA, pp. 908–913, March 2016
5. Ferreira, J.F., Amorim, I., Rocha, R.P., Dias, J.: T-SLAM: registering topological and geo-metric maps for robot localization in large environments. In: IEEE International Conference on Multisensor Fusion and Integration for Intelligent Systems MFI, pp. 392–398, August 2008
6. Lee, W., Chung, W.: Position estimation using multiple low-cost GPS receivers for outdoor mobile robots. In: 12th International Conference on Ubiquitous Robots and Ambient Intelligence, pp. 460–461. IEEE, October 2015
7. Diop, M., Ong, L.Y., Lim, T.S.: A computer vision-aided motion sensing algorithm for mobile robot's indoor navigation. In: 14th International Workshop on Advanced Motion Control, pp. 400–405. IEEE, April 2016
8. Song, Z., Jiang, G., Huang, C.: A survey on indoor positioning technologies. In: Zhou, Q. (ed.) ICTMF 2011. CCIS, vol. 164, pp. 198–206. Springer, Heidelberg (2011). doi:10.1007/978-3-642-24999-0_28
9. Bonin-Font, F., Ortiz, A., Oliver, G.: Visual navigation for mobile robots: a survey. J. Intell. Rob. Syst. **53**(3), 263–296 (2008)
10. Bezerra, G.L., Ferreira, D.S., Rebouças, P.P., Sombra, F.N.: Rotation-invariant feature extraction using a structural co-occurrence matrix. Measurement **94**(2), 406–415 (2016)
11. Haralick, R.M., Shanmugam, K.S., Dinstein, I.: Textural features for image classification. IEEE Trans. Syst. Man Cybern. **3**(6), 610–621 (1973)
12. Ojala, T., Pietikäinen, M., Mäenpää, T.: Multiresolution gray-scale and rotation invariant texture classification with local binary patterns. IEEE Trans. Pattern Anal. Mach. Intell. **24**(7), 971–987 (2002)
13. Flusser, J., Zitova, B., Suk, T.: Moments and Moment Invariants in Pattern Recognition. Wiley Publishing, Chichester (2009)
14. Gonzalez, R.C., Woods, R.E.: Digital Image Processing, 3rd edn. Pearson Prentice Hall, New Jersey (2010)

15. Lowe, D.G.: Distinctive image features from scale-invariant keypoints. Int. J. Comput. Vision **60**(2), 91–110 (2004)
16. Bay, H., Ess, A., Tuytelaars, T., Van Gool, L.: Speeded-up robust features (surf). Comput. Vis. Image Underst. **110**(3), 346–359 (2008)
17. Theodoridis, S., Koutroumbas, K.: Pattern Recognition, 4th edn. Academic Press, USA (2008)
18. Haykin, S.: Neural Networks and Learning Machines. Prentice Hall, McMaster University, Canada (2008)
19. Vapnik, V.N.: Statistical Learning Theory. Wiley, Nova Jersey (1998)
20. Crammer, K., Singer, Y.: On the algorithmic implementation of multiclass kernel-based vector machines. J. Mach. Learn. Res. **2**, 265–292 (2001)
21. Duan, K.-B., Keerthi, S.S.: Which is the best multiclass SVM method? An empirical study. In: Oza, N.C., Polikar, R., Kittler, J., Roli, F. (eds.) MCS 2005. LNCS, vol. 3541, pp. 278–285. Springer, Heidelberg (2005). doi:10.1007/11494683_28
22. Furnari, A., Farinella, G.M., Battiato, S.: Recognizing personal locations from egocentric videos. IEEE Trans. Hum.-Mach. Syst. **47**(1), 6–18 (2017)
23. Marinho, L.B., Almeida, J.S., Souza, J.W.M., Albuquerque, V.H.C., Rebouças Filho, P.P.: A novel mobile robot localization approach based on topological maps using classification with reject option in omnidirectional images. Expert Syst. Appl. **72**, 1–17 (2016)
24. Nunes, T.M., Coelho, A.L., Lima, C.A., Papa, J.P., de Albuquerque, V.H.C.: EEG signal classification for epilepsy diagnosis via optimum path forest - a systematic assessment. Neurocomputing **136**, 103–123 (2014)

Robust Features for Snapshot Hyperspectral Terrain-Classification

Christian Winkens[✉], Volkmar Kobelt, and Dietrich Paulus

Active Vision Group, Institute for Computational Visualistics,
University of Koblenz-Landau, Koblenz, Germany
{cwinkens,vkobelt,paulus}@uni-koblenz.de

Abstract. Hyperspectral imaging increases the amount of information incorporated per pixel in comparison to normal RGB color cameras. Conventional spectral cameras as used in satellite imaging use spatial or spectral scanning during acquisition which is only suitable for static scenes. In dynamic scenarios, such as in autonomous driving applications, the acquisition of the entire hyperspectral cube at the same time is mandatory. We investigate the eligibility of novel snapshot hyperspectral cameras which capture an entire hyperspectral cube without requiring moving parts or line-scanning. Captured hyperspectral data is used for multi class terrain classification utilizing machine learning techniques. Prior to classification, the data is segmented using Superpixel segmentation which is modified to work successfully on hyperspectral data. We further investigate a simple approach to normalize the hyperspectral data in terms of illumination, which yields vast improvements in classification accuracy, preventing most errors caused by shading and other influences. Furthermore we utilize Gabor texture features which add spatial information to the feature space without increasing the data dimensionality in an excessive fashion. The multi-class classification is evaluated against a novel hyperspectral ground truth dataset specifically created for this purpose.

Keywords: Hyperspectral imaging · Terrain classification · Spectral analysis · Autonomous vehicles

1 Introduction and Motivation

Spectral imaging is defined by acquiring light intensity for pixels in an image. Each pixel stores a vector of intensity values, which corresponds to the incoming light over a defined wavelength range. Typically, researchers utilize sensors like these on Landsat, SPOT satellites or the Airborne Visible Infrared Imaging Spectrometer (AVIRIS) systems. These line scanning sensors provide information of the Earth's surface and allow static analysis. This area has been firmly established for many years and is essential for several applications like earth observation, inspection and agriculture. Additionally onboard realtime hyperspectral image analysis for autonomous navigation is an exciting and promising application scenario. But this topic is relatively unexplored because the established hardware is only capable of capturing static scenes. This is due to the scanning requirements for constructing a full 3-D hypercube of a scene. Using line-scan

© Springer International Publishing AG 2017
M. Felsberg et al. (Eds.): CAIP 2017, Part I, LNCS 10424, pp. 16–27, 2017.
DOI: 10.1007/978-3-319-64689-3_2

cameras, multiple lines need to be scanned, while with cameras using special filters, several frames have to be captured to construct an spectral image of the scene. The slow acquisition time is responsible for motion artifacts which impede the observation of dynamic scenes. Therefore, new sensor techniques and procedures are needed here. This drawback can be overcome with novel highly compact, low-cost, snapshot mosaic (SSM) imaging cameras, which are able to capture a whole spectral cube in one shot. The capture time is considerably shorter than that of filter wheel solutions allowing to capture a hyperspectral cube at one discrete point in time. Utilizing these sensors, it is possible to use hyperspectral camera systems on unmanned land vehicles and utilize them for continuous terrain classification while moving. Most classifiers for spectral classification treat hyperspectral data as a set of spectral measurements and do not consider spatial dependencies. So the data is classified only based on their spectral information. These approaches discard information associated with correlations among neighboring pixels. Joint spectral and spatial classification techniques seem reasonable to address these disadvantages.

In this paper we investigate the use of snapshot mosaic hyperspectral cameras on unmanned land vehicles for drivability analysis and evaluate different spectral and contextual features for hyperspectral classification based on the data we captured. We make use of established supervised classifiers to recognize different classes like drivable, rough and obstacle which can bee seen as terrain recognition or environmental perception based on spectral reflectances.

The remainder of this paper is organized as follows. In the following section an overview of common algorithms for feature extraction and spectral classification is given. Then our general setup is presented in Sect. 3. Our feature extraction and classification approach is described in detail in Sects. 4 and 5. And in Sect. 6 we present our results on our new hand-labeled dataset. Finally a conclusion of our work is given in Sect. 7.

2 Related Work

The standard procedure for image-based terrain classification is defined by capturing regular RGB images and trying to identify different classes, like Chetan et al. [9] did. They used color information and local binary patterns (LBP) in combination with different supervised classifiers. Additionally, in recent years, hyperspectral classification has been under active development. Hyperspectral data allows for a more detailed insight into the composition and nature of materials like plants and soil than standard RGB data. Although there are some unsupervised classification algorithms in literature, we focus on supervised classification for the moment, because it is more widely used as shown by Plaza et al. [21]. Most supervised classifiers suffer from the Hughes effect [14], especially when dealing with high-dimensional hyperspectral data. To deal with this issue, Melgani et al. [18] and Camps-Valls et al. [5] introduced support vector machines with adequate kernels for hyperspectral classifications. Supervised techniques are limited by the availability of labeled training data and suffer from the high dimensionality of the data. While recording data is usually quite straightforward, the precise and correct annotation of the data is very time-consuming and complicated. Therefore semi-supervised techniques have come up to fix this as proposed by Camps-Valls et al. [7]. Jun et al. [16] presented a semi-supervised classifier that selects non-annotated data based on its entropy and adds it to the training set. The classification of hyperspectral data reveals several important challenges. There is a great mismatch between the high dimensionality of the data in the spectral range, its strong correlation and the availability of annotated data, which is absolutely necessary for the training. Another challenge is the correct combination and integration of spatial and spectral information to take advantage of features from both these domains.

In various experiments by Li et al. [17] it was observed that classification results can be improved by investigating spatial information in parallel with the spectral data. Different efforts have been made to incorporate context-sensitive information in classifiers for hyperspectral data [21]. Fauvel et al. [10] fuse morphological and hyperspectral data to enhance classification results. As a consequence, it has now been widely accepted that the combined use of spatial and spectral information offers significant advantages. To integrate the context into kernel-based classifiers, a pixel can be simultaneously defined both in the spectral domain and in the spatial domain by applying a corresponding feature extraction. Contextual features are achieved, for example, by the standard deviation per spectral band. This leads to a family of new kernel methods for hyperspectral data classification reported by Camps-Valls et al. [6] and implemented using a support vector machine. Brown et al. [4] used principal component analysis for dimensionality reduction and proposed an extension of the well known SIFT descriptor, called multi-spectral SIFT (MSIFT) for scene category recognition. Salamati et al. [22] investigated different combinations of SIFT and spectral information to enhance recognition accuracy. An alternative approach to combining contextual and spectral information is the use of Markov random fields (MRFs). They exploit the probabilistic correlation of adjacent labels [23].

In comparison there is only little research in literature on hyperspectral classification utilizing terrestrial spectral imaging, where data was not captured from an earth orbit or an airplane but from cameras which where mounted on land-based vehicles. One example is the vegetation detection in hyperspectral images as demonstrated by Bradley et al. [2], who showed that the use of the Normalized Difference Vegetation Index (NDVI) improves classification accuracy. Namin et al. [20] proposed an automatic system for material classification in natural environments by using multi-spectral images consisting of six visual and one NIR band. The combination of RGB and hyperspectral data, using the same hyperspectral snapshot cameras we use, was evaluated by Cavigelli et al. [8] on data with static background and a very small dataset utilizing deep neural nets.

3   Sensor Setup

In this work we used the MQ022HG-IM-SM4X4-VIS (*VIS*) manufactured by Ximea with an image chip from IMEC [12] utilizing a snapshot mosaic filter which has a per-pixel design. The filters are arranged in a rectangular mosaic pattern of n rows and m columns, which is repeated w times over the width and h times over the height of the sensor. These sensors are designed to work in a specific spectral range which is called

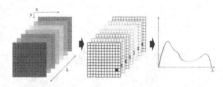

(a) Example raw image taken by the *VIS* camera.

(b) A schematic representation of a hypercube and an interpolated plot of a single data point.

Fig. 1. Raw image *VIS* camera with visible mosaic pattern. And a schematic representation of a hypercube.

the active range which is 470–620 nm for the current sensor. The *VIS* camera has a mosaic pattern with $n_{VIS} = 4, m_{VIS} = 4$. Ideally every filter has peaks centered around a defined wavelength spectrum with no response outside. However contamination is introduced into the response curve and the signal due to physical constraints. These effects can be summarized as a spectral shift, spectral leaking, and crosstalk and need to be compensated.

Therefore the raw data captured by the camera needs a special preprocessing. We need to construct a hypercube with spectral reflectances from the raw data. This step consists of cropping the raw-image to the valid sensor area, removing the vignette and converting to a three dimensional image, which we call a *hypercube*. Reflectance calculation is the process of extracting the reflectance signal from the captured data of an object. The purpose is to remove the influence of the sensor characteristics like quantum efficiency and the illumination source on the hyperspectral representation of objects. We define a hypercube as $\mathcal{H}: L_x \times L_y \times L_\lambda \rightarrow \mathbb{R}$ where L_x, L_y are the spatial domain and L_λ the spectral domain of the image. A visual interpretation of such a hypercube is displayed in Fig. 1b. The hypercube is understood as a volume, where each point $\mathcal{H}(x,y,\lambda)$ corresponds to a spectral reflectance. Derivated from the above definition a spectrum χ at (x,y) is defined as $\mathcal{H}(x,y) = \chi$, where $\chi \in \mathbb{R}^{|L_\lambda|}$ and $|L_\lambda| = n \cdot m$. The image with only one wavelength, called a spectral band $\mathcal{H}(z) = \mathcal{B}_{\lambda=z}$, is defined as follows: $\mathcal{B}_\lambda : L_x \times L_y \rightarrow \mathbb{R}$. This image contains $\mathbf{x} = (x,y)$ the wavelength sensitivity λ for each coordinate.

4 Classification Framework

For the evaluation of the different features, the image data is first segmented using the SLIC-Superpixels algorithm [1]. This technique joins pixels to a segment based on distance in color and image space. As it is proposed for RGB images, conversion to CIE-LAB color space is recommended to model human perception when measuring color similarity. As aesthetic properties aren't relevant for classification, the euclidean distance of the spectral vectors is used as the measure of similarity. Segmentation ensures homogeneous classification results and redundantizes post-processing. A representative

Fig. 2. RGB simulation of a hyperspectral image overlayed with the segmentation mask. The edge length is set to 15 pixels, resulting in segments covering approximately 225 pixels each.

segmentation result is shown in Fig. 2. The classifier used for evaluation is the Random Forest algorithm [3]. The image data was recorded using the *VIS* camera mounted on a car combined with several other sensors. For training and verification purposes the images need to be accordingly annotated. The other sensor's recorded data is supposed to be fused with the classification results in further works.

5   Extraction of Hyperspectral Features

Our main purpose in this work is the classification of hyperspectral data with k bands, utilizing spatial and spectral dimensions. To obtain good results and improve classification, features need to be extracted, which contain additional information to the raw spectra in the hypercube. A major source of error is the variable illumination of the scene, because our sensor measures reflectance values which change with illumination changes. So in practice classifiers are trained and may be used on data that shows different illumination situations. By using the normalized spectrum as a feature, we reduce the influence of scene illumination and other irregularities by making use of the hyperspectral counterpart to the log-chromaticity representation of RGB images [11]. In the RGB case, normalization of the values of pixel χ by the geometric mean

$$\chi^M(x,y) = \sqrt[3]{\prod_{i=1}^{3} \mathcal{B}_i(x,y)} \tag{1}$$

of it's components at position (x,y) is recommended. In the hyperspectral case, the number of bands is much higher, which can cause numerical instabilities, computation of the high-order roots being the cause. Instead the normalization by the sum of the spectrum's n values

$$\chi^S(x,y) = \sum_{i=1}^{n} \mathcal{B}_i(x,y) \tag{2}$$

can be used [13]. Logarithmizing isn't necessary for the feature extraction, as the particular axis in the resulting feature space wouldn't gain in variance. Hence the normalized spectrum at image position (x, y) is computed as

$$\mathcal{B}'_k(x,y) = \frac{\mathcal{B}_k(x,y)}{\chi^S(x,y)} \tag{3}$$

for each spectral band k. Additionally, the sum of all the spectrum's components χ^S is added to the feature vector to represent it's brightness.

To extend the feature space using the textures present in the images, a gabor filter bank is used. Each filter kernel is generated by modulating a gaussian by a sine and cosine term [19]. That way a set of uniformly spread kernels in orientation, scale and frequency are generated. Scale and frequency are inversely proportional. The higher the frequency of the kernel, the lower the scale, the bandwidth of the gaussian is chosen. This ensures the best possible trade-off between localization in frequency space and image space. For the available images, 6 orientations for the kernels with 6 combinations of scale and frequency each were used. The resulting filter bank can be seen in Fig. 3. In natural environments textures representing the same material are often oriented differently, for example blades of grass may be sloped to the side instead of standing straight. For that reason, the coefficients of the same frequency and scale but different directions are combined. Only the maximum coefficient of the multiple orientations for each frequency and kernel scale will be added to the feature vector. With 6 scales used for each orientation, 6 features per spectral band would amount to the feature vector. With the large number of available spectral components in hyperspectral images, it may still exhibit excessively high dimensionality. The dataset used for training and classification, which mainly shows dirt and tarmacked roads in rural areas, holds little difference in texture between the various channels. For that reason, a set of grayscale gabor-features is computed instead of applying the filter bank to each band individually. The underlying grayscale image representation is composed of the mean of the spectral components for each image position.

To obtain a more abstract texture feature, the ripple and granularity are taken into account. Based on the same filter bank and the same grayscale input image, for each combination of frequency and scale the maximum coefficient of the available orientations is extracted. Then the standard deviation between that maximum and the mean coefficient of the available orientations is computed and added to the feature vector. Surfaces with considerable ripple will result in a high value for the respective frequency. Low values occur when the pictured surface exhibits smooth or coarsely granular texture, so it shows no distinguishable orientation. In combination with the grayscale gabor features a wide range of possible texture situations can be represented in the feature space.

Additionally domain knowledge can be used to improve classification accuracy. The image situation for a driving vehicle shows constant conditions, for example sky is typically in the upper part of the image, while navigable terrain will not be found there. Including gravity-based information can support classification by modeling these conditions [15]. The probabilities of all possible class affiliations for each y-coordinate are extracted from the available ground truth images and stored in a lookup table. For feature

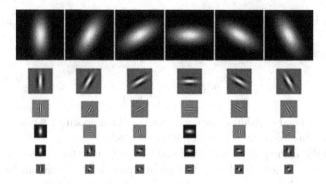

Fig. 3. Filter kernels computed for the gabor texture features.

extraction, the probabilities for individual class affiliations for the image position's y-coordinate is read off the lookup table and stored in the feature vector. This feature is restricted to the particular scenario described. Other navigation situations, like flight control of a drone, exhibit different vertical arrangements of scene components. Nevertheless it was used to show how it affects the other features' classification accuracy when paired with them. The proposed features are extracted from a segmented image as described in Sect. 4. The samples for the gabor based texture features and the gravity-based ones are picked from the center of the particular segment. The center p^c is defined using the 1st order moment as in

$$p^c = \frac{1}{n} \sum_{i=0}^{n} p_i \qquad (4)$$

for the n 2-dimensional coordinates p in image space that make up the segment. It can also be understood as the center of mass. For the normalized spectrum, the segment's sample isn't extracted from the center, but computed for each pixel and then the mean value for each band is returned. So we obtain an 32-dimensional feature vector for every segment if all features are used, the normalized spectra consisting of the 15 normalized bands and the sum of all bands. Both gabor features contribute six variables, as six scales and frequencies are available in the filter bank. The gravity-based feature offers a variable for each class to be recognized, totaling in 4 variables in our case.

6   Evaluation

As far as we know, there is no publicly available data set with hyperspectral data recorded by the MQ022HG-IM-SM4X4-VIS camera, which uses snapshot mosaic technique to acquire hyperspectral data. So we had to build a new dataset on our own, which will be published in the near future. We equipped a standard car with the camera manufactured by Ximea and collected a total of $\approx$200 GB of data driving through suburban and rural areas, from which we selected a subset for labeling hyperspectral data. The dataset was labeled in terms of drivability as illustrated in Fig. 4b. The main classes are drivable, rough and obstacle. In addition a sky class was introduced, because it is an important part of our scene and defines the border of the terrain. It consists of more than 1750000 sky, 1890000 drivable, 1070000 rough and 2910000 obstacle samples. During the labeling process not all image pixels have been assigned classes. This is due to the fact that border areas between materials are not unambiguously assignable. The feature extraction is realized as previously described in Sect. 5. The classifier used for evaluation is the Random Forest algorithm [3], which has proven to be successful in previous experiments [24]. For training, every fifth image in the available set was labeled and used to extract features. This way the training data represents the different possible image situations. The rest of the images was classified using all the different feature vectors. The overall accuracy is computed as h/p, with h being the sum of all correctly classified pixels in all images and p being the total population, made up of all labeled image pixels. Pixels that lack a label are classified as well, but for quantitative evaluation they need to be discarded. When merely using mean spectra of the segments as features, classification fails in some places due to different sources of error. In Fig. 4c the misinterpretation of shadows as obstacles and the sky's reflection on the engine hood as actual sky can be seen. The classification using the normalized mean spectra performs much better in the particular image showed in Fig. 4d. Shaded areas are recognized correctly in most cases and other details match the ground truth better as well. The overall accuracy of this classifier was 91.38%, while the one using the plain mean spectra of the segments achieved 88.92% mean accuracy for the available images. A lot of these images contain less challenging lighting situations, which explain the smaller difference in overall accuracy compared to the one displayed in Fig. 4.

Feature vectors as proposed in this work:

A Plain spectra
B Normalized spectra
C Grayscale gabor
D Ripple and granularity
E Normalized spectra and gravity-based feature
F All proposed features

Feature vectors as proposed by Namin et al. [20]:

G GLCM
H GLCM and plain spectra
I Plain spectra, std. deviation in segment, GLCM and fourier features

The other proposed features perform rather poor if they are used exclusively. The gravity-based feature can in no way represent the scene's content, while the grayscale gabor feature reaches an overall accuracy of 79.52%. The ripple and granularity feature performs similarly with 80.46% accuracy. Because the plain coefficients spring from image data that hasn't been normalized, they also depend on actual brightness. That causes the classification to misinterpret shaded areas just like the plain mean spectra. In combination with the normalized spectrum, better classification results can be achieved. The texture features offered only minor improvements, while the combination of normalized spectra and the gravity-based feature exceeded the other possible combinations at 94.82% accuracy. However, the classification results also show errors, that originate

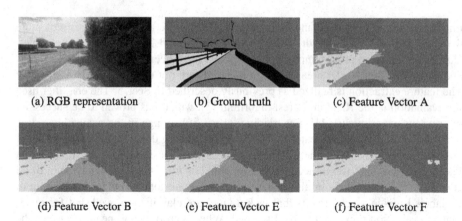

(a) RGB representation   (b) Ground truth   (c) Feature Vector A

(d) Feature Vector B   (e) Feature Vector E   (f) Feature Vector F

Fig. 4. Comparison of classification using plain mean spectra, normalized mean spectra, and combinations of features. For the plain mean spectra the accuracy for the shown image was 85.3%. Normalized mean spectra improved the accuracy to 95.8% in this case. The combination of normalized mean spectra and gravity-based feature offers 97.78%, the combination of all features 97.28% for the shown image. Annotations for navigability analysis are as follows: Green represents good and yellow fair navigability. Red shows obstacles and blue areas picture the sky. (Color figure online)

from the use of this feature. In particular, a lot of segments that are on similar height with the engine hood are classified as obstacles. The reason for this is that in training data the aforementioned engine hood is annotated as obstacle as well. The classification results in Fig. 5 all show that artifact on the right side of the engine hood. Generally, the gravity-based feature might translate poorly to other image situations.

For comparison, some of the features described in [20] were implemented and used for classification in the same fashion. The overall accuracy of 83.48% is worse than the one achieved with simple mean spectra. There are multiple reasons: Shadows weren't treated like in the original paper by annotating them in training. The results shown in Fig. 5 confirm that, as most shadowed areas are classified as obstacles. Furthermore, the feature used for vegetation recognition [2] isn't available because the *VIS* camera used in this work doesn't detect infrared light. These two approaches were essential for the great results that were attained. Also 15 instead of 7 bands in total were available and computing the GLCM for all of them might not add relevant information. Finally the features described in [20] are computed on neighborhood blocks for each pixel. As we used segmentation and created only one feature vector for each segment, for comparison, in this work the kernel sizes of the fourier features and the neighborhood to consider in GLCM were limited to the size of the segment. The scale of these was smaller than the block size proposed by Namin et al. [20], just as the input images are smaller than theirs. Especially the fourier features might yield better results on higher resolution images (Tables 1, 2 and 3).

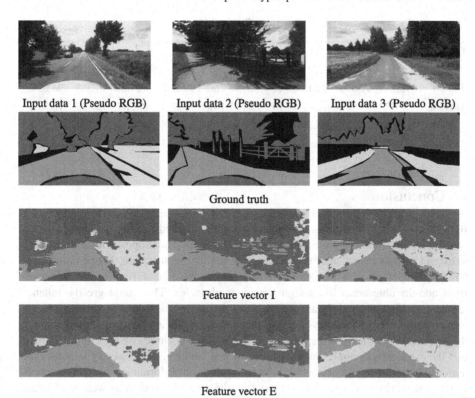

Input data 1 (Pseudo RGB)   Input data 2 (Pseudo RGB)   Input data 3 (Pseudo RGB)

Ground truth

Feature vector I

Feature vector E

Fig. 5. Comparison of classification results. The columns show an RGB representation of the hyperspectral input image, the ground truth, the feature vector proposed by [20] and the combination of normalized spectra and the gravity-based feature presented in this work.

Table 1. Comparison of accuracies achieved in classification using different feature vectors.

Feature set	A	B	C	D	E	F	G	H	I
Accuracy	88.92%	91.38%	79.52%	80.46%	94.82%	94.68%	81.15%	83.17%	83.48%

Table 2. Precision

Class	Feature vectors								
	A	B	C	D	E	F	G	H	I
Sky	82.02%	83.95%	75.24%	75.11%	89.17%	89.65%	80.45%	81.42%	81.94%
Drivable	81.18%	84.48%	69.53%	69.58%	86.09%	85.76%	67.39%	75.08%	74.70%
Rough	66.41%	66.31%	49.67%	52.34%	66.68%	64.48%	59.34%	54.78%	56.11%
Obstacle	67.97%	70.26%	65.39%	66.38%	74.01%	74.56%	63.44%	64.58%	64.52%

Table 3. Recall

Class	Feature vectors								
	A	B	C	D	E	F	G	H	I
Sky	97.22%	97.67%	94.18%	92.63%	98.26%	98.15%	88.42%	94.55%	94.22%
Drivable	89.63%	93.88%	70.65%	74.12%	97.10%	97.20%	84.21%	78.79%	78.58%
Rough	76.16%	79.72%	65.39%	67.02%	83.75%	83.64%	64.99%	73.76%	73.37%
Obstacle	88.03%	90.12%	81.39%	82.12%	95.41%	95.09%	80.36%	81.93%	83.23%

7 Conclusion

Both spectral and spatial information have been investigated for classification of hyperspectral images captured with snapshot mosaic cameras. The proposed features are easy to extract, which allows their use in real-time terrain classification. Based on the captured hyperspectral data we were able to precisely distinguish road or drivable areas from non-drivable areas like rough terrain or obstacles. This could greatly enhance terrain classification performance. Especially the use of the normalized mean spectra improved the overall accuracy in a way that would enable actual navigation. Although the proposed gabor texture features didn't improve the accuracy enough to justify their use, the extraction of ripple and granularity turned out be an efficient way to reduce the dimensionality in feature space while retaining relevant information for classification. In other scenarios and situations, or used on the normalized data as well, the texture features might be more powerful. The adaption of a segmentation algorithm that has been widely used on RGB images for hyperspectral data has been shown and yielded satisfying results.

Further work might validate the benefit of classification on segmented images, especially when the results need to be fused with data from other sensors. For better comparability with other approaches that extract features from hyperspectral data, both the features proposed here and the ones from referenced work need to be validated on a data set that is fully compatible.

References

1. Achanta, R., Shaji, A., Smith, K., Lucchi, A., Fua, P., Süsstrunk, S.: Slic superpixels compared to state-of-the-art superpixel methods. IEEE Trans. Pattern Anal. Mach. Intell. **34**(11), 2274–2282 (2012)
2. Bradley, D.M., Unnikrishnan, R., Bagnell, J.: Vegetation detection for driving in complex environments. In: 2007 IEEE International Conference on Robotics and Automation, pp. 503–508. IEEE (2007)
3. Breiman, L.: Random forests. Mach. Learn. **45**(1), 5–32 (2001)
4. Brown, M., Süsstrunk, S.: Multi-spectral sift for scene category recognition. In: IEEE Conference on Computer Vision and Pattern Recognition (CVPR), pp. 177–184. IEEE (2011)
5. Camps-Valls, G., Bruzzone, L.: Kernel-based methods for hyperspectral image classification. IEEE Trans. Geosci. Remote Sens. **43**(6), 1351–1362 (2005)

6. Camps-Valls, G., Gomez-Chova, L., Muñoz-Marí, J., Vila-Francés, J., Calpe-Maravilla, J.: Composite kernels for hyperspectral image classification. IEEE Geosci. Remote Sens. Lett. **3**(1), 93–97 (2006)
7. Camps-Valls, G., Tuia, D., Gómez-Chova, L., Jiménez, S., Malo, J.: Remote sensing image processing. Synth. Lect. Image Video Multimedia Process. **5**(1), 1–192 (2011)
8. Cavigelli, L., Bernath, D., Magno, M., Benini, L.: Computationally efficient target classification in multispectral image data with deep neural networks. In: SPIE Security+Defence. p. 99970L. International Society for Optics and Photonics (2016)
9. Chetan, J., Krishna, M., Jawahar, C.: Fast and spatially-smooth terrain classification using monocular camera. In: Pattern Recognition (ICPR), 2010 20th International Conference on. pp. 4060–4063. IEEE (2010)
10. Fauvel, M., Benediktsson, J.A., Chanussot, J., Sveinsson, J.R.: Spectral and spatial classification of hyperspectral data using SVMs and morphological profiles. IEEE Trans. Geosci. Remote Sens. **46**(11), 3804–3814 (2008)
11. Finlayson, G.D., Hordley, S.D., Lu, C., Drew, M.S.: On the removal of shadows from images. IEEE Trans. Pattern Anal. Mach. Intell. **28**(1), 59–68 (2006)
12. Geelen, B., Tack, N., Lambrechts, A.: A compact snapshot multispectral imager with a monolithically integrated per-pixel filter mosaic. In: SPIE Moems-Mems. p. 89740L. International Society for Optics and Photonics (2014)
13. Gevers, T., Stokman, H., Weijer, J.v.d.: Colour constancy from hyper-spectral data. In: Proceedings of the British Machine Vision Conference, pp. 30.1-30.10. BMVA Press (2000)
14. Hughes, G.: On the mean accuracy of statistical pattern recognizers. IEEE Trans. Inf. Theory **14**(1), 55–63 (1968)
15. Javanbakhti, S., Zinger, S., de With, P.H.N.: Context-based region labeling for event detection in surveillance video. In: 2014 International Conference on Information Science, Electronics and Electrical Engineering, vol. 1, pp. 94–98, April 2014
16. Li, J., Bioucas-Dias, J.M., Plaza, A.: Semisupervised hyperspectral image segmentation using multinomial logistic regression with active learning. IEEE Trans. Geosci. Remote Sens. **48**(11), 4085–4098 (2010)
17. Li, J., Bioucas-Dias, J.M., Plaza, A.: Spectral-spatial hyperspectral image segmentation using subspace multinomial logistic regression and markov random fields. IEEE Trans. Geosci. Remote Sens. **50**(3), 809–823 (2012)
18. Melgani, F., Bruzzone, L.: Classification of hyperspectral remote sensing images with support vector machines. IEEE Trans. Geosci. Remote Sens. **42**(8), 1778–1790 (2004)
19. Naghdy, G.A., Wang, J., Ogunbona, P.O.: Texture analysis using gabor wavelets, vol. 2657, pp. 74–85 (1996)
20. Namin, S.T., Petersson, L.: Classification of materials in natural scenes using multi-spectral images. In: 2012 IEEE/RSJ International Conference on Intelligent Robots and Systems (IROS), pp. 1393–1398. IEEE (2012)
21. Plaza, A., Benediktsson, J.A., Boardman, J.W., Brazile, J., Bruzzone, L., Camps-Valls, G., Chanussot, J., Fauvel, M., Gamba, P., Gualtieri, A., et al.: Recent advances in techniques for hyperspectral image processing. Remote Sens. Environ. **113**, S110–S122 (2009)
22. Salamati, N., Larlus, D., Csurka, G.: Combining visible and near-infrared cues for image categorisation. In: Proceeding of the 22nd British Machine Vision Conference (BMVC 2011), No. EPFL-CONF-169247 (2011)
23. Tarabalka, Y., Fauvel, M., Chanussot, J., Benediktsson, J.A.: SVM-and MRF-based method for accurate classification of hyperspectral images. IEEE Geosci. Remote Sens. Lett. **7**(4), 736–740 (2010)
24. Winkens, C., Sattler, F., Paulus, D.: Hyperspectral terrain classification for ground vehicles. In: 12th International Conference on Computer Vision Theory and Applications (VISAPP) (2017)

Improved Stixel Estimation Based on Transitivity Analysis in Disparity Space

Noor Haitham Saleem[1](✉), Hsiang-Jen Chien[1], Mahdi Rezaei[2], and Reinhard Klette[1]

[1] EEE Department, School of Engineering, Computer, and Mathematical Sciences, Auckland University of Technology, Auckland, New Zealand
nalani@aut.ac.nz
[2] CE Department, Faculty of Computer and Information Technology Engineering, Qazvin Islamic Azad University, Qazvin, Iran

Abstract. We present a novel method for stixel construction using a calibrated collinear trinocular vision system. Our method takes three conjugate stereo images at the same time to measure the consistency of disparity values by means of the transitivity error in disparity space. Unlike previous stixel estimation methods that are built based on a single disparity map, our proposed method introduces a multi-map fusion technique to obtain more robust stixel calculations. We also apply a polynomial curve fitting approach to detect an accurate road manifold, using the v-disparity space which is built based on a confidence map, which further supports accurate stixel calculation. Comparing the depth information from the extracted stixels (using stixel maps) with depth measurements obtained from a highly accurate LiDAR range sensor, we evaluate the accuracy of the proposed method. Experimental results indicate a significant improvement of 13.6% in the accuracy of stixel detection compared to conventional binocular vision.

1 Introduction

Vision-based driver assistance systems (VB-DAS) contribute to the current transition process towards autonomous vehicles. They are already widely used in current modern cars [1]. Cameras are one type of sensors that are commonly installed in modern cars. In particular, stereo vision contributes to systems that aim at distance measurements, surface modelling, or object detection [2]. This is important, for example, for scene analysis [3], feature descriptors [4], optimising learning time [5], or for reducing processing efforts in general [6].

In 2009 a novel "super-pixel representation" has been proposed for urban road scenes. The method is known as *stixel* (from "stick elements"). It groups vertically space cubes which belong to an on-road object [7]. The representation yields a highly efficient modelling of scene objects in urban traffic environments [9]. Recently, joint stixel representations, combining semantic data and depth, are proposed to integrate both categories in terms of a joint optimized scene model [10].

© Springer International Publishing AG 2017
M. Felsberg et al. (Eds.): CAIP 2017, Part I, LNCS 10424, pp. 28–40, 2017.
DOI: 10.1007/978-3-319-64689-3_3

Fig. 1. A stixel world for a scene in KITTI's *residential* dataset [11]. *Top-left*: Disparity map using an SGM-variant visualized by applying a color key. *Top-right*: Improved disparity map. *Bottom-left*: Stixels on a ground plane using binocular vision (red rectangles indicate missing stixels). *Bottom-right*: Proposed stixel estimation. (Color figure online)

To construct a "stixel world" (see Fig. 1), multiple independent techniques may have to be cascaded.[1] These may include mapping disparities into occupancy grids, ground manifold estimation, object height detection, and finally stixel extraction.

The *free space* is a region in the ground manifold "without any obstacle" [13], i.e. regions ahead of the ego-vehicle where this vehicle may potentially drive in, for example, in the next few seconds. Free-space and stixel calculations are closely related to each other; the existence of a stixel excludes free space at this place; stixels are "sitting" on the ground manifold, and the free space is a subset of the ground manifold. The detection of free-space is important for intelligent transportation control [14]. It is also crucial for collision avoidance for the *ego-vehicle* (i.e. the vehicle in which the system is operating in) and assisting a blind pedestrian [32].

Having VB-DAS as a core component over other active sensors, many *advanced driver assistance systems* (ADAS) demonstrate prominent developments in this area (e.g. [15]). An ADAS provides a better understanding of the environment in order to improve traffic safety and efficiency [16].

Accuracy of stixels requires a disparity signal of "good" quality; this quality often decreases in cases of occlusions or textureless image patches [17]. Unfortunately, these issues are common in traffic scenes, thus more efforts are needed to improve disparity signals, also aiming at more reliable free-space estimation and stixel calculations.

A binocular vision system depends on calculated disparity values which are calculated by implementing stereo matching algorithms [12,18] on images obtained by a left and right camera.

Since noisy 3D points have a considerable impact on free-space detection, it is very important to identify unreliable disparity values before they are transformed into 3D space and used for stixel estimation. Therefore, we consider the use of

---

[1] We adopt a *semi-global matching* (SGM) algorithm [12] for disparity calculation.

confidence maps (see [19] for different options for such maps) with the aim of improving stixel segmentation.

The remainder of this paper is structured as follows. Section 2 addresses work closely related to stixel estimation. In Sect. 3, the proposed approach is described in detail. In Sect. 4, experimental results are given and discussed. Section 5 concludes.

2   Related Work

We briefly discuss work on stixel extraction. Stixels are a compact representation towards semantic segmentation of traffic scenes; space elements above neighbouring pixels at the same depth are vertically grouped [20], according to an estimated object height at those pixels. Apparently, stixels are like rectangular thin columns on the ground manifold defined on a regular grid. A stixel starts at the top at a detected object surface and ends at the bottom on the level of the ground manifold. Free space (for the ego-vehicle) is a subset of the ground manifold not covered by stixels.

Rapid stixels describe techniques which enhance stixel extraction by reducing computational costs. In [23], a direct stixel computation is proposed by changing the parametrization from disparity space into pixel-wise cost volumes for speed improvement. In [21], the authors use deep convolution neural networks for free-space detection using monocular vision, while obstacle detection and stixel estimation are done using stereo vision. Fast stixel computation without depth maps is proposed in [22]; it allows high-speed pedestrian detection up to 200 fps.

Color fusion models compute stixels using stereo images, and also involve a combination of color appearance and depth cues for free-space and obstacle detection. Such methods have been presented in the stixel segmentation literature [17, 24, 25]. Their implementation can be done by using low-level fusion of depth or semantic information in the stixel generation process. Scharwächter et al. employed pixel classification by random decision forests [24], while in [25] semantic information via object detectors is used for a suitable set of classes. Yet another method to improve stixels is by using low-level appearance models in an on-line self-supervised framework; see [17].

Stereo confidence-based methods, on the other hand, use confidence estimation within the stereo-matching process to replace spurious disparity matches by interpolating surrounding disparity values at these locations; see [26–28] for examples. In [26], the authors incorporate three confidence measures, namely the naïve peak-ratio (PKRN), the maximum-likelihood measure (MLM), and local curve (LC) information into stixel representations. The stereo confidence measures use stereo confidence cues based on an extended Bayesian approach. In [28], an ensemble learning classifier is adopted to increase accuracy in stereo-error detection. In [27], histogram-sensor models are explored to model on a real-world application using a global formulation of 3D reconstruction through an occupancy grid.

Rapid-stixel methods may have some drawbacks; they may suffer from low-depth accuracy which affects stixel extraction negatively. In order to perform stixel segmentation, an adopted colour fusion model might not be suitable due to shortages highlighted in [10]. With promising results achieved by adopting confidence information, this paper proposes the following:

1. altogether a low-cost architecture for reducing false-positives in stixel estimation,
2. in particular the use of a confidence measure derived from trinocular stereo matching, and
3. a method for performance evaluation of stixel estimation assuming the availability of LiDAR data.

3   Stixels in Trinocular Stereo Vision

We consider a trinocular calibrated video recording system which allows us to perform stereo matching on one of the three possible camera pairs. Thus we may have up to three different left-right disparity maps; they may be fused and warped to a selected reference camera (one of the three). Based on the fused (and thus enhanced) disparity map, the *ground manifold* (i.e. a generalisation from a plane) is estimated using a *v-disparity* technique. This is followed by detections of base- and top-points of stixels applying means of membership voting and a cost image. In a final step, base- and top-points are used for extracting the stixels.

3.1   Transitivity Error in Disparity Space

Given a collinear m-camera configuration, we have $m(m-1)/2$ left-right disparity maps. It has been shown that the accumulative transitivity error among these maps can be effectively used as a confidence indicator on a stereo matcher [29]. Let $(u,v) \in \mathbb{R}^2$ denote a pixel location in left-image coordinates. A disparity map $\delta : \mathbb{R}^2 \to \mathbb{R}_0^+$ finds its corresponding pixel in right-image coordinates $(u - \delta(u,v), v)$. A disparity map can therefore be used to define the warping of a function $\mathcal{M} : \mathbb{R}^2 \to \mathbb{R}$ as follows:

$$\phi(\mathcal{M}, \delta)(u,v) = \mathcal{M}(u - \delta(u,v), v) \tag{1}$$

The warping function ϕ is used to construct the concatenation of two disparity maps

$$\tau(\delta_{01}, \delta_{12})(u,v) = \delta_{01}(u,v) + \phi(\delta_{12}, \delta_{01})(u,v) \tag{2}$$

where δ_{01} and δ_{12} are the disparity maps with respect to camera pairs $(0,1)$ and $(1,2)$, respectively, in a trinocular configuration.

Let $\bar{\delta}_{02} = \tau(\delta_{01}, \delta_{12})$ be a *combined* disparity map, and δ_{02} the explicitly computed one for camera pair $(0,2)$. We define our new, say, *trinocular confidence measure* by

$$\Gamma(u,v) = \frac{1}{\|\delta_{02}(u,v) - \bar{\delta}_{02}(u,v)\| + 1} \tag{3}$$

Fig. 2. Trinocular confidence and free space. *Top row*: Trinocular stereo pair from the KITTI *road* dataset. *Bottom left*: TED-based disparity. *Bottom middle*: Red and blue pixels indicate high and low confidence values, respectively. *Bottom right*: Calculated free-space (using *v*-disparity, confidence map, and polynomial curve fitting). (Color figure online)

where the absolute difference $\|\delta_{02}(u,v) - \bar{\delta}_{02}(u,v)\|$ is the *transitivity error in disparity space* (TED). See Fig. 2 for an example of this confidence indicator; $\bar{\delta}_{02}$ is also called the *TED-based disparity*.

3.2 Detection of Base-Points of Stixels

Our stixel calculation works on TED-based disparities. We propose a new polynomial curve-fitting technique to identify the lower envelop in the common *v*-disparity space. This identification supports the base-point calculation of stixels. We consider base-points $b_1, b_2, ..., b_{N_{\mathrm{col}}}$ of obstacles in row v.

The *v*-disparity map is computed by accumulating pixels with the same disparity value in one row v, $1 \leq v \leq N_{\mathrm{row}}$, of the disparity map:

$$V(v,d) = \mathrm{card}\{u : 1 \leq u \leq N_{\mathrm{col}} \wedge \mathrm{int}(\delta(u,v)) = d\} \tag{4}$$

where $0 \leq d \leq d_{\max}$ defines the quantized disparity range for δ in the $N_{\mathrm{row}} \times N_{\mathrm{col}}$ disparity map; int is the nearest integer.

In Eq. (4), each element in the disparity map is considered equally. In this work we propose to use a weighted sum of our trinocular confidence values:

$$V(v,d) = \sum_{1 \leq u \leq N_{\mathrm{col}} \wedge \mathrm{int}(\delta(u,v)) = d} \Gamma(u,v) \tag{5}$$

Here, elements with higher TED-based confidence become more influential.

The next step is to extract a ground manifold from the generated *v*-disparity map. The ground manifold is identified with an approximated lower envelope in the *v*-disparity space.

Assuming a ground plane, a Hough transform is used in [30,31] to detect a lower envelop function in form of a straight line in the *v*-disparity space. In order to construct this envelop function, the method starts at first with a lower and upper envelop. The envelop estimation is based on calculating the intensity sum of all pixels along a considered curve in the *v*-disparity image, and then selecting the envelop for which this sum becomes a minimum.

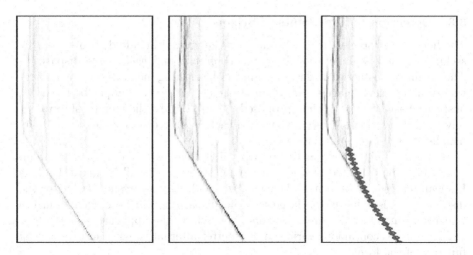

Fig. 3. Demonstration of v-disparity. *Left*: Common cardinality-based v-disparity map. *Middle*: Novel TED-based v-disparity map. *Right*: Detected curve using polynomial fitting.

Considering that a road surface is not a perfect plane, and possibly also more irregular in slope changes than a continuous curve, we consider polynomial curve fitting for extracting best fits to the v-disparities, defining a polynomial as being our envelop function; see Fig. 3 for an example.

In general, there is always room for improvements in curve fitting. (A ground manifold also remains to be approximated to some degree only when assuming identical height across one image row v.) We apply a polynomial curve fitting technique to find the coefficients of a polynomial $P(x)$ of degree n that best fits the lower envelop in the v-disparity image:

$$y = P(x) = a_n x^n + a_{n-1} x^{n-1} + \ldots + a_1 x + a_0 \qquad (6)$$

where $a_0, a_1, \ldots, a_n$ are the coefficients, and the degree n is selected according to accuracy requirements for the algorithm. In order to generate the coefficients of the polynomial according to the degree specified, we need to compute a least-square polynomial for a given set of data. Following the least-square principle, we obtain the parameters $a_0, a_1, \ldots, a_n$, which minimize the total square error:

$$E(a_0, a_1, \ldots, a_n) = \sum_{i=1}^{m} \left[y_i - (a_n x^n + a_{n-1} x^{n-1} + \ldots + a_1 x + a_0) \right]^2 \qquad (7)$$

where $m \geq n$ is the number of samples. The optimal coefficients can be solved linearly.

The computed curve then defines a value $d_R(v)$ for row v, and function d_R altogether estimates for "on-road disparities". The profile is used to find the base points b_u of obstacles in column u, following [31].

3.3 Detection of Top-Points of Stixels

The height of obstacles (which "stand" on the ground manifold) is obtained by seeking an ideal segmentation between *foreground* and *background* disparities. The height-of-obstacle calculation begins with selecting membership votes. Next we estimate a cost image to approximate $t_1, t_2, ..., t_{N_{col}}$, the upper boundary of obstacles based on the method proposed by [20]. Briefly, the membership values rely on the selection of every disparity of each column from the disparity for its member to the foreground obstacle.

A membership value can be positive if it does not exceed the maximum distance of the expected obstacle disparity; otherwise it will be negative. This Boolean representation brings the challenge to identify a threshold value for the distance; if this value is too large then all disparities will be chosen from the foreground membership, and vice-versa. Therefore, the application of Boolean membership in continuous variation is a better alternative with an exponential function of the form

$$M(u, v) = 2^{1-\varepsilon(u,v)} - 1 \qquad (8)$$

where

$$\varepsilon(u, v) = \left[\frac{\hat{d}_u - \delta(u, v)}{\hat{d}_u - Z^{-1}\left(Z(\hat{d}_u) + \triangle Z\right)} \right]^2 \qquad (9)$$

where $\hat{d}_u = \delta(u, b_u)$ is the disparity of an obstacle's base point in column u, and Z is the disparity-to-depth conversion function; $\triangle Z$ as a defined soft constraint range in depth.

Fig. 4. Stixel world on KITTI data. *Left:* Membership votes. *Middle:* Cost image (data term). *Right:* Extracted stixels. (Color figure online)

A visualization of membership votes is illustrated in Fig. 4. Green represents true positives (belonging to an object), pale-blue shows free-space, and blue shows true negatives (background).

From the membership values, the *cost image* is computed as follows:

$$C(u, v) = \sum_{j=1}^{v-1} M(u, j) - \sum_{j=v}^{b_u} M(u, j) \qquad (10)$$

Note that row 1 is on top of the image, and row N_{row} at the bottom. Thus, both sums evaluate in each column u membership above b_u only. The function C essentially decides the cost of dividing $[1, b_u]$ into two intervals, namely $[1, t_u - 1]$ (i.e. the background) and $[t_u, b_u]$ (i.e. the foreground), in each column.

A result of the membership cost image, used for the height segmentation, is shown in Fig. 4, middle. The figure shows the height cost of foreground and background disparities. As can be seen, there are bright values which show a high likelihood for performing a foreground-background separation.

The obstacles' top-points $t_1, t_2, ...t_{N_{col}}$ are obtained from the computed cost image C following the approach proposed in [20], i.e. dynamic programming is used for approaching a minimum when all the pixels in $[t_u, b_u]$ are close to the target disparity while the remaining in $[1, t_u - 1]$ are far away from the target.

3.4   Stixel Extraction

By combining base-points $b_1, b_2, ...b_{N_{col}}$ found in Sect. 3.2 and top-points $t_1, t_2, ...t_{N_{col}}$ found in Sect. 3.3, we extract the stixels.

In this paper we adopt a column grouping technique proposed in [7,8]. Given $w \in \mathbb{N}^+$ as a predefined width of stixels, every w neighbouring columns are grouped across the whole image, resulting in $\lfloor \frac{N_{col}}{w} \rfloor$ non-overlapping stixels in one row. For the i-th stixel we have a set of w base-points $B_i = \{b_{u_i}, b_{u_i+1}, ..., b_{u_i+w-1}\}$ and a set of w top-points $T_i = \{t_{u_i}, t_{u_i+1}, ..., t_{u_i+w-1}\}$ where $u_i = (i-1)w + 1$.

The rectangle spanned from column $u = u_i$ to $u = u_i + w - 1$ and row $v = \min(T_i)$ to $v = \max(B_i)$ defines the *scope of a stixel*. Instead of using only base-points' disparities, we integrate all the disparities within the scope to yield a more robust estimate of the stixel's depth z_i, by means of a histogram-based regression technique proposed in [7] with w set to 5 pixels.

Figure 5 shows resulting stixels. The colours of the stixels encode the distance to the ego-vehicle. Red-scale colours represent objects farther away. Stixels of

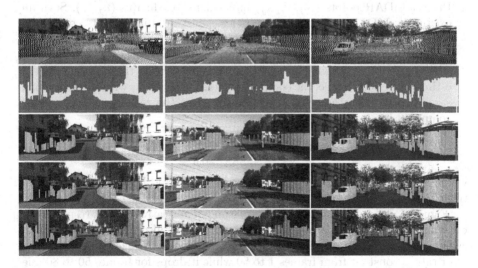

Fig. 5. Qualitative results using KITTI *residential* (first column), *road* (second column), and *city* (third column) data. *First row*: LiDAR projection. *Second row*: Stixel map. *Third row*: Stixels estimated using binocular stereo and ground plane. *Fourth row*: Stixels estimated using trinocular stereo and ground plane. *Fifth row*: Stixels estimated using TED-based disparities and polynomial ground manifold approximation.

"minor height" have been ignored. The figure illustrates that a stixel represents the height of the first "substantial" obstacle facing the ego-vehicle along a viewing direction.

4   Experimental Results

This section evaluates the performance of three methods for stixel detection: binocular-based occupancy grid (i.e. assuming a plane as ground manifold) stixels, trinocular-based occupancy grid stixels, and stixels using TED-based disparities on a polynomial manifold.

We selected 655 stereo images from KITTI's city, residential, and road datasets which include cars, pedestrians, trees, and traffic signals. Previous literature states challenges in evaluating stixels using KITTI data. The challenges are given by a lack of annotated road images, or a lack of stixel ground truth. It is also of limited relevance to evaluate the quality of the 3D reconstruction subjectively based on manually observed disparity images.

Since 3D laser scanners are accurate as reference sensors, we employ the Velodyne LiDAR data obtained by a 3D laser scanner which are publicly available [11].

We evaluate all stixels in every frame individually for understanding the efficiency of the proposed trinocular stixels in terms of distance errors. This comprises several processes:

1. Generate a stixel map which forms stixels above the ground manifold, as shown in Fig. 5 (third to fifth row) for "dominating" stixels.
2. Project LiDAR points (X_j, Y_j, Z_j) into image coordinates (u_j, v_j). Such an exemplary LiDAR point projection is also illustrated in Fig. 5 (second row). The projections are used to build a LiDAR-stixel correspondence function β_{ij}, where $\beta_{ij} = 1$ if LiDAR point j hits stixel i, otherwise $\beta_{ij} = 0$.
3. The degree of correspondence of these images verifies the accuracy of the estimated stixels. Hence, the comparison of LiDAR depths with corresponding stixel depths form the error measurement using the root-mean-square error computed by

$$\text{RMSE} = \sqrt{\frac{\sum_{i=1}^{N_{\text{stx}}} \sum_{j=1}^{N_{\text{pts}}} \beta_{ij}(z_i - Z_j)^2}{N_{\text{hit}}}} \tag{11}$$

where N_{stx} and N_{pts} represent the number of stixels and LiDAR points, respectively, and N_{hit} is the number of non-zero elements in β.

Figures 5 and 6 demonstrate qualitative and quantitative results, respectively. city and residential data differ from road data by also showing pedestrians, cyclists, and sometimes cyclists having a baby stroller. As shown in Fig. 6, the error rate is constant from frames 1 to 50 while it drops for frames 50 to 80 due to unexpected interference by a cyclist. A similar pattern is noted after frame 80 till frame 150. After that, the error rate fluctuates roughly at the first road junction (frame 155) shows a different pattern of traffic while it increases again at the end of the data sequence due to a round-about.

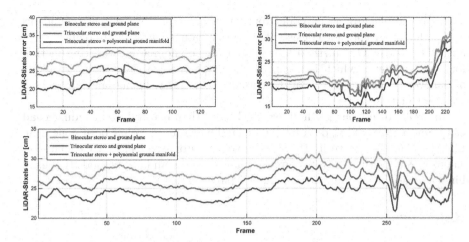

Fig. 6. The error rates represent the difference of distances between LiDAR data and stixels, shown for the three methods. *Top-left*: Error rate on `residential` data. *Top-right*: Error rate on `city` data. *Bottom*: Error rate on `road` data.

This occurs for all the three methods. Detection using binocular stereo on a plane shows the highest error rate and the highest false-positive rate too, due to degrading disparities. Detection using trinocular stereo on a plane performs better than the binocular stereo method. Our proposed method (TED-based trinocular and polynomial ground manifold) covers more valid disparities compared to others, and appears to be insensitive to slope changes. It outperforms others regarding a smaller rate of false alarms.

On the `residential` data, the error rate for all three methods was less compared to the `city` data. The used data show a car parked on the side of the road, houses, and road junctions. The accuracy of the stixels detected via the trinocular+plane method in frame 60 is very close to the one using trinocular-TED+polynomial. For the two standard methods (binocular or trinocular on plane), there are a few stixels that were not detected at the visible end of the road; see Fig. 5. They were successfully detected via the proposed trinocular-TED+polynomial. The `road` data show the effectiveness of the proposed method. It can perform better than the other two methods when dealing with open road situations. In summary, the experimental analysis illustrates an improved robustness of the proposed trinocular-TED+polynomial method across various data sets. Low texture, or changes in the slope of road surfaces are in particular cases where our method is more robust in detecting stixels.

5   Conclusions

This paper proposed a novel method for stixel construction. The stixels, built using TED-based disparities provided by trinocular vision, have been found to provide better accuracy over conventional binocular ones, especially when also using polynomial ground-manifold approximation. Our method uses a confidence map, which can vote for consistent disparity values within a trinocular stereo analysis process. Our method also includes a polynomial curve-fitting method for road geometry which is insensitive to slope changes. The main advantage of our work is to produce a low-cost architecture for reducing false-positives in stixel estimation.

In order to test our method, we used more than 600 frames including `road`, `city`, and `residential` data from KITTI. The verification has been done using LiDAR range data to verify the accuracy of the proposed method. We compared the proposed method with two ground-plane-based standard methods (i.e. using binocular or simply unified trinocular disparities).

References

1. Klette, R.: Vision-based driver assistance. In: Wiley Encyclopaedia Electrical Electronics Engineering, pp. 1–15. Wiley (2015)
2. Rezaei, M., Klette, R.: Computer Vision for Driver Assistance. CIV, vol. 45. Springer, Cham (2017)
3. Kaaniche, K., Demonceaux, C., Vasseur, P.: Analysis of low-altitude aerial sequences for road traffic diagnosis using graph partitioning and Markov hierarchical models. In: Proceeding International Multi-conference Systems Signals Devices, pp. 656–661 (2016)
4. Wu, J., Cui, Z., Sheng, V.S., Zhao, P., Su, D., Gong, S.: A comparative study of SIFT and its variants. Measur. Sci. Rev. **13**(3), 122–131 (2013)
5. Farabet, C., Couprie, C., Najman, L., Lecun, Y.: Learning hierarchical features for scene labeling. IEEE Trans. Pattern Anal. Mach. Intell. **35**(8), 1915–1929 (2013)
6. Anders, J., Mefenza, M., Bobda, C., Yonga, F., Aklah, Z., Gunn, K.: A hardware/software prototyping system for driving assistance investigations. J. Real-Time Image Process. **11**(3), 559–569 (2016)
7. Badino, H., Franke, U., Pfeiffer, D.: The stixel world - a compact medium level representation of the 3D-world. In: Denzler, J., Notni, G., Süße, H. (eds.) DAGM 2009. LNCS, vol. 5748, pp. 51–60. Springer, Heidelberg (2009). doi:10.1007/978-3-642-03798-6_6
8. Pfeiffer, D., Franke, U.: Efficient representation of traffic scenes by means of dynamic stixels. In: Processing Intelligent Vehicles Symposium, pp. 217–224 (2010)
9. Scharwächter, T., Enzweiler, M., Franke, U., Roth, S.: Stixmantics: a medium-level model for real-time semantic scene understanding. In: Fleet, D., Pajdla, T., Schiele, B., Tuytelaars, T. (eds.) ECCV 2014. LNCS, vol. 8693, pp. 533–548. Springer, Cham (2014). doi:10.1007/978-3-319-10602-1_35
10. Schneider, L., Cordts, M., Rehfeld, T., Pfeiffer, D., Enzweiler, M., Franke, U., Pollefeys, M., Roth, S.: Semantic stixels: depth is not enough. In: Proceeding Intelligent Vehicles Symposium, pp. 110–117 (2016)
11. Geiger, A., Lenz, P., Urtasun, R.: Are we ready for autonomous driving? The KITTI vision benchmark suite. In: Proceeding of Computer Vision Pattern Recognition, pp. 3354–3361 (2012)
12. Hirschmüller, H.: Stereo processing by semiglobal matching and mutual information. IEEE Trans. Pattern Anal. Mach. Intell. **30**, 328–341 (2008)
13. Shin, B.-S., Xu, Z., Klette, R.: Visual lane analysis and higher-order tasks: a concise review. Mach. Vis. Appl. **25**(6), 1519–1547 (2014)

14. Onoguchi, K., Takeda, N., Watanabe, M.: Obstacle location estimation using planar projection stereopsis method. Syst. Comput. Japan **32**(14), 67–76 (2001)
15. Saleem, N.H., Klette, R.: Accuracy of free-space detection for stereo versus monocular vision. In: Proceeding of Image Vision Computing New Zealand, pp. 48–53 (2016)
16. Seo, J., Oh, C., Sohn, K.: Segment-based free space estimation using plane normal vector in disparity space. In: Proceeding of Connected Vehicles Expo, pp. 144–149 (2015)
17. Sanberg, W.P., Dubbelman, G., de With, P.H.N.: Color-based free-space segmentation using online disparity-supervised learning. In: Proceeding Intelligent Transportation Systems, pp. 906–912 (2015)
18. Spangenberg, R., Langner, T., Adfeldt, S., Rojas, R.: Large scale semi-global matching on the CPU. In: Proceeding IEEE Intelligent Vehicles Symposium, pp. 195–201 (2014)
19. Klette, R.: Concise Computer Vision. Springer, London (2014). doi:10.1007/978-1-4471-6320-6
20. Pfeiffer, D.: The stixel world. Doctoral Thesis, Humboldt Universität Berlin (2011)
21. Levi, D., Garnett, N., Fetaya, E.: StixelNet: a deep convolutional network for obstacle detection and road segmentation. In: Proceeding British Machine Vision Conference, vol. 1, p. 12 (2015)
22. Benenson, R., Timofte, R., Van Gool, L.: Stixels estimation without depth map computation. In: Proceeding International Conference Computer Vision Workshops, pp. 2010–2017 (2011)
23. Benenson, R., Mathias, M., Timofte, R., Van Gool, L.: Fast stixels estimation for fast pedestrian detection. In: Proceeding of European Conference Computer Vision, pp. 11–20 (2012)
24. Scharwächter, T., Franke, U.: Low-level fusion of color, texture and depth for robust road scene understanding. In: Proceeding of Intelligent Vehicles Symposium, pp. 599–604 (2015)
25. Cordts, M., Schneider, L., Enzweiler, M., Franke, U., Roth, S.: Object-level priors for stixel generation. In: Proceeding of German Conference on Pattern Recognition, pp. 172–183 (2014)
26. Pfeiffer, D., Gehrig, S., Schneider, N.: Exploiting the power of stereo confidences. In: Proceeding of Conference Computer Vision, Pattern Recognition, pp. 297–304 (2013)
27. Brandao, M., Ferreira, R., Hashimoto, K., Takanishi, A., Santos-Victor, J.: On stereo confidence measures for global methods, evaluation, new model and integration into occupancy grids. In: Proceeding of Pattern Analysis Machine Intelligence, pp. 116–128 (2016)
28. Haeusler, R., Nair, R., Köndermann, D.: Ensemble learning for confidence measures in stereo vision. In: Proceeding of Conference Computer Vision, Pattern Recognition, pp. 305–312 (2013)
29. Chien, H.-J., Geng, H., Klette, R.: Improved visual odometry based on transitivity error in disparity space: a third-eye approach. In: Proceeding Image Vision Computing, New Zealand, pp. 72–77 (2014)
30. Iloie, A., Giosan, I., Nedevschi, S.: UV disparity based obstacle detection and pedestrian classification in urban traffic scenarios. In: Proceeding of Intelligent Computer Communication Processing, pp. 119–125 (2014)

31. Labayrade, R., Aubert, D., Tarel, J.: Real time obstacle detection in stereovision on non flat road geometry through v-disparity representation. In: Proceeding of IEEE Intelligent Vehicle Symposium, pp. 646–651 (2002)
32. Keller, C., Dang, T., Fritz, H., Joos, A., Rabe, C., Gavrila, D.: Active pedestrian safety by automatic braking and evasive steering. In: Proceeding of Intelligent Transportation Systems, pp. 1292–1304 (2011)

Motion and Tracking

Mixing Hough and Color Histogram Models for Accurate Real-Time Object Tracking

Antoine Tran[✉] and Antoine Manzanera

U2IS, ENSTA ParisTech, Université de Paris-Saclay,
828, Bd des Maréchaux, 91762 Palaiseau Cedex, France
{antoine.tran,antoine.manzanera}@ensta-paristech.fr

Abstract. This paper presents a new object tracking algorithm, which does not rely on offline supervised learning. We propose a very fast and accurate tracker, exclusively based on two complementary low-level features: gradient-based and color-based features. On the first hand, we compute a Generalized Hough Transform, indexed by gradient orientation. On the second hand, a RGB color histogram is used as a global rotation-invariant model. These two parts are processed independently, then merged to estimate the object position. Then, two confidence maps are generated and combined to estimate the object size. Experiments made on VOT2014 and VOT2015 datasets show that our tracker is competitive among all competitors (in accuracy and robustness, ranked in the top 10 and top 15 respectively), and is one of the few trackers running at more than 100 fps on a laptop machine, with one thread. Thanks to its low memory footprint, it can also run on embedded systems.

1 Introduction

Object tracking is a very popular task in computer vision. Basically, the goal is to accurately localize one defined object (the target) in a video. Among difficulties, we can cite object deformation, motion change, rotation, scaling or those linked to the context (illumination change, occlusion, camouflage, camera motion). Applications such as human-computer interaction or augmented reality require reactive algorithms, so the computational cost may be a critical issue.

For efficiency, our tracker is based on very light methods, and combines low-level shape and color features. It is a model-free tracker, meaning that the offline training is done only at the first frame of the sequence. The gradient orientation is computed and used as an index of a Generalized Hough Transform (GHT) [2]. A RGB histogram is used to represent the color aspect of the target object, and distinguish it from the background. These two parts are processed independently, and merged to finally estimate the object location. They are complementary: the original GHT is robust to illumination, but weak against scale and rotation, unlike the color histogram model.

By testing and evaluating our tracker on VOT2014 and VOT2015 [15,16], we show that this combination leads to high performance in terms of accuracy and robustness. Moreover, by associating low-level features and light algorithms,

© Springer International Publishing AG 2017
M. Felsberg et al. (Eds.): CAIP 2017, Part I, LNCS 10424, pp. 43–54, 2017.
DOI: 10.1007/978-3-319-64689-3_4

our tracker can run at more than 100 fps on a laptop machine, without explicit multithreading. It is one of the fastest among all competitors, while being ranked second among real-time trackers in terms of accuracy and robustness criteria from VOT2015. Its lightness also makes it suitable to embedded systems.

This paper is organized that way: after a short state-of-the-art of object tracking, we will explain our method. Finally, we will show some results obtained in the academic datasets.

2 Related Work

Given a sequence, object tracking consists in estimating the state of one target object at each frame. This state can be its center, its bounding box or its silhouette. Many works have addressed the tracking problem, and we refer to Yilmaz' survey for a coverage of the task [25].

Structurally, our tracker belongs to the class of trackers combining different methods [1,3,9,18,24]. Among recent works, STAPLE [3] is related to our technique: it combines correlation and color histogram model to provide an effective and fast tracker. As we are using the GHT in its original form, our template-part tracker is simpler and lighter than Bertinetto's, as it only requires gradient computation (instead of HOG features [5]). Duffner's PixelTrack [9] is also close to our tracker: it combines a GHT with a foreground/background color model into a very fast tracking algorithm (above 100 fps).

The VOT committee annually proposes a dataset to evaluate and rank trackers [14–16]. In this challenge, the most accurate trackers are based on Convolutional Neural Network [19,20] and correlation filters [3,7]. If we focus on the fastest algorithms of VOT15 challenge, Vojir's tracker [23] based on the Meanshift [4] proposes decent accuracy, with a speed far beyond real-time. Maresca [18] proposes a fast tracker based on estimation of object motion, obtained by the combination of several light trackers.

Our proposed tracker is based on very low-level features and lightweight operations: color histogram and GHT. Color histogram is popular in object tracking [4,21,23], since it is fast and robust to scale and in-plane rotation changes. The GHT is an extension of the Hough Transform [8], used to detect arbitrary shapes. It consists in considering some elements of this shape (pixels, patches) and, according to their local appearances, make them vote for potential positions of the shape center. The estimated center is then determined by the location that has accumulated the highest number of votes. The GHT is robust to illumination changes and to camouflage issue, compensating some weaknesses of the color-based model. In tracking context, the main issue of the GHT is the robustness to scale and rotation changes. However, several authors proposed Hough-based trackers [9–11,17] by modifying the original GHT, or using complementary methods. Amongst Hough-based trackers, only Hua [13] outperforms our algorithm, but is 100 times slower. However, his algorithm essentially relies on a HOG-based detector, which estimates several candidates locations, and the Hough Transform is only used to discard wrong candidates. PixelTrack [9] is at a same order of magnitude in speed, but much less accurate and robust.

In the last VOT challenges [14,15], few trackers belong to the category of real-time trackers (21 competitors among 132), with diverse performances in accuracy and robustness. In this category, our method is one of the fastest (more than 100 fps), but also one of the most accurate and robust (only beaten by [23]). Compared to slower tracker, ours is still competitive: ranked in the top 15 for both criteria in VOT2015 [15] and ranked in the top 10 in VOT2014 [16].

Our state representation is based on a bounding box. Given a frame t, the aim is to estimate the bounding box $B_t = (c_t, w_t, h_t)$ of an object $\mathcal{O}$, where each parameter is respectively: bounding box's center, width and height.

3   Our Contribution

First, we explain how to initialize our tracker. Second, we deal with state estimation. As position is estimated before scale, we will explain these two steps in different sections. Third, we explain the model updating process.

3.1   Initialization

Our tracker is initialized at the first frame $\mathbf{I}_0$, with a manually set bounding box B_0. Let $\mathbf{I}(B)$ be the restriction of an image $\mathbf{I}$ to any subset of pixels B.

First, the target RGB histogram H_0 of $\mathbf{I}_0(B_0)$, with $n_c \times n_c \times n_c$ bins, is generated ($n_c = 12$). H_t will denote the target RGB histogram at the frame t.

Second, for the geometrical model, let $\mathbf{M}_0$ and $\mathbf{\Phi}_0$ be the gradient magnitude and orientation of $\mathbf{I}_0(B_0)$. The goal is to initialize the *R-Table R* (which will be updated over the time), indexed by $n_o = 16$ orientations. It consists in considering all pixels $p \in \mathbf{I}_0(B_0)$, for which $\mathbf{M}_0(p) > \epsilon_M$ ($\epsilon_M = 70.0$) and whose quantified gradient orientation is θ_p, then to store in $R(\theta_p)$ the couple $(\vec{u} = \vec{pc_0}, \omega_{\vec{u}})$, which is the displacement from p to the bounding box center c_0, and $\omega_{\vec{u}} = 1.0$ a default weight value.

3.2   Estimation of Position

Given an image $\mathbf{I}_t$ from a sequence, our tracker first estimates the object center, then its scale. We will explain these two steps independently.

On the one hand, we perform a basic GHT. As during initialization, it requires to evaluate gradient image from $\mathbf{I}_t$, keep pixels p whose gradient magnitude is above ϵ_M, and quantify its orientation as θ_p. Then, for each couple $(\vec{u}, \omega_{\vec{u}}) \in R(\theta_p)$, p votes for all displacements $p + \vec{u}$ with the weight $\omega_{\vec{u}}$. More formally:

$$\mathbf{HT}_t(p) = \sum_q \sum_{(\vec{u},\omega) \in R(\theta_q)} \omega \cdot \delta(p, q + \vec{u}) \qquad (1)$$

with δ the Kronecker function. Finally, when all pixels have voted, the created map (the Hough Transform) $\mathbf{HT}_t$ emphasizes the most probable locations of the object center, with respect to the geometrical model. The GHT is performed

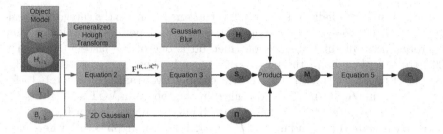

Fig. 1. Position tracking diagram.

under its simplest form, unlike [9] who indexes their R-Table using gradient and color features and [11,17] who use machine learning classifiers.

We then blur $\mathbf{HT}_t$ (3×3 discrete Gaussian filter), in order to add robustness to deformation. The GHT is intrinsically robust to illumination change (conservation of the orientation), but not to scaling (pixels votes spread away from the center) nor to rotation (pixels vote according to the wrong list in the R-Table).

On the other hand, to exploit the color model, we define foreground and background areas. We use the color model histogram H_{t-1}, and build a background histogram H_t^{bck}. These features have the advantage to compensate for the GHT weaknesses, and vice-versa. Given the last estimated bounding box B_{t-1}, let S_t be the background area, defined by all pixels inside the bounding box of center c_{t-1}, and dimension $(\alpha \cdot w_{t-1}, \alpha \cdot h_{t-1})$ ($\alpha = 2.0$), excluding B_{t-1}. From S_t, we build H_t^{bck}. Then, for every pixel $p \in (B_{t-1} \cup S_t)$, let q_t^p be its quantified color in $\mathbf{I}_t$. As proposed in [21], we define the foregroundness with respect to the object $\mathcal{O}$, knowing the object color histogram H_{t-1} and the background color model:

$$\mathbf{F}_{\mathcal{O}}^{(H_{t-1}, H_t^{bck})}(p) = \begin{cases} \frac{H_{t-1}(q_t^p)}{H_{t-1}(q_t^p) + H_t^{bck}(q_t^p)} & \text{if } p \in (B_{t-1} \cup S_t) \\ 0 & \text{otherwise} \end{cases} \tag{2}$$

$\mathbf{F}_{\mathcal{O}}^{(H_{t-1}, H_t^{bck})}(p)$ indicates how probably p belongs to the target. Unlike [21], we do not combine Eq. 2 with a distractor-aware model, as we aim to remain as simple as possible. Compared to PixelTrack [9], this method only needs a model histogram and no prior information about background. Then, given a bounding box B of size (w_{t-1}, h_{t-1}) inside $(B_{t-1} \cup S_t)$, let $\mathbf{S}_{c,t}(B)$ be the normalized score evaluating whether the target is inside B:

$$\mathbf{S}_{c,t}(B) = \frac{\sum_{p \in B} F_{\mathcal{O}}^{(H_{t-1}, H_t^{bck})}(p)}{w_{t-1} \cdot h_{t-1}} \tag{3}$$

This formulation is simpler than Possegger's [21], who proposed a method to discard distractors. We also use a prediction map, that indicates the likelihood to find c_t at x, given B_{t-1}:

$$\mathbf{\Pi}_t^{B_{t-1}}(x) = \exp\left(\frac{-(x - c_{t-1})^2}{2 \cdot \min(w_{t-1}, h_{t-1})^2}\right) \tag{4}$$

Finally, for all pixels x, we define $\mathbf{M}_t$ such as $\mathbf{M}_t(x) = \mathbf{HT}_t(x) \cdot \mathbf{S}_{c,t}(B_x) \cdot \mathbf{\Pi}_t^{B_{t-1}}(x)$ with B_x the rectangle centered in x of size (w_{t-1}, h_{t-1}). From $\mathbf{M}_t$, we finally estimate the object position c_t as follows:

$$c_t = \begin{cases} \underset{x}{\operatorname{argmax}}(\mathbf{M}_t(x)) & \text{if } \max_x(\mathbf{M}_t(x)) \neq 0 \\ c_{t-1} + \overrightarrow{c_{t-2}c_{t-1}} & \text{otherwise} \end{cases} \tag{5}$$

The second case of Eq. 5 assumes that, when the support of $\mathbf{HT}_t$ and $\mathbf{S}_{c,t}$ are disjoint, the target is translating with a vector $\overrightarrow{c_{t-2}c_{t-1}}$. We choose a pixel-wise multiplication to merge our two trackers, unlike STAPLE [3] who used a weighted average. In that way, we do not have to deal with the difference of magnitude of $\mathbf{HT}_t$ and $\mathbf{S}_{c,t}$, and to adjust a weight. We also differ from PixelTrack [9], where the final position is obtained by linear combination of the centers estimated by the GHT on one hand and the color segmentation map on the other hand. Figure 1 describes all steps for our position estimation.

3.3 Estimation of Scale

The second step consists in estimating object scale.

On the one hand, from the GHT, let us define the backprojection map $\mathbf{BP}_t$, for all pixels p fulfilling conditions to be stored in the R-Table:

$$\mathbf{BP}_t(p) = \frac{\sum\limits_{(\overrightarrow{u},\omega) \in R(\theta_p)} \mathbf{M}_t(p + \overrightarrow{u})}{|R(\theta_p)|} \tag{6}$$

with $|R(\theta_p)|$ the cardinality of $R(\theta_p)$. The approach is similar to Duffner's one [9]. However, Duffner only backprojects the peak of the GHT, while we consider the sum of the voted positions for all pixels. In both cases, the made assumption is that the higher the backprojection is, the more likely it belongs to the target.

On the other hand, we consider $\mathbf{F}_{\mathcal{O}}^{(H_{t-1}, H_t{}^{bck})}$, defined Eq. 2, as a color confidence map. Then, let $\mathbf{BF}_t$ be the final confidence map:

$$\mathbf{BF}_t = 0.5 \cdot (\mathbf{BP}_t + \mathbf{F}_{\mathcal{O}}^{(H_{t-1}, H_t{}^{bck})}) \tag{7}$$

Figure 2 illustrates these maps. Then, inspired by Possegger [21], we consider the set of object pixels $OP_t = \{p | \mathbf{BF}_t(p) > 0.5\} \cup R_t$, with R_t a safe foreground area defined as the rectangle centered on c_t, of size $(\beta \cdot w_{t-1}, \beta \cdot h_{t-1})$ ($\beta = 0.20$). From OP_t, we only retain the connected component that contains c_t, to discard isolated pixels that could generate scale overestimation. Finally, we estimate a potential bounding box by computing the bounding box $\bar{B}_t$ of this connected component. Then, we reject bounding box sizes whose relative area variation with respect to B_{t-1} is above 5%. Otherwise, we update object's size using the same aspect ratio, as follows:

$$X_t = \lambda_t \cdot X_{t-1} \tag{8}$$

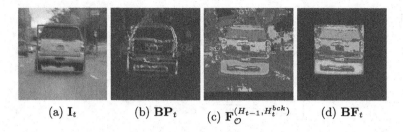

(a) $\mathbf{I}_t$ (b) $\mathbf{BP}_t$ (c) $\mathbf{F}_{\mathcal{O}}^{(H_{t-1}, H_t^{bck})}$ (d) $\mathbf{BF}_t$

Fig. 2. Cropped frame from *car1* from VOT2015, with the ground truth in blue. Back-projection map $\mathbf{BP}_t$ and pixel color likelihood map $\mathbf{F}_{\mathcal{O}}^{(H_{t-1}, H_t^{bck})}$ are complementary: while the first one indicates which border pixels are more likely to belong to the target, the second one gives high results for pixels of car's back. (Color figure online)

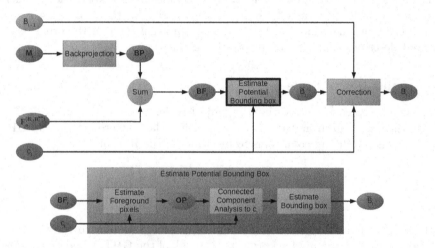

Fig. 3. Scale estimation diagram.

with $X \in \{w, h\}$ and $\lambda_t = \min(1.05, \max(0.95, \frac{\mathcal{A}(\bar{B}_t)}{\mathcal{A}(B_{t-1})}))$ ($\mathcal{A}$ being the area operator). Figure 3 summarizes scale estimation operations. Finally, to prepare the updating process, let $\mathbf{SG}_t$ be the shape and color-based confidence map such that $\mathbf{SG}_t(p) = \mathbf{BF}_t(p)$ if $p \in B_t$, and 0 otherwise.

3.4 Model Update

The last step consists in updating the model, knowing the estimated bounding box B_t. To update the color model, let H_t^o be the color histogram of $\mathbf{I}_t(B_t)$ and $\mu_c = 0.05$ the color updating rate:

$$H_t = (1 - \mu_c) \cdot H_{t-1} + \mu_c \cdot H_t^o. \tag{9}$$

To update the R-Table R, we start by reducing all displacement weights:

$$\forall \theta, \forall (\overrightarrow{u}, \omega_{\overrightarrow{u}}) \in R(\theta), \omega_{\overrightarrow{u}} \leftarrow (1 - \mu_g) \cdot \omega_{\overrightarrow{u}} \tag{10}$$

Then, considering the confidence map $\mathbf{SG}_t$, and the object center c_t, for all pixels $p \in B_t$ with gradient orientation θ_p, we consider the displacement $\vec{v} = \vec{pc_t}$, with a weight equal to $\mu_g \cdot \mathbf{SG}_t(p)$. Then, we consider two cases:

- if $\vec{v}$ is in $R(\theta_p)$, we increment its weight by $\mu_g \cdot \mathbf{SG}_t(p)$ ($\mu_g = 0.05$), in order to reinforce the most relevant elements of the R-Table
- otherwise we add an entry into $R(\theta_p)$ with the weight $\mu_g \cdot \mathbf{SG}_t(p)$

Finally, for each index of the R-Table, we keep only the $N_R = 200$ displacements with the strongest weights, to limit computational and memory cost.

4 Experiments

Before dealing with experiments on academic datasets, we will detail our implementation setup.

4.1 Implementations Details

Our algorithm is developed using C++ and the OpenCV 2.4.9 library, and tested on a laptop at 2.4 GHz, without explicit multithreading. In terms of implementation, iterative pixel access is done using image pointers, and image histogram computation is done using Look Up Table. At frame t, our tracker only processes

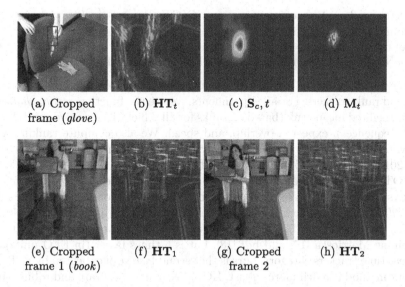

(a) Cropped frame (*glove*)   (b) $\mathbf{HT}_t$ (c) $\mathbf{S}_{c,t}$ (d) $\mathbf{M}_t$

(e) Cropped frame 1 (*book*)   (f) $\mathbf{HT}_1$ (g) Cropped frame 2   (h) $\mathbf{HT}_2$

Fig. 4. The first line illustrates failures due to the transparency of the glove (ground truth in blue, tracker output in red), making its color similar to the chair. The GHT still performs correctly (sharp peak on the second image). On the second line the GHT fails due to the rotation of the book: the peak disappears from $\mathbf{HT}_1$ to $\mathbf{HT}_2$. (Color figure online)

area centered in c_{t-1}, and of size two times the last bounding box area. To evaluate the map $\mathbf{S}_{c,t}$, we use integral images. Otherwise, no major algorithmic optimization has been done.

In terms of memory footprint, target's informations, composed of the color histogram ($n_c{}^3 = 12^3$ floating point numbers), the R-Table ($n_0 \cdot N_R \cdot 2 \cdot 4$ integers for displacements and $n_0 \cdot N_R \cdot 2$ floats for weights) and the two last states (4 integers for coordinates and scales), resulting, with our set of parameters, in a memory footprint of about 45 ko. This quantity is independent of target' size. It is however negligible compared to the number of temporary images: 2 8-bit images (gradient orientation and magnitude maps), 1 RGB image (the sub-image in which we are tracking the target) and 5 32-bit images (float) ($\mathbf{HT}_t$, $\mathbf{BP}_t$, $\mathbf{S}_{c,t}$, $\mathbf{F}_{\mathcal{O}}^{(H_{t-1}, H_t^{bck})}$ and $\mathbf{BF}_t$), which depends on object's size (and its associated search window): for a 100×100 object's size, the footprint will reach 2 Mo. Considering the speed obtained experimentally, we are convinced that our method is suitable embedded systems. All parameters have been tuned for the best trade-off between performance and speed on VOT2015 [15]. For experiments case, we will denote as CHT the position tracker only, and CHTs our complete tracker.

4.2   Results on VOT 2014 and VOT 2015

Each year, the VOT committee proposes a dataset to test and evaluate trackers. Each frame from each sequence is labeled according to its difficulty (occlusion, camera motion, size, motion or illumination change). Evaluation criteria are:

- Accuracy: based on overlap measure $O(GT_t, B_t) = \frac{GT_t \cap B_t}{GT_t \cup B_t}$, with GT_t the ground truth at the frame t
- Robustness: given by the number of failures (frames where $O(GT_t, B_t) = 0$)
- Speed: based on a normalized speed (EFO units, see [15])

The VOT committee provides results of all competitors, and a toolkit to evaluate and rank trackers. For all experiments, we use the function *report_challenge* to get weighted mean rank (based on ranks for all difficulties), pooled rank (based on all sequences), expected overlap, and speed. We also compute ranking with the whole set of results, but only display those of relevant trackers. Results for VOT2014 and VOT2015 are summarized Table 1.

VOT2014 [16] is a dataset composed of 25 sequences, and results for 40 trackers are available. We choose to show our results compared to DSST [6] (VOT2014 winner), Hough-based trackers [9,17,18] and real-time trackers [9,12, 18,22]. Amongst Hough-based trackers, our method is as well ranked as Matflow, combining Matrioska [17] and bdf [18], but is 6 times faster (in EFO units). In the real-time trackers category, ours is the second fastest one, beaten by FoT [22] but our method is much more robust. KCF [12] is more accurate and robust than our method, but is slower. Globally, our tracker is well ranked (top 10 among 40 competitors) but is one of the tracker proposing the best trade-off between effectiveness and speed. Surprisingly, our position-only tracker (CHT) and our complete one have similar performance in terms of accuracy and robustness, but the complete one is faster (probably due to object size reduction).

Table 1. Results in VOT2014 and VOT2015 for different trackers. The smaller the rank, the better the tracker. Real-time trackers appear in bold.

VOT2014							
Tracker	Weighted mean rank		Pooled rank		Expected overlap	Speed	
	Accuracy	Robustness	Accuracy	Robustness		(EFO)	(fps)
CHTs	8.83	4.33	6	9	0.2960	129.77	159.21
CHT	8.42	4.08	6	9	0.2916	109.75	134.65
DSST [6]	1.83	4.33	1	3	0.3693	5.80	13.07
Matrioska [17]	9.83	12	6	9	0.2671	10.20	21.88
bdf [18]	10.67	7.50	10	9	0.3097	46.82	100.45
Matflow [17,18]	8.67	3.17	6	4	0.3120	19.08	40.94
FoT [22]	7	18.17	6	22	0.2859	114.64	306.52
PTp [9]	25.33	11.17	30	9	0.2519	49.89	127.87
KCF [12]	2	4.67	1	5	0.3641	24.23	63.42
VOT2015							
CHTs	12.67	13.67	13	17	0.2606	103.89	111.22
CHT	13.83	15.67	13	20	0.2615	101.91	109.10
Staple [3]	1	4.33	1	5	0.345		
ASMS [23]	7.50	11	2	13	0.2353	115.09	142.26
bdf [18]	29.33	32	27	43	0.2054	200.24	78.43
FoT [22]	19.50	42.50	16	53	0.1934	143.62	177.53
DSST [6]	4.0	23.67	1	38	0.2707	3.29	4.47
DAT [21]	13.73	17.33	6	20	0.2428	9.82	14.87
HT [11]	20	28.50	13	43	0.2045	0.91	0.56
Matflow [17,18]	22.17	27.33	23	43	0.2098	81.34	31.86
MDNET [20]	1	1.33	1	1	0.3789	0.87	0.97
sPST [13]	1.67	4.50	1	5	0.3134	1.03	1.16

VOT2015 challenge is composed of 60 sequences and results for 62 competitors are available. Compared to MDNET [20], VOT2015 winner based on CNNs, and STAPLE [3] (results from author's website, without speed results) our tracker is less effective, but faster (being lighter, and knowing that the author mentioned 80 fps on a 4.0 GHz CPU, we expect our method to be faster than STAPLE). Compared to STAPLE, our shape-based tracker relies on the GHT, with gradient computation only, while STAPLE requires a more complex algorithm (correlation-based tracker using HOG features). Amongst real-time trackers [18,22,23], our tracker is one of the few above 100 EFO, and is only beaten by Vojir's extension of Meanshift [23] in rankings and speed, but with slightly higher expected overlap. Otherwise, we are better ranked than other real-time trackers. We perform better than other Hough-based trackers [11,17], including Matflow, which was on par with our tracker in VOT2014, but excluding sPST [13], which is 100 times slower and relies on an object detector. Compared to Possegger [21],

Fig. 5. Results from VOT2014 and 2015. White bounding boxes are ground truth, red ones are obtained with our tracker. The two first images are cropped from *sphere* and *torus* sequences from VOT2014 (PixelTrack in blue, and MatFlow in green). The two last are cropped from *birds2* and *motocross* sequences from VOT2015 (DAT in blue, and STAPLE in green). (Color figure online)

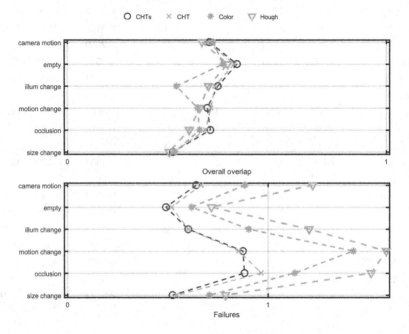

Fig. 6. Expected overlap and number of failures for our complete tracker, and for partial versions (position tracking only, Hough only and Color only). (Color figure online)

from which our color part tracker is inspired, we demonstrate the usefulness of the Hough part, since, in weighted mean ranking, our method is slightly more robust. Some results are shown Figs. 4 and 5. We also used VOT2015 to see performances of each part of the tracker (Hough and color ones). In both cases, the loss of performance is dramatic, as we can see on Fig. 6, where we show the results of different versions in terms of expected overlap (the higher, the better) and failures (the lower, the better) for several difficulties, both obtained on the set of frames concerned by the difficulties.

5   Conclusion

In this paper, we proposed a tracker working without offline training and tracking arbitrary objects. Unlike most state-of-the-art trackers, our method is based on a very low level representation. First, our geometrical model is only based on gradient, through a GHT. Then, an object color histogram is used to generate a map indicating the likeliness of the pixel to belong to the object. These two operations, done independently, are combined to estimate object center. Second, from the two features, we generate two confidence maps, and merge them in order to estimate the object size. The final confidence map is built from the fusion of these two maps and used to update the object geometrical and color models. Our whole tracker relies on computationally efficient operations, and performs tracking task beyond real-time (about 100 fps) with a low memory footprint. Experiments were done on recent academic dataset [15,16], for which our tracker is ranked in the first third for accuracy and robustness. It is also one of the fastest from VOT2014 and VOT2015 challenges. Thanks to its speed and low memory footprint, our algorithm can be implemented on embedded systems, combined with other trackers to improve accuracy and robustness or with object detection or background subtraction to automatically initialize tracked regions.

Acknowledgments. We gratefully acknowledge financial support from the French Government Defense procurement and technology agency (DGA/MRIS).

References

1. Babenko, B., Yang, M.H., Belongie, S.: Robust object tracking with online multiple instance learning. IEEE Trans. Pattern Anal. Mach. Intell. **33**(8), 1619–1632 (2011)
2. Ballard, D.H.: Generalizing the Hough transform to detect arbitrary shapes. Pattern Recogn. **13**(2), 111–122 (1981)
3. Bertinetto, L., Valmadre, J., Golodetz, S., Miksik, O., Torr, P.: Staple: complementary learners for real-time tracking. arXiv preprint arXiv:1512.01355 (2015)
4. Comaniciu, D., Ramesh, V., Meer, P.: Real-time tracking of non-rigid objects using mean shift. In: IEEE Conference on Computer Vision and Pattern Recognition, Proceedings, vol. 2, pp. 142–149. IEEE (2000)
5. Dalal, N., Triggs, B.: Histograms of oriented gradients for human detection. In: 2005 IEEE Computer Society Conference on Computer Vision and Pattern Recognition (CVPR 2005), vol. 1, pp. 886–893. IEEE (2005)
6. Danelljan, M., Häger, G., Khan, F., Felsberg, M.: Accurate scale estimation for robust visual tracking. In: British Machine Vision Conference, Nottingham, 1–5 September 2014. BMVA Press (2014)
7. Danelljan, M., Robinson, A., Shahbaz Khan, F., Felsberg, M.: Beyond correlation filters: learning continuous convolution operators for visual tracking. In: Leibe, B., Matas, J., Sebe, N., Welling, M. (eds.) ECCV 2016. LNCS, vol. 9909, pp. 472–488. Springer, Cham (2016). doi:10.1007/978-3-319-46454-1_29
8. Duda, R.O., Hart, P.E.: Use of the Hough transformation to detect lines and curves in pictures. Commun. ACM **15**(1), 11–15 (1972)

9. Duffner, S., Garcia, C.: PixelTrack: a fast adaptive algorithm for tracking non-rigid objects. In: 2013 IEEE International Conference on Computer Vision (ICCV), pp. 2480–2487. IEEE (2013)
10. Gall, J., Yao, A., Razavi, N., Van Gool, L., Lempitsky, V.: Hough forests for object detection, tracking, and action recognition. IEEE Trans. Pattern Anal. Mach. Intell. **33**(11), 2188–2202 (2011)
11. Godec, M., Roth, P.M., Bischof, H.: Hough-based tracking of non-rigid objects. Comput. Vis. Image Underst. **117**(10), 1245–1256 (2013)
12. Henriques, J.F., Caseiro, R., Martins, P., Batista, J.: High-speed tracking with kernelized correlation filters. IEEE Trans. Pattern Anal. Mach. Intell. **37**(3), 583–596 (2015)
13. Hua, Y., Alahari, K., Schmid, C.: Online object tracking with proposal selection. In: Proceedings of the IEEE International Conference on Computer Vision, pp. 3092–3100 (2015)
14. Kristan, M., et al.: The visual object tracking VOT2016 challenge results. In: Hua, G., Jégou, H. (eds.) ECCV 2016. LNCS, vol. 9914, pp. 777–823. Springer, Cham (2016). doi:10.1007/978-3-319-48881-3_54
15. Kristan, M., Matas, J., Leonardis, A., et al.: The visual object tracking VOT2015 challenge results. In: Proceedings of the IEEE International Conference on Computer Vision Workshops, pp. 1–23 (2015)
16. Kristan, M., Pflugfelder, R., Leonardis, A., Matas, J., et al.: The visual object tracking vot2014 challenge results (2014). http://www.votchallenge.net/vot2014/program.html
17. Maresca, M.E., Petrosino, A.: MATRIOSKA: a multi-level approach to fast tracking by learning. In: Petrosino, A. (ed.) ICIAP 2013. LNCS, vol. 8157, pp. 419–428. Springer, Heidelberg (2013). doi:10.1007/978-3-642-41184-7_43
18. Maresca, M.E., Petrosino, A.: Clustering local motion estimates for robust and efficient object tracking. In: Agapito, L., Bronstein, M.M., Rother, C. (eds.) ECCV 2014. LNCS, vol. 8926, pp. 244–253. Springer, Cham (2015). doi:10.1007/978-3-319-16181-5_17
19. Nam, H., Baek, M., Han, B.: Modeling and propagating CNNs in a tree structure for visual tracking. arXiv preprint arXiv:1608.07242 (2016)
20. Nam, H., Han, B.: Learning multi-domain convolutional neural networks for visual tracking. In: Proceedings of the IEEE Conference on Computer Vision and Pattern Recognition, pp. 4293–4302 (2016)
21. Possegger, H., Mauthner, T., Bischof, H.: In defense of color-based model-free tracking. In: Proceedings of the IEEE Conference on Computer Vision and Pattern Recognition, pp. 2113–2120 (2015)
22. Vojíř, T., Matas, J.: The enhanced flock of trackers. In: Cipolla, R., Battiato, S., Farinella, G.M. (eds.) Registration and Recognition in Images and Videos. SCI, vol. 532, pp. 113–136. Springer, Heidelberg (2014). doi:10.1007/978-3-642-44907-9_6
23. Vojir, T., Noskova, J., Matas, J.: Robust scale-adaptive mean-shift for tracking. In: Kämäräinen, J.-K., Koskela, M. (eds.) SCIA 2013. LNCS, vol. 7944, pp. 652–663. Springer, Heidelberg (2013). doi:10.1007/978-3-642-38886-6_61
24. Yang, F., Lu, H., Yang, M.H.: Robust visual tracking via multiple kernel boosting with affinity constraints. IEEE Trans. Circuits Syst. Video Technol. **24**(2), 242–254 (2014)
25. Yilmaz, A., Javed, O., Shah, M.: Object tracking: a survey. ACM Comput. Surv. (CSUR) **38**(4), 13 (2006)

DCCO: Towards Deformable Continuous Convolution Operators for Visual Tracking

Joakim Johnander[✉], Martin Danelljan,
Fahad Shahbaz Khan, and Michael Felsberg

Computer Vision Laboratory, Department of Electrical Engineering,
Linköping University, Linköping, Sweden
joakimjohnander@gmail.com

Abstract. Discriminative Correlation Filter (DCF) based methods have shown competitive performance on tracking benchmarks in recent years. Generally, DCF based trackers learn a rigid appearance model of the target. However, this reliance on a single rigid appearance model is insufficient in situations where the target undergoes non-rigid transformations. In this paper, we propose a unified formulation for learning a deformable convolution filter. In our framework, the deformable filter is represented as a linear combination of sub-filters. Both the sub-filter coefficients and their relative locations are inferred jointly in our formulation. Experiments are performed on three challenging tracking benchmarks: OTB-2015, TempleColor and VOT2016. Our approach improves the baseline method, leading to performance comparable to state-of-the-art.

Keyword: Visual tracking

1 Introduction

Generic visual object tracking is the computer vision problem of estimating the trajectory of a target throughout an image sequence, given only the initial target location. Visual tracking is useful in numerous applications, including autonomous driving, smart surveillance systems and intelligent robotics. The problem is challenging due to large variations in appearance of the target and background, as well as challenging situations involving motion blur, target deformation, in- and out-of-plane rotations, and fast motion.

To tackle the problem of visual tracking, several paradigms exist in literature [13]. Among different paradigms, approaches based on the Discriminative Correlation Filters (DCF) based framework have achieved superior results, evident from recent the Visual Object Tracking (VOT) challenge results [13,14]. This improvement in performance, both in terms of precision and robustness, is largely attributed to the use of powerful multi-dimensional features such as HOG, Colornames, and deep features [5,10,20], as well as sophisticated learning models [8,9].

Despite the improvement in tracking performance, the aforementioned state-of-the-art DCF based approaches employ a single rigid model of the target.

© Springer International Publishing AG 2017
M. Felsberg et al. (Eds.): CAIP 2017, Part I, LNCS 10424, pp. 55–67, 2017.
DOI: 10.1007/978-3-319-64689-3_5

However, this reliance on a single rigid model is insufficient in situations involving rotations and deformable targets. In such complex situations, the rigid filters fail to capture information of the target parts that move relative to each other. This desired information can be retained by integrating deformability in the DCF filters. Several recent works aim at introducing part-based information into the DCF framework [16,18,19]. These approaches introduce an explicit component to integrate the part-based information in the learning. Different to these approaches, we investigate a deformable DCF model, which can be learned in unified fashion (Fig. 1).

In many real-world situations, such as a running human or a rotating box, different regions of the target deform relative to each other. Ideally, such information should be integrated in the learning formulation by allowing the regions of the appearance model to deform accordingly. This flexibility in the tracking model reduces the need of highly invariant features, thereby increasing the discriminative power of the model. However, increasing the flexibility and complexity of the model introduces the risk of over-fitting and complex inference mechanisms, which degrades the robustness of the tracker. In this paper, we therefore advocate a unified formulation, where the deformable filter is learned by optimizing a single joint objective function. Additionally, this unified strategy enables the careful incorporation of regularization models to tackle the risk of over-fitting.

Fig. 1. Example tracking results of our deformable correlation filter approach on three challenging sequences. The circles mark sub-filter locations and the green box is the predicted target location. The red boxes (in the middle and lower rows) show the baseline predictions. The sub-filter locations deform according to the appearance changes of the target in the presence of deformations. (Color figure online)

Contribution. We propose a unified framework for learning a deformable convolution filter in a discriminative fashion. The deformable filter is represented as a linear combination of sub-filters. The deformable filter is learned by jointly optimizing the sub-filter coefficients *and* their relative locations. To avoid overfitting, we propose to regularize the sub-filter locations with an affine deformation model. We further derive an efficient online optimization procedure to infer the parameters of the model. Experiments on three challenging tracking benchmarks suggest that our method improves the performance in challenging situations.

2   Related Work

In recent years, Discriminative Correlation Filters (DCF) based tracking methods have shown competitive performance in terms of accuracy and robustness on tracking benchmarks [13,22]. In particular, the success of DCF based methods is evident from the outcome of the Visual Object Tracking (VOT) 2014 and 2016 challenges [13] where the top-rank trackers employ variants of the DCF framework. In DCF framework, a correlation filter is learned from a set of training samples to discriminate between the target and background appearance. The training of the filter is performed in a sliding-window manner by exploiting the properties of circular correlation. The original DCF based tracking approach by Bolme et al. [3] was restricted to a single feature channel and was later extended to multi-channel feature maps [10–12]. Most recent advancement in DCF based tracking performance is attributed to including scale estimation [6,15], deep features [7,20], spatial regularization [8], and continuous convolution filters [9].

Several recent works have shown that integrating the part-based information improve the tracking performance. The work of [18] introduces a part-based approach where each part utilizes the kernalized correlation filter (KCF) tracker and argues that partial occlusions can effectively be handled by adaptive weighting of the parts. The work of [16] tracks several patches, each with a KCF, by fusing the information using a particle filter to estimate position, width and height. Lukezic et al. [19] introduces a sophisticated model with several parts held together by a spring-like system by minimizing an energy function based on the part-filter responses.

Our approach: Different to aforementioned approaches, we propose a theoretical framework by designing a single deformable correlation filter. In our approach, the coefficients and locations of all sub-filters are learned jointly in a *unified* framework. Additionally, we integrate our deformable correlation filter in a recently introduced state-of-the-art DCF tracking framework [9].

3   Continuous Convolution Operators for Tracking

In this work, we propose a deformable correlation tracking formulation. As a starting point, we use the recent Continuous Convolution Operator Tracker (C-COT) formulation [9] due to two main advantages compared to current template based correlation filter trackers. Firstly, the continuous reformulation of the learning problem benefits from a natural integration of multi-resolution deep features and continuous-domain score map predictions. Secondly, it provides an efficient optimization framework based on the Conjugate Gradient method. For efficiency, we also employ components of its descendant tracker *ECO* [4].

For a given target object in a video, the C-COT discriminatively learns a convolution filter f that acts as an instance-specific object detector. Different from previous approaches, the filter f is viewed as a continuous function represented by its Fourier series coefficients. The detection scores are computed by first extracting a D-dimensional feature map x from the local image region of interest. Typically, the sample x consists of HOG or multi-resolution deep convolutional features. We let $x_d[n_1, n_2]$ denote the value of the d-th feature channel at the spatial location (n_1, n_2) in the feature map. The continuous scores in the corresponding image region are determined by the convolution operation $S_f\{x\} = \sum_{d=1}^{D} f_d * J^d\{x_d\}$, where $J^d\{x_d\}$ is an interpolation operator mapping the samples from the discrete to the continuous domain.

The filter f is trained in a supervised fashion, given a set of sample feature maps $\{x^1, x^2, \ldots, x^C\}$ and corresponding label score maps $\{y^1, y^2, \ldots, y^C\}$, by minimizing the objective,

$$\epsilon(f) = \sum_{c=1}^{C} \alpha^c \|S_f\{x^c\} - y^c\|^2 + \sum_{d=1}^{D} \|w^d \cdot f_d\|^2. \tag{1}$$

The first term penalizes classification errors of each sample using the squared L^2-norm. The sample c is weighted by the positive weight factor α^c, which is typically set using a learning rate parameter. The second term deploys a continuous spatial regularization function w^d, that penalizes high magnitude filter coefficients to alleviate the periodic boundary effects. Element-wise multiplication is denoted as $\cdot$. The label score function y^c is generally set to a Gaussian function with a narrow peak at the target center. Note that a sample feature map x^c contains both target appearance and the surrounding background. The filter is hence trained to predict high activation scores at the target center and low scores at the neighboring background. In practice, training and detection is performed directly in the Fourier domain, utilizing the FFT algorithm and the convolution properties of the Fourier series.

As related methods, the C-COT method works in two main steps. (i) When a new sample is received, the target position and scale are estimated, i.e. $S_f\{x\}$ is calculated using the estimated filter f for different scales using a scale pyramid. The new target state is then estimated as the position and scale that maximizes the detection score. (ii) To update the model, a sample (x^c, y^c) is first added to the training set, where x^c is extracted in the estimated target scale. The filter is then refined by minimizing the objective (1). This is done by using conjugate gradient to solve the arising normal equations. We refer to [9] for further details. To enhance the efficiency of the tracker, we further deploy the factorized convolution approach and update strategy recently proposed in [4].

4 Method

Here, we introduce a deformable correlation filter tracking model. A classic DCF contains an assumption that the target is rigid and will not rotate. The filter can handle violations to this assumption if a significant part of the target still fulfills it, or by using features with sufficient invariance. Examples of such model violations are sequences showing humans running or a change of perspective. By dividing the filter into sub-filters which can move relative to each other, they can fit more accurately onto a smaller part of the target. A standard DCF may choose to discard or weigh down information about a moving part whereas our approach allows one sub-filter to focus on this information explicitly, and move with that part. By writing the filter as a linear combination of sub-filters we can optimize a joint loss over all the sub-filter coefficients and the sub-filter positions jointly.

4.1 Deformable Correlation Filter

We construct a deformable convolution filter as a linear combination of trainable sub-filters. The filter becomes deformable by allowing the relative locations of the filters to change along to the target transformations. Formally, we denote the sub-filter with f^m and let $p^{c,m} = (p_1^{c,m}, p_2^{c,m})$ be its relative location in the frame c. The filter f at frame c is obtained as a linear combination of the shifted sub-filters,

$$f(t_1, t_2) = \sum_{m=1}^{M} f^m(t_1 - p_1^{c,m}, t_2 - p_2^{c,m}).$$ (2)

We jointly learn both the sub-filter coefficients f^m and their locations $p^{c,m}$ by minimizing a joint loss,

$$\epsilon(f, p) = \epsilon_1(f, p) + \epsilon_2(f) + \epsilon_3(p),$$ (3)

where each term is described below.

Classification Error. The loss for the discrepancy between the desired response and the filter response for sample x^c is

$$\epsilon_1(f, p) = \sum_{c=1}^{C} \alpha^c \|S_f\{x^c\} - y^c\|^2,$$ (4)

where α^c is the weight for sample c. From the translation invariance of the convolution operation and the definition (2), the classification scores can be computed as,

$$S_f\{x^c\}(t_1, t_2) = \sum_{m=1}^{M} S_{f^m}\{x^c\}(t_1 - p_1^{c,m}, t_2 - p_2^{c,m}).$$ (5)

The score operator $S_{f^m}\{x^c\}$ is defined as described in Sect. 3.

Spatial Regularization. A spatial regularization of the filters enforces low filter coefficients close to the edges,

$$\epsilon_2(f) = \sum_{m=1}^{M} \sum_{d=1}^{D} \|w^{m,d} \cdot f_d^m\|^2, \tag{6}$$

where $w^{m,d}$ is the continuous spatial regularization function for filter m. We assume different spatial regularization functions for the different sub-filters as it may be desireable for the sub-filters to track regions of different size. In our experiments, by using two different spatial regularizations where one is much tighter, we let one sub-filter track the whole target while the others track smaller patches. Please note that $\epsilon_2(f)$ does not depend on the sub-filter positions.

Regularization of Sub-filter Positions. To regularize the sub-filter positions, we add a deformable model that incorporates prior information of typical target deformations. In this work, we use a simple yet effective model, namely that the current sub-filter positions are related to their initial positions by a linear mapping. The resulting regularization term is thus given by,

$$\epsilon_3(p) = \lambda_p \sum_{m=1}^{M} \|p^{c,m} - Rp^{1,m}\|^2. \tag{7}$$

Here, $p^{c,m}$ is the position of sub-filter m in frame c, and $R \in \mathbb{R}^{2 \times 2}$ is a transformation matrix. In our experiments we use a full linear transform, which is optimized jointly during the learning. λ_p is a parameter determining the regularization impact. This part of the loss does not depend on the sub-filter coefficients.

4.2   Fourier Domain Formulation

The optimization is performed in the Fourier domain using Parseval's formula. This results in a finite representation of the continuous filters using truncated Fourier series.

Let $\hat{\cdot}$ denote the Fourier coefficients for any given, sufficiently nice function. By linearity of the Fourier transform

$$\widehat{S_f\{x^c\}}[k_1, k_2] = \sum_{m=1}^{M} \beta[k_1, k_2]\widehat{S_{f^m}\{x^c\}}[k_1, k_2] \tag{8}$$

where

$$\beta[k_1, k_2] = e^{-i2\pi p_1^{c,m} k_1 / T_1} e^{-i2\pi p_2^{c,m} k_2 / T_2} \tag{9}$$

and

$$\widehat{S_{f^m}\{x^c\}}[k_1, k_2] = \left(\sum_{d=1}^{D} \hat{f}_d^m[k_1, k_2]\widehat{J^d\{x^c\}}[k_1, k_2] \right). \tag{10}$$

Given C samples, we optimize the filter in the C-COT framework. The objective 3 is minimized by using Parseval's formula. We get the corresponding objective

$$\epsilon(f,p) = \sum_{c=1}^{C} \alpha^c \|\widehat{S_f\{x^c\}} - \hat{y}^c\|^2 + \sum_{m=1}^{M} \sum_{d=1}^{D} \|\hat{w}^{m,d} * \hat{f}_d^m\|^2 + \lambda_p \sum_{m=1}^{M} \|p^{c,m} - Rp^{1,m}\|^2 \tag{11}$$

which will be minimized by an alternate optimization strategy where we iteratively update the sub-filter coefficients and positions.

4.3   Updating the Filter Coefficients

The Fourier coefficients are truncated such that for feature dimension d only the K^d first coefficients are used (resulting in $2K^d + 1$ coefficients in total for that dimension). Also define $K = \max_d K^d$. To minimize the functional we rewrite it as a least squares problem which can be solved via its normal equations. The normal equations are then solved using conjugate gradient. Let $\cdot^H$ be the conjugate transpose. We define a block matrix with $C \times MD$ blocks

$$A = \begin{pmatrix} A^1 \\ \vdots \\ A^C \end{pmatrix}, \quad A^c = \begin{pmatrix} A^{c,1} \ldots A^{c,M} \end{pmatrix}, \quad A^{c,m} = \begin{pmatrix} A^{c,m,1} \ldots A^{c,m,D} \end{pmatrix} \tag{12}$$

where $A^{c,m,d}$ is a diagonal matrix of size $K \cdot K \times K^d \cdot K^d$

$$A^{c,m,d} = \text{diag} \begin{pmatrix} \beta[-K^d, -K^d]\widehat{J^d\{x^c\}}[-K^d, -K^d] \\ \vdots \\ \beta[-K^d, K^d]\widehat{J^d\{x^c\}}[-K^d, K^d] \\ \vdots \\ \beta[K^d, K^d]\widehat{J^d\{x^c\}}[K^d, K^d] \end{pmatrix}. \tag{13}$$

Further define

$$\hat{\mathbf{f}} = \begin{pmatrix} \hat{\mathbf{f}}^1 \\ \vdots \\ \hat{\mathbf{f}}^M \end{pmatrix}, \quad \hat{\mathbf{f}}^m = \begin{pmatrix} \hat{\mathbf{f}}_1^m \\ \vdots \\ \hat{\mathbf{f}}_D^m \end{pmatrix}, \quad \hat{\mathbf{f}}_d^m = \begin{pmatrix} f_d^m[-K^d, -K^d] \\ \vdots \\ f_d^m[-K^d, K^d] \\ \vdots \\ f_d^m[K^d, K^d] \end{pmatrix} \tag{14}$$

and

$$\hat{\mathbf{y}} = \begin{pmatrix} \hat{y}^1 \\ \vdots \\ \hat{y}^C \end{pmatrix}. \tag{15}$$

Lastly, let Γ denote a diagonal matrix containing the learning rate α^c, of size $CK \times CK$; and W denote a Toeplitz matrix corresponding to summation of the convolutions with $w^{m,d}$. Using these definitions the objective becomes

$$\epsilon(f,p) = \sum_{c=1}^{C} \alpha^c \|A^c \hat{\mathbf{f}} - \hat{\mathbf{y}}^c\|^2 + \|W\hat{\mathbf{f}}\|^2 + \epsilon_3(p). \tag{16}$$

We discard $\epsilon_3(p)$ while minimizing the objective over f, as it will be addressed in the next step. The objective is then minimized by solving

$$(A^H \Gamma A + W^H W)\hat{\mathbf{f}} = A^H \hat{\mathbf{y}} \tag{17}$$

using the method of conjugate gradient.

4.4 Displacement Estimation of the Sub-filters

The sub-filters are moved by minimizing the objective with respect to the sub-filter positions. This problem is not convex, and we resort to gradient descent utilizing Barzilai-Borwein's method [1]. The perk of their method is that the steplength is adaptive. The gradient is found as

$$\frac{d}{dp^{c,m}}\epsilon(f) = \frac{d}{dp^{c,m}}\epsilon_1(f) + \frac{d}{dp^{c,m}}\epsilon_3(p) \tag{18}$$

where

$$\frac{d}{dp^{c,m}}\epsilon_1(f) = 2(\widehat{S_f\{x^c\}} - \hat{y}^c)e^{-i2\pi p_1^{c,m}k_1/T_1}e^{-i2\pi p_2^{c,m}k_2/T_2}\widehat{S_{f^m}\{x^c\}}\begin{pmatrix} -i2\pi k_1/T \\ -i2\pi k_2/T \end{pmatrix} \tag{19}$$

and

$$\frac{d}{dp^{c,m}}\epsilon_3(p) = 2\lambda_p(p^{c,m} - Rp^{1,m}). \tag{20}$$

Note that $\epsilon_2(f)$ does not depend on the sub-filter positions, and hence the derivative with respect to the sub-filter positions is zero. In our experiments we let R be either the identity matrix, or an affine transform. The translation part of the affine transform is handled during the target position estimation described in Sect. 3. Hence the affine transform can be considered equivalent to a linear transform. The linear transform is estimated in each step of gradient descent using a closed form expression. This is done by rewriting the problem as an over-determined linear system of equations and solve it via its normal equations.

5   Experiment and Results

We validate our approach by performing comprehensive experiments on three tracking benchmarks: OTB-2015 [22], TempleColor [17] and VOT2016 [13].

5.1 Implementation Details

In our experiments we employ two types of features: Color Names, and "Deep Features" extracted from the Convolutional Neural Network (CNN). We use the network VGG-m and extract features from the layers Conv-1 and Conv-5. We use different number of sub-filters depending on the target size. We employ a "root-filter" which is a subfilter that is always centered around the target and utilizes both shallow features and deep features from a CNN. The locations of the sub-filters are continuously updated and has a strong regularization to enforce locality. We test different feature sets for these sub-filters. The sub-filters are initialized in the first frame where they are placed in a grid. We use $\lambda_P = 3 \cdot 10^{-6}$ on VOT2016 and TempleColor datasets, and use $\lambda_P = 3 \cdot 10^{-4}$ on the OTB-2015 dataset. We use the same set of parameters for all videos in each dataset.

5.2 Baseline Comparison

We perform baseline comparisons on the OTB-2015 dataset with 100 videos. We compare different features for the sub-filters, and different regularization for their positions. We evaluate the tracking performance in terms of mean overlap precision (OP) and area-under-the-curve (AUC). The overlap precision (OP) is calculated as the fraction of frames in the video where the intersection-over-union (IoU) overlap with the ground truth exceeds a threshold of 0.5 (PASCAL criterion). The area-under-the-curve (AUC) is calculated from the success plot where the mean OP is plotted over the range of IoU thresholds over all videos.

Table 1 shows the results of the baseline and proposed approach with the sub-filter positions regularized either with an affine transform, or the identity transform (Sect. 4.4). The proposed approach based on an affine transform provides improved tracking performance. This shows that regularization of the sub-filter positions is important and using an affine transform is superior compared to

Table 1. Baseline comparison on the OTB-2015 dataset with the two different regularizations of the sub-filter positions. The affine transform provides the best results.

	Baseline, no deformability	Affine	Identity
Mean OP	83.2	83.9	83.4
Mean AUC	68.4	69	68.5

Table 2. Baseline comparison on the OTB-2015 dataset when using different set of features for the sub-filters.

	Baseline	Shallow + CN	Shallow	Shallow + Deep	Deep	CN
Mean OP	83.2	83.6	83.5	83.6	83.9	83.9
Mean AUC	68.4	69	68.9	68.9	69	68.8

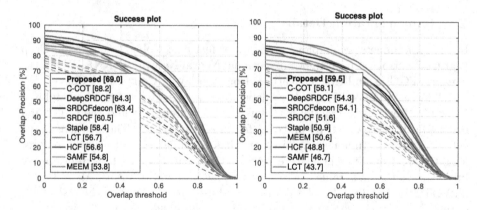

Fig. 2. Success plots on the OTB-2015 (left) and TempleColor (right) datasets, compared to state-of-the-art. The AUC score of each tracker is shown in the legend. We show slight performance increases on both datasets.

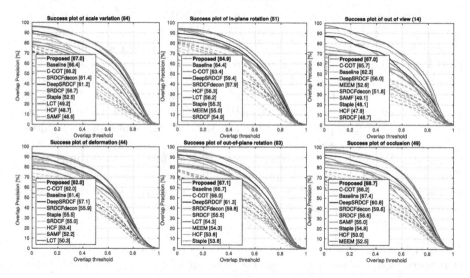

Fig. 3. Attribute-based comparison on the OTB-2015 dataset. Success plots are shown for six attributes. Our approach achieves improved performance compared to existing trackers in these scenarios.

an identity transform. Table 2 shows the baseline comparison when using different set of features. The deep features provide improved performance. However, performance comparable to deep features is also achieved by using colornames.

5.3   State-of-the-Art Comparison

OTB-2015. Figure 2 (on the left) shows the success plot for the OTB-2015 dataset which consists of 100 videos. The area-under-the-curve (AUC) score for each tracker is represented in the legend. Among existing approaches, the C-COT tracker [9] achieves an AUC score of 68.2%. It is worth to mention that the recently introduced ECO tracker [4] achieves the best results with an AUC score of 70.0%. However, the ECO tracker also employs HOG features together with colornames (CN) and deep features. Instead, our deformable convolution filter approach achieves competetive performance without using HOG features, with an AUC score of 69.0%. Figure 3 shows the attribute based comparison on the OTB-2015 dataset. All videos in the OTB-2015 dataset are annotated with 11 different attributes. Our approach provides the best results on 7 attributes.

5.4   TempleColor

Figure 2 (on the right) shows the success plot for the TempleColor dataset consisting of 128 videos. The SRDCF tracker [8] and its deep features variant (Deep-SRDCF) [7] achieve AUC scores of 51.6% and 54.3% respectively. The C-COT tracker yields an AUC score of 58.1%. Our approach improves the performance by 1.4% compared to the C-COT tracker.

5.5   VOT2016

The VOT2016 which consists of 60 videos compiled from a set of more than 300 videos. On the VOT2016 dataset, the tracking performance is evaluated both in terms of accuracy (average overlap during successful tracking) and robustness (failure rate). The overall tracking performance is calculated using Expected Average Overlap (EAO) which takes into account both accuracy and robustness. For more details, we refer to [14]. Table 3 shows the comparison on the VOT2016 dataset. We present the results in terms of EAO, failure rate, and accuracy. Our approach provides competetive performance in terms of accuracy and provides the best results in terms of robustness, with a failure rate of 0.70.

Table 3. State-of-the-art in terms of expected area overlap (EAO), robustness (failure rate), and accuracy on the VOT2016 dataset. The proposed approach show a slight decrease in EAO but a slight improvement to failure rate.

	SRBT [13]	EBT [23]	DDC [13]	Staple [2]	MLDF [13]	SSAT [13]	TCNN [21]	C-COT [9]	ECO [4]	Proposed Our
EAO	0.290	0.291	0.293	0.295	0.311	0.321	0.325	0.331	0.374	0.368
Fail. rt	1.25	0.90	1.23	1.35	0.83	1.04	0.96	0.85	0.72	0.70
Acc	0.50	0.44	0.53	0.54	0.48	0.57	0.54	0.52	0.54	0.54

6   Conclusions

We proposed a unified formulation to learn a deformable convolution filter. We represented our deformable filter as a linear combination of sub-filters. Both the coefficients and locations of all sub-filters are learned jointly in our framework. Experiments are performed on three challenging tracking datasets: OTB-2015, TempleColor and VOT2016. Our results clearly suggest that the proposed deformable convolution filter provides improved results compared to the baseline, leading to competitive performance compared to state-of-the-art trackers.

Acknowledgments. This work has been supported by SSF (SymbiCloud), VR (EMC^2, starting grant 2016-05543), SNIC, WASP, and Nvidia.

References

1. Barzilai, J., Borwein, J.M.: Two-point step size gradient methods. IMA J. Numer. Anal. **8**(1), 141–148 (1988)
2. Bertinetto, L., Valmadre, J., Golodetz, S., Miksik, O., Torr, P.H.S.: Staple: complementary learners for real-time tracking. In: CVPR (2016)
3. Bolme, D.S., Beveridge, J.R., Draper, B.A., Lui, Y.M.: Visual object tracking using adaptive correlation filters. In: CVPR (2010)
4. Danelljan, M., Bhat, G., Shahbaz Khan, F., Felsberg, M.: Eco: efficient convolution operators for tracking. In: CVPR (2017)
5. Danelljan, M., Häger, G., Khan, F., Felsberg, M.: Adaptive decontamination of the training set: a unified formulation for discriminative visual tracking. In: CVPR (2016)
6. Danelljan, M., Häger, G., Shahbaz Khan, F., Felsberg, M.: Accurate scale estimation for robust visual tracking. In: BMVC (2014)
7. Danelljan, M., Häger, G., Shahbaz Khan, F., Felsberg, M.: Convolutional features for correlation filter based visual tracking. In: ICCV Workshop (2015)
8. Danelljan, M., Häger, G., Shahbaz Khan, F., Felsberg, M.: Learning spatially regularized correlation filters for visual tracking. In: ICCV (2015)
9. Danelljan, M., Robinson, A., Shahbaz Khan, F., Felsberg, M.: Beyond correlation filters: learning continuous convolution operators for visual tracking. In: Leibe, B., Matas, J., Sebe, N., Welling, M. (eds.) ECCV 2016. LNCS, vol. 9909, pp. 472–488. Springer, Cham (2016). doi:10.1007/978-3-319-46454-1_29
10. Danelljan, M., Shahbaz Khan, F., Felsberg, M., van de Weijer, J.: Adaptive color attributes for real-time visual tracking. In: CVPR (2014)
11. Galoogahi, H., Sim, T., Lucey, S.: Multi-channel correlation filters. In: ICCV (2013)
12. Henriques, J.F., Caseiro, R., Martins, P., Batista, J.: High-speed tracking with kernelized correlation filters. TPAMI **37**(3), 583–596 (2015)
13. Kristan, M., et al.: The visual object tracking VOT2016 challenge results. In: Hua, G., Jégou, H. (eds.) ECCV 2016. LNCS, vol. 9914, pp. 777–823. Springer, Cham (2016). doi:10.1007/978-3-319-48881-3_54
14. Kristan, M., Matas, J., Leonardis, A., Felsberg, M., Čehovin, L., Fernández, G., Vojír, T., Nebehay, G., Pflugfelder, R., Häger, G.: The visual object tracking VOT2015 challenge results. In: ICCV workshop (2015)
15. Li, Y., Zhu, J.: A scale adaptive kernel correlation filter tracker with feature integration. In: Agapito, L., Bronstein, M.M., Rother, C. (eds.) ECCV 2014. LNCS, vol. 8926, pp. 254–265. Springer, Cham (2015). doi:10.1007/978-3-319-16181-5_18
16. Li, Y., Zhu, J., Hoi, S.C.: Reliable patch trackers: robust visual tracking by exploiting reliable patches. In: CVPR (2015)

17. Liang, P., Blasch, E., Ling, H.: Encoding color information for visual tracking: algorithms and benchmark. TIP **24**(12), 5630–5644 (2015)
18. Liu, T., Wang, G., Yang, Q.: Real-time part-based visual tracking via adaptive correlation filters. In: CVPR (2015)
19. Lukežič, A., Čehovin, L., Kristan, M.: Deformable parts correlation filters for robust visual tracking. arXiv preprint (2016). arXiv:1605.03720
20. Ma, C., Huang, J.B., Yang, X., Yang, M.H.: Hierarchical convolutional features for visual tracking. In: ICCV (2015)
21. Nam, H., Baek, M., Han, B.: Modeling and propagating CNNS in a tree structure for visual tracking. CoRR abs/1608.07242 (2016)
22. Wu, Y., Lim, J., Yang, M.H.: Object tracking benchmark. TPAMI **37**(9), 1834–1848 (2015)
23. Zhu, G., Porikli, F., Li, H.: Tracking randomly moving objects on edge box proposals. arXiv preprint (2015). arXiv:1507.08085

Two-Letters-Key Keyboard for Predictive Touchless Typing with Head Movements

Adam Nowosielski[✉]

Faculty of Computer Science and Information Technology,
West Pomeranian University of Technology, Szczecin,
Żołnierska 52, 71-210 Szczecin, Poland
anowosielski@wi.zut.edu.pl

Abstract. Head operated interfaces offer touchless interaction with electronic devices for physically challenged people who are unable to operate standard input equipment. The great challenge of such interfaces is text entry. Most existing approaches are based on camera mouse where the on-screen keyboard is operated by the pointer controlled with head movements. Movements of the head are also employed to cycle through keys or their groups to access the intended letter. While the process of direct selection requires substantial precision the traverse procedure is time consuming. The main contribution of this paper is proposition of the Two-Letters-Key Keyboard for touchless typing with head movements. The solution offers substantial acceleration in accessing the desired keys. The typing proceeds with directional head movements and only two consecutive moves are required to reach the expected key. No additional mechanisms (like eye blink or mouth open) are required for head typing.

Keywords: Touchless typing · Typing with head movements · Virtual keyboard · Gesture interaction · Human-computer interaction

1   Introduction

New assistive technologies are required for physically challenged people to provide information access and means for operation in the electronic world. Through advanced interfaces people with motion impairments can achieve computer-mediated communication with others. They can increase own independence in their daily lives. One of the option here is the head operated interface. Other include: eye tracking [1], speech recognition [2], brain computer interfaces [3]. The price level of these solutions is far too high for many individuals. However, despite the existence of low-cost alternatives [4], the choice of the appropriate interface is largely dependent on the form of disability. In this paper, the attention is focused on head operated interfaces and computer vision techniques.

Head operated interfaces largely focus on conventional mouse replacement. User's head movements are captured and translated into the motion of a pointer in the Graphical User Interface (GUI). Rotation and translation of the head are

© Springer International Publishing AG 2017
M. Felsberg et al. (Eds.): CAIP 2017, Part I, LNCS 10424, pp. 68–79, 2017.
DOI: 10.1007/978-3-319-64689-3_6

denoted as rigid motions [5]. The second group, the non-rigid motions, includes actions like [5]: eye winks, cheeks twitch, mouth movements (opening, closing, stretching) etc. The non-rigid motions are predominantly used for the selection process. With the help of the above means the complete pointer manipulation can be achieved and the special term - *camera mouse* - has been adopted here [5,6].

Based on the above concept some interfaces have already been reported in the scientific literature. In [5] and [6] the mouse cursor navigation is obtained by 3D head pose. Some of the proposed solutions concentrated on tracking only user's facial features. The position of user's nostrils related to the face region has been used for mouse movements in [7,8]. The mouse control based on the image plane position of the eyes can be found in [9,10]. Clicking events, in turn, are obtained with different non-rigid motions and examples include: eye blinks [6,8, 9], mouth shape changes [5,10–12] or brows movements [12]. Interesting approach is presented in [6] where the clicking events are obtained on the base of the distance between the user's head and a camera.

The process of typing with head movements is possible by the combination of the camera mouse and an on-screen keyboard. Modern operating systems offer soft keyboards (usually with the QWERTY layout) as the substitution for a physical keyboard. The typing is achieved through direct pointing of the key followed by the confirmation procedure. Another possibility is provided by the traverse procedure where consecutive keys or groups are accessed after the rigid motion detection. In the first case, the process of direct selection is difficult and requires substantial precision. The main drawback with the cycle method is the time required to reach subsequent characters.

In this paper, a novel technique for touchless typing with head movements is introduced. It allows to reach the base key with only two steps. This is substantial improvement to existing solutions and it directly derives from the 3-Steps Keyboard proposed in [13]. The main concepts of the 3-Step Keyboard are quoted in Sect. 3 after the overview of reduced interaction keyboards presented in Sect. 2. In Sect. 4 the main concepts of the Two-Letters-Key Keyboard for predictive touchless typing with head movements are proposed. The details of user interactions and their recognition with computer vision methods are provided in Sect. 5. The summary and final conclusions are included in Sect. 6.

2   Limited Interaction Typing

Human-computer interaction based on head movements may be the only option for some people. Free head movements suit well for camera mouse approach. However, when typing with direct pointing of the keys using the on-screen keyboard the precision problems arise. For this reason, free and unconstrained movements are reduced to only four primary directions: up (U), down (D), left (L) and right (R). First pair of directions are achieved with the upward and downward nod gestures. The horizontal direction can be obtained with rotary head movements or shift movements. Limiting the number of possible moves to the four main directions forces the change in the approach to typing. Captured head movements are used for traversing through individual keys or groups (rows or columns). Such approach requires many steps to reach the intended letter. This concept is popular and some solutions have already been reported in the literature [14] or made available for public as open source projects [15,16].

Typing with head movements limited to the four main directions resembles in principle the four-key text entry method. This form of typing is applied with game controllers, TV remotes and not so long ago has been a standard in pagers. The input mechanism requires from the user to press physical directional keys to navigate the on-screen keyboard. The directional keys are usually used to move some kind of visual indicator. Additional key for pressing is compulsory here. It should be noticed that different combinations of directions lead to specific letters depending on the initial position (previously entered character). Another approach, however, is possible. The idea is to assigning codes (specific combination of directions) to symbols. Such procedure has two advantages. First, no additional key for pressing is necessary. Second, each character is accessed in the same way regardless of the initial position resulting from previously typed letter. The main question concerns the most efficient coding system which reduces the number of key presses. The extensive analysis of this problem appears in [17] where a family of reduced-key keyboards is proposed.

Considering the typing process with directional movements on four-key keyboard, four alternatives can be accessed in the first step. In the two-steps procedure there are 16 possibilities (4×4) and with another succeeding step - 64 (16×4). The English alphabet consists of 26 characters. There are also supplementary symbols (e.g.: space, punctuation marks) and the typing requires additional special-function keys (like backspace). Nonetheless, considering the 4-choice base and the lack of any predictive technique or dictionary support the typing of any character requires at least 3 steps. This is the case when fixed length codes are employed. It is however possible to minimize the average number of steps (key-presses identified with directional movements) per character using the variable code length. This way, frequent symbols get shorter codes and those occurring less frequently get longer codes. [17] uses Huffman coding to assign minimized key sequences to letters and average value of 2.321 key presses is achieved in the proposed H4-Writer method. The H4-Writer uses four directions as the base and the shortest combinations consists of 2 directions but the longest require 5 directions. The E character, for example, is achieved with UU (double up movement) in the H4-Writer technique, while the Z is accessed with the DRLLL combination.

Different number of combinations of motion might be unusual. However, such approach has been popular and the multi-tap on a mobile phone keyboard is the best example. It requires consecutive taps to reach successive letters on the same key. Finally, this technique has been replaced with the T9 approach where dictionary support allowed a single press on the ambiguous key. This example shows that significant acceleration of typing speed is possible with the usage of predictive techniques and the dictionary support. For this reason, in this paper we propose to improve the recently proposed the 3-Steps Keyboard (a reduced interaction interface for touchless typing with head movements) with the dictionary support. The contribution allows further reduction in the number of head movements from 3 to 2.

3   3-Steps Keyboard Overview

In [13] new concept to touchless text typing with head movements has been proposed. The keyboard has the form of a single row with vertically translated alphabetically arranged characters presented in Fig. 1.

Fig. 1. The layout of the reduced interaction 3-Steps Keyboard [13]

The letters form two-levels distinctive groups. They are arranged in four main groups. The first and fourth group are located on the same level. The two middle groups are shifted, one upward and the other downward [13]. The displacement indicates the direction which has to be selected to reach each group. This way, the first group is selected with the left-direction movement, the second with the up movement, the third with the down movement, and the fourth with the movement to the right [13]. After the choice of an individual group, other groups are deactivated. Further interactions continue within the selected group and employ the same procedure. Each subgroup is translated in the same manner as main groups and consist of only two letters. Hereby, the second directional movement leads to the pair of characters. All remaining pairs are deactivated and the user has to select the appropriate letter from the pair with left or right directional movement. After that, the keyboard returns to the initial appearance and all the keys are activated again. The procedure of letter selection includes three steps. Figure 2 quotes an example where the A character is typed with three movements (i.e. LUL).

The base character can be reached with only three steps in the 3-Steps Keyboard. The access to other symbols or control keys is also provided. The extreme buttons represents backspace and space. Dot, comma and enter are also available for direct selection. Numbers and other symbols are accessed by switching the keyboard state (accomplished by the second key and LLR combination of movements) [13].

4   Two-Letters-Key Keyboard Interface

In the 3-Steps Keyboard presented in [13] the main concept is the 3-steps interaction which allow to reach any base character. Figure 3 presents the modified

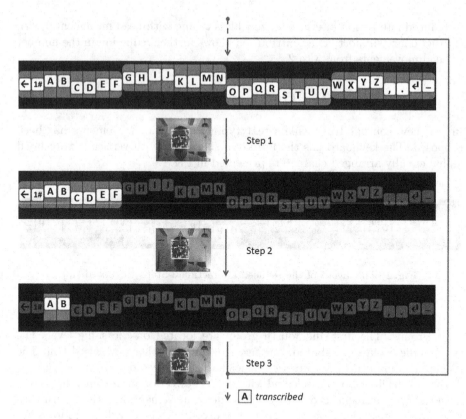

Fig. 2. Examplary interaction: letter 'A' transcription with the LUL combination of movements

version where two letters have been combined in individual keys. Similar concept has already been applied in the QWERTY layout and the RIM BlackBerry 7100t is an example of finished product. The adjacent letters in RIM BlackBerry 7100t (i.e. Q and W, E and R, and so on) are merged into one physical button. Such approach introduces ambiguity since same combinations might produce different correct words. It must be noticed, however, that the ambiguity level in keyboards containing two letters per key is much smaller compared to well-known T9 predictive text technology introduced for mobile phones. In mobile phone keyboard each key represents three or four letters. Three key-presses on the mobile phone keyboard produces at least 27 ($3 \times 3 \times 3$) unique combinations of initial strings while the keyboard including two letters on the key produces only 8 combinations.

User interaction with Two-Letters-Key Keyboard for touchless typing with head movements is very similar to the one present in the 3-Steps Keyboard (described in brief in Sect. 3). The main difference is the exclusion of the third step. Only two consecutive movements are required to reach the expected key. In the first step, with directional head movement, the user selects one of the four

Fig. 3. Layout of the proposed Two-Letters-Key Keyboard with 2-steps access

groups of letters. Then, with another directional movement a subgroup of two letters is selected and this finishes the procedure of key selection. User proceeds to next letter selection and at the same time the word generator engine provides suitable suggestions. If the expected word appears it can be selected without typing subsequent characters. The interaction details are provided further.

Comparing the layout of the modified version with the original 3-Steps Keyboard less special keys are available. With the RD sequence of movements in the Two-Letters-Key Keyboard the '1#' key is accessed. It switches the keyboard state and allows typing other symbols and numbers. This issue, however, is not considered in this paper. The extreme left button represents backspace. It is accessible with two consecutive movements to the left LL. The extreme right button - accessed with the RR sequence - has a special meaning. It moves the interaction to the suggestion panel.

Figure 4 presents the window of created prototype. The keyboard is located at the bottom part (it is deactivated). The middle part contains a panel of suggested words and the text box is placed in the upper part. The screenshot presents the state of the interface after the user moved head in the following sequence: LDDLURDLDR. This sequence denotes subsequent keys: 'CD', 'OP', 'MN', 'OP', 'UV'. As the result of this interactions the intended word *computer* appeared as the third suggestion, visible in the middle panel. The user could have continued typing next letters. However, with two movements to the right RR the user has been transferred to the suggestion panel where with the down movement D the proper suggestion has been selected. The suggestion panel contains four elements which are accessible with one of the directional movements. Hence the mutual displacement of the elements suggesting the proper direction of movement. After the user makes a choice the appropriate suggestion is briefly highlighted and the current string placed between vertical bars (in the upper text box) is replaced with that suggestion, space is added automatically and a single vertical bar presenting cursor appears.

The inference of words, in the created prototype, is based on the dictionary analysis. During typing each key provides the possibility of two letters. Consecutive keys generate new strings which are dictionary validated. If a given combination does not create a proper word it is removed. As the writing progresses the number of potential words decreases. All suggestions are sorted according to the length and the shortest are presented for the user. Finally, when the word suggestion engine fails, it is possible to switch to the original 3-Steps Keyboard (through the '1#' key accessed with RD movements). The introduction of sentence'

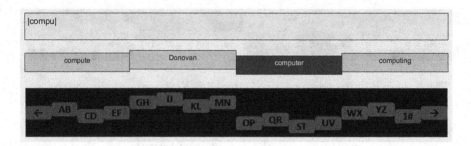

Fig. 4. Created prototype interface of Two-Letters-Key Keyboard

level suggestions appears now as the most valuable improvement here. It requires more complex prediction model and this issue is left as the future work.

For the task of dictionary analysis the United States (US) English dictionary provided for the OpenOffice suite has been utilized. The employed version of *en_US.dic* consisted of 52890 entries. This is typical dictionary and statistics on length of words are provided in Fig. 5. The seven-letter words are the most common (7944 entries). Words containing from 6 to 9 letters constitute more than half of the dictionary (29156 entries).

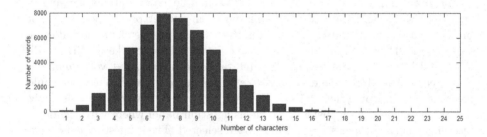

Fig. 5. Word length statistics

To validate the proposed approach some experiments have been performed. The set of short phrases provided in [18] have been used to check the efficiency of the dictionary support mechanism in typical everyday life texts. No problems have been reported and intended words appeared properly as suggestions. The typing speed was measured during experiments conducted by 9-person group of computer science students on the human-computer interaction course. People with disabilities are the main target of this interface, however in the initial experiments there were no handicapped participants. After the brief acquaintance, the fastest typist were able to type with speed reaching 20 CPM (chars per minute). On average the acceleration of 19.23% compared to the original 3-Steps Keyboard has been achieved.

5   The Control Routine

The interaction with the proposed keyboard interface is based on recognition of the user actions performed with the head. The fundamental task in the control routine is the face detection followed by a tracking procedure. The detection of faces has been an active research area for the past two decades and despite the existence of very good solutions it still remains interesting and challenging. It evolved from relatively simple approaches like those based on skin colour [19], through classifiers based on AdaBoost learning applied to different types of low-level descriptors (Haarlike features, Histogram of Oriented Gradients, Local Binary Patterns, etc.) [20,21], to methods based on deep learning [22]. New imaging techniques, beyond the visible spectrum, are also considered in the context of face detection [21].

The procedure of face detection and tracking in the presented solution does not differ from the one employed in [13]. It consists of well known Viola and Jones approach [20] in the first step. Then, the face region is analyzed and distinctive feature points are selected for tracking using the minimum eigenvalue algorithm. Those points are then tracked with the Kanade-Lucas-Tomasi (KLT) feature-tracking procedure. The counterpart for each point is searched in the new frame. Outliers are excluded with the MSAC algorithm [23]. The matched pairs are employed to calculate the geometric transformation which is then applied to the bounding box of the previous face localization. Figure 6 on the left presents a user in typical office environment with the result of face detection and distinctive feature points selected for tracking. Face centre is denoted with horizontal and vertical black lines.

 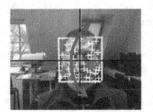 

Fig. 6. Result of face and distinctive feature points detection (left), schematic representation of vertical (middle) and horizontal (right) movements

The centre of the bounding box serves for the steering purposes. Four directional head movements have to be recognized. It is assumed that the user makes up and down directional movements with the upward or downward rotary nod gesture, presented schematically in the middle image of Fig. 6. These gestures are natural and can be performed with ease. More possibilities exist with horizontal movements (the right image of Fig. 6). Similarly as with vertical moves the left and right directions can be achieved with the rotary movements. Our investigations, supported by the feedback from participants who tested the interface, proved that horizontal rotary head movements are the fastest and the easiest to perform. Some users, however, complained that during this form of movement they are forced look at the screen at an angle which is uncomfortable. Alternatively, horizontal head movements can be accomplished with head shift or tilt. Shift movements are the hardest to perform and the tilt in this group is best balanced. Finally, it is the user who chooses the preferred form of interaction. All variants of movements are available in the interface. From the 9-person group of participants 4 have selected the rotary movements, 2 preferred the tilt operation and 1 has chosen the shift variant. Two remaining participants could not made a decision and one of them used alternately tilt with rotary movements while the second opted for tilt and shift movements.

When applying the face detection and tracking procedure for the steering purposes some issues have to be addressed. The first one concerns free movements. In head operated interfaces it is difficult to differentiate the steering movements from those casual. When user remains in the neutral position the reference centre is continuously adapting to small involuntary head movements. The typing is triggered when the motion parameters exceed specified thresholds. However, when no other action follows it is assumed that no steering was intended, the user has taken a more comfortable position, and new reference centre is calculated.

It must be noticed that during horizontal head movements with the tilt approach the change in the abscissa coordinate is expected. Unfortunately, the ordinate coordinate of the head centre also changes in the tilt case. The problem has been resolved with the simple procedure where the motion detection in a given direction automatically locks the competitive direction (the direction with a higher value of shift wins).

In the 3-Steps Keyboard after each directional movement the return to the center position was expected. In the Two-Letters-Key Keyboard another improvement is proposed. To simplify and accelerate the interaction the return to the center is not compulsory and the diagonal movements are allowed. For the example, the standard procedure for accessing the 'AB' key requires the left gesture with the return to the centre followed by the up gesture with the return to the centre. In the improved procedure, after performing the left gesture, user can directly move diagonally to the up position (the procedure, where the direction with a higher value of shift wins, still applies here). After the second gesture the return to the centre is compulsory. Hereby, each key requires two directional movements after which the head should return to the approximate centre position.

When new key is selected the face detection and tracking procedure is restarted and new reference centre is calculated. The whole process is imperceptible by the user and there are following reasons for justification of such operation. First, there are some variations in the head centre location after the return to the neutral area. Second, there is a problem of accumulative error during tracking. Finally, some points can be lost during tracking a face which is not a rigid object and can change its shape. The forced face detection, new feature points selection and the update of the reference centre coordinates prevent from the above threats.

6 Conclusions

The main contribution of this paper is the proposition of the Two-Letters-Key Keyboard for touchless typing with head movements. Most existing approaches requires substantial precision when operated on the camera mouse routine or suffer from time consuming process when operated with the traverse procedure. The proposed interface offers substantial acceleration in accessing the desired keys. Only two consecutive moves are required to reach the expected key. The proposed interface is based on the recently proposed the 3-Steps Keyboard and improves the original by the elimination of the third step. In the Two-Letters-Keys Keyboard each key consists of two letters and is accessed with two directional movement. The problem of ambiguity has been solved with the dictionary support which provides the predictive typing. The suggestion panel consists of four suggestions achievable with a directional moves. Since the switch to the suggestion panel requires two movements (RR), the whole process of suggested word selection requires three directional movements. It is worth noting that no additional mechanisms (like eye blink or mouth open) are required for head typing. Finally, the minor improvement in the proposed solution is the introduction of diagonal movements which eliminates the need of return to the center in most cases.

The main target of the presented interface are people with disabilities. In the initial experiments there were no handicapped participants. We are planning to perform evaluations with target users in the future. Another study, we want to perform, is the use of eyetracker to analyze the subjective performance and perception of the interface. The last but not least, left as the future work, is the introduction of sentence level suggestions. We hope that further improvements will make the interface better for those in need.

References

1. Jacob, R.J.K., Karn, K.S.: Eye tracking in human-computer interaction and usability research: ready to deliver the promises. In: The Minds Eye: Cognitive and Applied Aspects of Eye Movement Research, pp. 573–605. Elsevier Science, Oxford (2003)
2. Rebman, C.M., Aiken, M.W., Cegielski, C.G.: Speech recognition in the human-computer interface. Inf. Manage. 40(6), 509–519 (2003)
3. Graimann, B., Allison, B.Z., Pfurtscheller, G.: Brain-computer interfaces: a gentle introduction. In: Graimann, B., Pfurtscheller, G. (eds.) Brain Computer Interfaces: Revolutionizing Human-Computer Interaction. The Frontiers Collection. Springer, Heidelberg (2010)

4. Mantiuk, R., Kowalik, M., Nowosielski, A., Bazyluk, B.: Do-it-yourself eye tracker: low-cost pupil-based eye tracker for computer graphics applications. In: Schoeffmann, K., Merialdo, B., Hauptmann, A.G., Ngo, C.-W., Andreopoulos, Y., Breiteneder, C. (eds.) MMM 2012. LNCS, vol. 7131, pp. 115–125. Springer, Heidelberg (2012). doi:10.1007/978-3-642-27355-1_13

5. Tu, J., Tao, H., Huang, T.: Face as mouse through visual face tracking. Comput. Vis. Image Underst. **108**(2007), 35–40 (2007)

6. Nabati, M., Behrad, A.: 3D head pose estimation and camera mouse implementation using a monocular video camera. Signal Image Video Process. **9**(1), 39–44 (2015)

7. Morris, T., Chauhan, V.: Facial feature tracking for cursor control. J. Netw. Comput. Appl. **29**(1), 62–80 (2006)

8. Varona, J., Manresa-Yee, C., Perales, F.J.: Hands-free vision-based interface for computer accessibility. J. Netw. Comput. Appl. **31**(4), 357–374 (2008)

9. Santis, A., Iacoviello, D.: Robust real time eye tracking for computer interface for disabled people. Comput. Methods Progr. Biomed. **96**(1), 1–11 (2009)

10. Shin, Y., Ju, J.S., Kim, E.Y.: Welfare interface implementation using multiple facial features tracking for the disabled people. Pattern Recognit. Lett. **29**(2008), 1784–1796 (2008)

11. Bian, Z.-P., Hou, J., Chau, L.-P., Magnenat-Thalmann, N.: Facial position and expression-based human-computer interface for persons with tetraplegia. IEEE J. Biomed. Health Inform. **20**(3), 915–924 (2016)

12. Gizatdinova, Y., Spakov, O., Surakka, V.: Face typing: vision-based perceptual interface for hands-free text entry with a scrollable virtual keyboard. In: IEEE Workshop on Applications of Computer Vision, Breckenridge, CO, USA, pp. 81–87 (2012)

13. Nowosielski, A.: 3-steps keyboard: reduced interaction interface for touchless typing with head movements. In: Kurzynski, M., Wozniak, M., Burduk, R. (eds.) CORES 2017. AISC, vol. 578, pp. 229–237. Springer, Cham (2018). doi:10.1007/978-3-319-59162-9_24

14. Nowosielski, A., Chodyła, Ł.: Touchless input interface for disabled. In: Burduk, R., Jackowski, K., Kurzynski, M., Wozniak, M., Zolnierek, A. (eds.) Proceedings of the 8th International Conference on Computer Recognition Systems CORES 2013. AISC, vol. 226. Springer, Heidelberg (2013)

15. Assistive Context-Aware Toolkit (ACAT). Project page (2017). https://01.org/acat

16. QVirtboard. Project page (2017). http://qvirtboard.sourceforge.net

17. MacKenzie, I.S., Soukoreff, R.W., Helga, J.: 1 thumb, 4 buttons, 20 words per minute: design and evaluation of H4-writer. In: Proceedings of the 24th Annual ACM Symposium on User Interface Software and Technology (UIST 2011), Santa Barbara, California, USA, pp. 471–480 (2011)

18. Komninos, A., Dunlop, M.: Text input on a smart watch. IEEE Pervasive Comput. **13**, 50–58 (2014)

19. Kukharev, G., Nowosielski, A.: Fast and efficient algorithm for face detection in colour images. Mach. Graph. Vis. **13**(4), 377–399 (2004)

20. Viola, P., Jones, M.: Robust real-time face detection. Int. J. Comput. Vis. **57**(2), 137–154 (2004)

21. Forczmański, P.: Performance evaluation of selected thermal imaging-based human face detectors. In: Kurzynski, M., Wozniak, M., Burduk, R. (eds.) CORES 2017. AISC, vol. 578, pp. 170–181. Springer, Cham (2018). doi:10.1007/978-3-319-59162-9_18
22. Farfade, S.S., Saberian, M., Li, L.-J.: Multi-view face detection using deep convolutional neural networks. In: Proceedings of the 5th ACM on International Conference on Multimedia Retrieval (ICMR 2015), pp. 643–650 (2015)
23. Torr, P.H.S., Zisserman, A.: MLESAC: a new robust estimator with application to estimating image geometry. Comput. Vis. Image Underst. **78**(1), 138–156 (2000)

Recognizing Interactions Between People from Video Sequences

Kyle Stephens and Adrian G. Bors$^{(\boxtimes)}$

Department of Computer Science, University of York, York YO10 5GH, UK
adrian.bors@york.ac.uk

Abstract. This research study proposes a new approach to group activity recognition which is fully automatic. The approach adopted is hierarchical, starting with tracking and modelling local movement leading to the segmentation of moving regions. Interactions between moving regions are modelled using Kullback-Leibler (KL) divergence. Then the statistics of such movement interactions or as relative positions of moving regions is represented using kernel density estimation (KDE). The dynamics of such movement interactions and relative locations is modelled as well in a development of the approach. Eventually, the KDE representations are subsampled and considered as inputs of a support vector machines (SVM) classifier. The proposed approach does not require any intervention by an operator.

Keywords: Group activity recognition · Streaklines · Moving regions · Kullback-Leibler divergence · Kernel density estimation · SVM

1 Introduction

Human activity recognition has received considerable attention, by modelling and identifying the movement of isolated individuals. Nevertheless, many human activities take place in a social context of interaction with other people. Most human activity recognition methods start with extracting local features from video sequences which are then modelled either syntactically or statistically and the resulting modelling data is fed into a machine learning classifier. More recently, this area evolved towards detecting anomalies in the videos representing human activity, such as by using dynamic texture models [1], and Markov random fields [2]. An observational approach, detecting new activities in the scene, by using the Kullback-Leibler (KL) divergence from a dictionary of pre-observed events was proposed in [3,4].

Following behaviour studies resulting from the complexity of modern life lead to the requirement of contextual modelling of human activities instead of that of simple movements by individual persons. Group activity requires more complex descriptions of how people interact with each other and with their surroundings. In the study by Ni *et al.* [5] group activities are recognized using manually initialized tracklets while a heat-map based algorithm was used for

© Springer International Publishing AG 2017
M. Felsberg et al. (Eds.): CAIP 2017, Part I, LNCS 10424, pp. 80–91, 2017.
DOI: 10.1007/978-3-319-64689-3_7

modelling human trajectories when recognising group activities in videos in [6]. A statistical approach of modelling data acquired by a multi-camera system was used in [7] and a hierarchical semantic granularity approach was employed for group activity in [8]. Movement trajectories have been represented as either histograms of features extracted from tracklets [10] or as Gaussian processes modelling time-series of movement trajectories [11]. Such approaches rely on either the training of a pedestrian detector for each scene, or on the manual annotation of trajectories.

This research study describes an automatic method for group activity recognition by modelling the inter-dependant relationships between human activity characteristic features over time. Features representing medium-term tracking of moving regions are extracted using the method from [12] leading to the segmentation of compactly moving regions. The interdependency between moving regions is represented by evaluating the relative movement and location between pairs of segmented moving regions. Kernel Density Estimation (KDE) is then used to model the statistics of the movement, location, as well as their evolution in time, representing the dynamics of such interactions between moving regions. The group interaction model keeps track of stationary pedestrians by automatically marking the locations where these stop and then when they start an activity again. Section 2 describes the features used for representing moving regions, while the statistical modelling is provided in Sect. 3. Section 4 describes the classification approach. Section 5 provides the experimental results on two group activity datasets while Sect. 6 draws the conclusions.

2   Modelling Human Interactions

The proposed methodology for group activity recognition has three main processing stages: estimating streaklines of movement, modelling moving regions and their dynamics and group activity recognition. Optical flow estimation leads to tracking of regions of movement in the image [13,14]. Streaklines [12], similarly to the approach from [14], represent the smooth movement of particles of fluid. Modelling streaklines relies on the Lagrangian framework for fluid dynamics, ensuring the robustness and the continuity of movement estimation. Unlike in the approach from [12], where streaklines are computed for each pixel, in this research study each streakline is associated with a block of pixels of fixed size by computing the marginal median of all streakline vectors located in a specific region. A streakline consists of several vectors head-to-tail located along a localized trajectory of movement which is then fit by a first degree polynomial for smoothing.

The general assumption is that movement in the scene corresponds to moving people, but interactions with other moving objects such as vehicles is accounted for in this model as well. Firstly, we begin by segmenting the streakflow field into distinctly moving regions. The Expectation-Maximization (EM) algorithm, assuming Gaussian Mixture Models (GMM) is used for segmenting and modelling each inter-connected region. The number of clusters and the centers of the Gaussian functions are initialized using the modes of the streakline flow histograms. A two-step approach is adopted for movement segmentation in order to address the effects of perspective projection, which are mostly observed in the case of video sequences acquired with wide-angle lens cameras located at low heights. The assumption is that in the upper part of the video frames, objects and their motion is smaller than in the lower part, due to the perspective of the scene. In the first step, the segmentation is performed in order to estimate the height of the moving objects, which is used to derive a scaling factor. In the second step, the segmentation is repeated by considering this scaling factor, applied to the movements estimated from the video sequence, according to the location of its corresponding moving region in the scene. The motion $\mathbf{M}_i$ of region i is then scaled by a factor s_i:

$$\mathbf{M}'_i = s_i \mathbf{M}_i, \tag{1}$$

where s represents the perspective projection scaling factor estimated for the given scene from the video sequence. Each moving region is therefore represented by a GMM, defined by its mean and variance.

3 Modelling Interactions Between Moving Regions

The key characteristics of group activities are often present in the interdependent relationship between the people present in the scene as well as between them and the surroundings. The general assumption is that moving regions correspond to human activities and in the following we model the relationship between such regions. In the first instance, we compute statistical differences between streakflow distributions $\mathcal{A}_{I(t)}$ and $\mathcal{A}_{J(t)}$, corresponding to two moving regions $I(t)$ and $J(t)$ at time t by

$$M(I(t), J(t)) = \mathrm{e}^{-\frac{D_{SKL}(\mathcal{A}_{I(t)} \| \mathcal{A}_{J(t)})}{\sigma_m}} \tag{2}$$

where $D_{SKL}(\mathcal{A}_{I(t)} \| \mathcal{A}_{J(t)})$ is the symmetrized KL divergence between the local statistics of streaklines corresponding to the moving regions $I(t)$ and $J(t)$ at time t, [3] and σ_m is a scaling factor for movement differences. The background is considered as one of the regions as well. The calculation of equation (2) results in a value within the range $[0, 1]$ which models the inter-dependancy between regions $I(t)$ and $J(t)$. For example, individuals moving in completely opposite directions will have $M(I(t), J(t)) = 0$, whilst individuals moving in the same direction and at the same speed will have $M(I(t), J(t)) = 1$. These are then concatenated to form a vector representing the inter-dependant group relationships of the streakflows at a particular time t.

A similar approach is adopted for the locations of the moving regions by forming distributions of location coordinates corresponding to each moving region, including the background. The distributions of relative locations for the people from the scene, both moving or stationary, is modelled as well. The characteristic parameters of GMMs in this case correspond to the location, size and approximative size and shape of each moving region. Similarly to Eq. (2), we model the interaction between two GMMs $\mathbf{C}_{I(t)}$ and $\mathbf{C}_{J(t)}$ representing the moving regions $I(t)$ and $J(t)$ at time t, as:

$$D(I(t), J(t)) = e^{-\frac{D_{SKL}(^{C}I(t)\|^{C}J(t))}{\sigma_l}} \qquad (3)$$

where σ_l represents the characteristic scale parameter for locations. Similarly to the streakflow model, this provides a value in the range $[0,1]$ representing the spatial relationship between the two moving regions. For example, individuals characterised by moving regions $I(t)$ and $J(t)$ at time t, located far apart, will have $D(I(t), J(t)) = 0$, whilst individuals located closer together will have $D(I(t), J(t)) = 1$. A vector, representing all the inter-relationships of locations for the group activity at time t, is then formed as shown in Fig. 1(a).

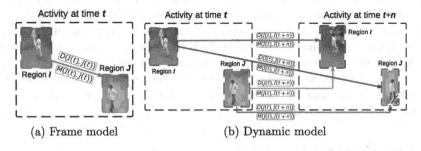

(a) Frame model   (b) Dynamic model

Fig. 1. Modelling the inter-dependencies of moving regions in both space and time.

We also model the dynamic changes of relative differences between moving regions over subsequent frames by computing the differences between all streakflow models $M(I(t), J(t)) = 1$ at time t and those identified at other times $t + n$. These are computed as in Eq. (2), except that the models are now calculated across subsequent sets of frames. A vector of streakflow differences representing all the inter-dependant relationships of streakflow models between the time instances t and $t + n$ is then formed. The same modelling of dynamic changes is applied for changes in the relative distances between the locations, sizes and shapes of the moving regions by using inter-distances $D(I(t), J(t)) = 1$ from (3) at times t and $t + n$. Another issue addressed in this research study is the modelling of people who become stationary after they have moved through the scene. If there is no movement detected in a particular area and its neighbouring surroundings of the scene where motion was previously detected, during p consecutive frames, this indicates that a previously moving region ceased to move. Such stationary regions are characterised by their location and by zero motion. Finally, when movement occurs again in the region of a stationary person, then such regions are considered to be moving again as components of the group activity model. The dynamic model is illustrated in Fig. 1(b).

4   Classifying Types of Interactions Between People

Kernel Density Estimation (KDE) is a non-parametric representation which provides a good model for complex data such as those defining human interactions. On the other hand KDE smoothes the data representation reducing the uncertainty when compared to assuming a certain parametric statistical model. The bandwidth parameters of the bi-variate Gaussian kernel are used to help control the smoothing effects of the kernel density estimator. In this study, we use the bivariate KDE method employing diffusions on data representations, proposed in [16], which considers a Gaussian kernel, and uses an automatic bandwidth selection method.

A discrete representation of the resulting KDE's for each set of features is represented on a grid of fixed size $K \times K$. By using a fixed grid size for representing the movement in the scene, the locations of the regions of movement, dynamics of movement and their region locations, we implicitly apply a data normalization, because such data representations do not depend on the frame size or on the actual number of frames. Such KDE's are then sampled and used as a feature vector representing the characteristics of the group activity taking place in the given video sequence. The feature vectors are then used to train a Support Vector Machine (SVM) algorithm, having K^2 inputs, while the outputs separate each group activity.

5   Experimental Results

In the following we provide the experiments when considering two databases containing group interaction videos: NUS-HGA [5], and Colective [9] datasets. This first data set consists of six different group activities collected in five different sessions containing 476 video sequences, each session representing staged actions. Initially, streaklines are extracted for blocks of size 14×14 over 10 consecutive frames. The motion is segmented and each moving region is represented by the Gaussian Mixture Model (GMM) of streakflows vectors and their locations GMM. Figure 2 shows an example of the estimated streakflows, motion histograms, and the moving region segmentation for the fight activity from the NUS-HGA dataset. In this particular activity, movement is intense and chaotic. In Fig. 2b the solid green bars correspond to peaks of the histogram, while the solid red bars are entries with the height below 15% of the maximum bar height which are eventually removed for not being significant enough in the context of the scene' movement. The moving regions are well segmented and the small regions obtained in region 1 of Fig. 2c help characterize the smaller atomic events performed in the group, for example pushing or kicking which usually happens during the fighting activity.

We account for the perspective projection effects, where smaller movements in smaller segmented regions would correspond to movements detected from farther away in the scene. The segmentation is done in two stages, where during the first segmentation stage a scaling factor is calculated and then the motion is scaled

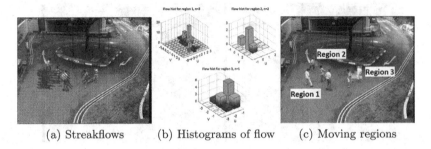

(a) Streakflows   (b) Histograms of flow   (c) Moving regions

Fig. 2. Example of streakflows, histograms of flow and the moving regions before and after segmentation on a fight sequence from the NUS-HGA dataset. In (b) "n" refers to the number of histogram peaks. (Color figure online)

accordingly and the scene resegmented. The detection of the stationary regions detector is applied considering the number consecutive frames for estimating the streaklines as $p = 25$. Two examples of detecting stationary pedestrians are shown in Fig. 3 for the Talking and Gathering activities. In Figs. 3a and c the pedestrians are still moving and therefore their corresponding moving regions are properly detected. In Figs. 3b and d the individuals have stopped and their stationary regions are properly detected.

The streakflow movement model, streakflow dynamics, location and location dynamics relationship differences are computed as described in Sect. 3, considering the scaling parameters $\sigma_m = 15$, $\sigma_l = 550$ for motion and location differences respectively, and $\sigma_m = 17.5$, $\sigma_l = 650$ for the motion and location dynamics. The number of frames, considered for the dynamic window from Sect. 3, is set to $n = 13$. The bivariate kernel density estimation from [16] is computed over a fixed grid size of 16×16. Representations of the KDEs using Gaussian kernels for various human interaction activities are shown in Figs. 4a–f when considering motion estimation and segmentation, and in Figs. 4g–l when modelling the locations of moving regions. The gathering motion shown in Fig. 4f displays a diversity of differences in movement, which is expected as some individuals are gathering coming from different directions. The Walking activity location differences, shown in Fig. 4e, are all close to 1. This implies that the individuals are tightly grouped, which is expected in the Walk in Group activity. The Gather activity location differences shown in Fig. 4l display clear transitions between locations situated far apart leading to closer-together locations. This is expected, as the gathering activity involves individuals coming from far away towards gathering in a tight group at the end of the activity.

For classification purposes, the density estimations are subsampled and fed into the classifier. The motion and location features represent complimentary information and can be combined for the final activity classification. We use SVM with the RBF kernel as a classifier, considering the parameters $C = 2.83$ and $\gamma = 0.00195$ for the SVM margin and kernel bandwidth. For all experiments,

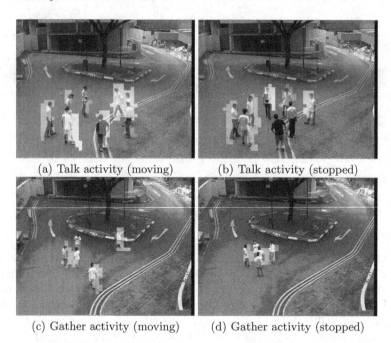

(a) Talk activity (moving)   (b) Talk activity (stopped)

(c) Gather activity (moving)   (d) Gather activity (stopped)

Fig. 3. Identifying when pedestrians stop during the video frames showing gathering and talking activities from the NUS-HGA dataset.

we follow the evaluation protocol described in [5], where the NUS-HGA dataset is split into 5-fold training and testing.

The Collective dataset [9] consists of 6 different activities: Gathering, Talking, Dismissal, Walking Together, Chasing and Queueing. The dataset consists of 32 video sequences, where each video sequence contains multiple examples of each activity. The video sequences are recorded using a hand-held camera, and therefore the perspective distortion is quite strong in the scenes from this dataset. The spatio-temporal segmentation of these video sequences takes place into blocks of 20×20 pixels by 10 frames, where the streaklines are extracted for each block of 10 frames. Examples of the streakflows and movement segmentation are shown in Fig. 5 for the Chasing and Gather activities. In both cases, the moving regions are well segmented, particularly in the chasing example where the chaser and chasee are segmented separately despite forming one connected region moving in the same direction. The next step involves applying the stationary pedestrian detector as in Sect. 2, assuming the number of prior frames used as $p = 25$. The videos from the Collective dataset show different activities, displaying transitions from one activity to another, including times when people are stationary. Such situations are identified and an example of transitions through activities is shown in Fig. 6. Initially, as in Fig. 6a, the pedestrians are moving towards each other performing the gathering activity. People are eventually gathered together towards the end of this activity, and the transition to

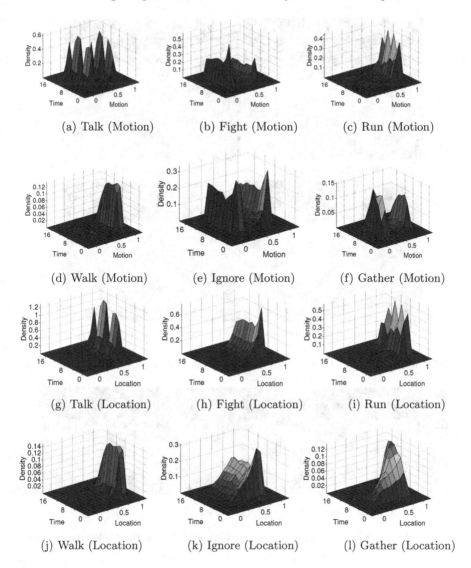

(a) Talk (Motion)   (b) Fight (Motion)   (c) Run (Motion)

(d) Walk (Motion)   (e) Ignore (Motion)   (f) Gather (Motion)

(g) Talk (Location)   (h) Fight (Location)   (i) Run (Location)

(j) Walk (Location)   (k) Ignore (Location)   (l) Gather (Location)

Fig. 4. KDEs and histograms representing motion dynamics in (a)–(f) and for location in (g)–(l) for the NUS-HGA dataset.

the talking activity is evident in Fig. 6b. The stationary people detection has successfully recorded the locations of the individuals when stopping, as seen in Fig. 6b. Finally, after a period of time, the individuals begin to move again performing the dispersing activity shown in Fig. 6c. In Fig. 6c, the new moving regions are detected replacing the previously identified stopped regions which are no longer present.

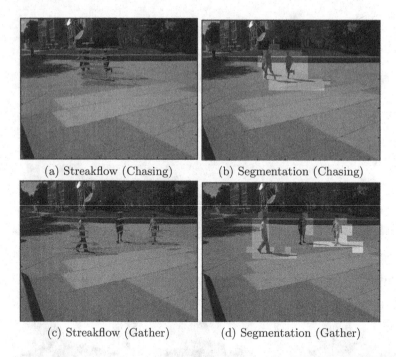

(a) Streakflow (Chasing)   (b) Segmentation (Chasing)

(c) Streakflow (Gather)   (d) Segmentation (Gather)

Fig. 5. Examples of streakflow and segmentation from the Collective dataset.

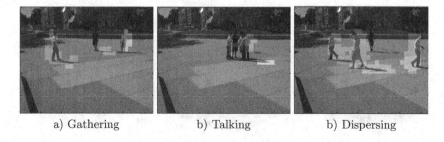

a) Gathering   b) Talking   b) Dispersing

Fig. 6. Pedestrians transitioning through various activities in the Collective dataset.

In the following, the human activity features, representing the streakflow differences, streakflow dynamics, location differences and location dynamics are computed for each moving region as described in Sect. 3. The scaling parameters are $\sigma_m = 15$ and $\sigma_l = 450$ for motion features and location features, respectively, while the size of the dynamic window for the motion dynamics and location dynamics is $n = 5$. Then, the data is represented over time using KDE, as described in Sect. 4, over a grid size of 8×8, using the 2-column feature matrices as input data. The grid-based representation of the KDE is then used as input to the SVM classifier with the RBF kernel.

For the tests on the Collective dataset we divide the dataset into 3 subsets for 3-fold training and testing according to the tests in [9]. We split the sequences

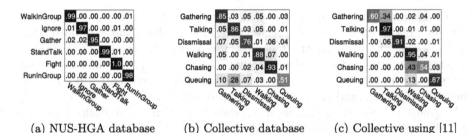

(a) NUS-HGA database   (b) Collective database   (c) Collective using [11]

Fig. 7. Confusion matrices for the recognition results of the proposed method when combining all features modelling movement, location distribution and their dynamics as well on (a) NUS-HGA, resulting in 90% classification accuracy and (b) Collective dataset, resulting in 79.7% accuracy. (c) The confusion matrix for Colective dataset when using [11], resulting in 80.3% classification accuracy.

Table 1. Group activity recognition results on the NUS-HGA and Collective datasets

Method	NUS-HGA dataset (%)	Collective dataset (%)
Localized causalities [5]	74.2	–
Group interaction zone [17]	96.0	–
Multiple-layered model [11]	96.2	80.3
Monte Carlo Tree Search [18]	–	77.7
Collective activities [8]	–	79.2
Motion	86.2	75.4
Location	87.1	64.3
Motion dynamics	91.6	76.8
Location dynamics	92.6	71.6
Motion+Location	94.5	76.5
Motion Dynamics+Location Dynamics	97.1	78.4
Motion+Location+Motion Dynamics+Location Dynamics	98.0	79.7

during training and testing into short sequences of 60 frames for the evaluation and then calculate the average recognition accuracy across all classes. Confusion matrices for all features combined are compared to the approach from [11] as shown in Fig. 7. The results for the Queuing activity are not that good because that stationary pedestrians forming queues are not moving at all for the duration of the sequence, and therefore are not detected. However, it can be observed from Fig. 7 that the results of the proposed methods show a greater consistency across all the other activities then other approaches.

Comparative results are provided in Table 1 for NUS-HGA and Collective datasets. The location features provide a better recognition result than the motion features while the results for the dynamics models for motion and location emphasise their importance for the Group activity recognition. The combination of all features account for movement, location, as well as the dynamics of both movement and location, and gives the best result of 98% for the NUS-HGA dataset. The group interaction method from [17] does not evaluate the results using the 5-fold training and testing as suggested in [5] for the NUS-HGA dataset. The proposed methodology, which is fully automated, provides a clear improvement of about 2% over the best other approach for the NUS-HGA dataset. For the Collective dataset, the proposed method is comparative to the state-of-the-art and superior to the other methods when not considering the queuing activity. The motion and movement dynamics outperform the location inter-dependency features, while the dynamic features outperform their equivalent frame-by-frame features. Similarly to the results on the NUS-HGA database, the best results for the Collective dataset are achieved when combining all features. However, all the other comparative methods are not fully automatic and use some form of human intervention during the experiments. Meanwhile, the proposed methodology is completely automatic and does not require any human intervention.

6   Conclusion

A completely automatic approach for modelling interactions between people is proposed in this paper. Streakflows of localized movement along several frames are estimated from the video sequence. Statistical distributions of vectors forming streakflows, as well as their locations are represented using kernel density estimation (KDE) and are used in order to identify compactly moving regions. We also consider the dynamics of change in the streakflows and in the locations of the moving regions. The relative movement of each moving region with all the other moving regions, including the background, is then represented statistically. Scaling is used in order to mitigate the effects of perspective projection in the scene, while the dynamics of change in the moving regions considers the timing when people are stationary. Eventually, SVM with RBF kernels, considering sampled KDE representations of movement, location, and their dynamics, as inputs, is used as a classifier.

Acknowledgment. This research work was supported by DSTL grant DSTLX1000074616 "Human Activity Recognition."

References

1. Li, W., Mahadevan, V., Vasconcelos, N.: Anomaly detection and localization in crowded scenes. IEEE Trans. Pattern Anal. Mach. Intell. **32**(1), 18–32 (2014)
2. Nallaivarothayan, H., Fookes, C., Denman, S., Sridharan, S.: An MRF based abnormal event detection approach using motion and appearance features. In: Proceedings of IEEE International Conference on Advanced Video and Signal-Based Surveillance, pp. 343–348 (2014)
3. Stephens, K., Bors, A.G.: Observing human activities using movement modelling. In: Proceedings of IEEE International Conference on Advanced Video and Signal-based Surveillance, paper # 44, pp. 1–6 (2015)
4. Stephens, K., Bors, A.G.: Grouping multi-vector streaklines for human activity identification. In: Proceedings of IEEE Workshop on Image, Video and Multidimensional Signal Processing, Bordeaux, France (2016)
5. Ni, B., Yan, S., Kassim, A.: Recognizing human group activities with localized causalities. In: Proceedings of IEEE Conference on Computer Vision and Pattern Recognition, pp. 1470–1477 (2009)
6. Lin, W., Chu, H., Wu, J., Sheng, B., Chen, Z.: A heat-map-based algorithm for recognizing group activities in videos. IEEE Trans. Circuits Syst. Video Technol. **23**(11), 1980–1992 (2013)
7. Chang, M., Ge, W.: Probabilistic group-level motion analysis and scenario recognition. In: Proceedings of International Conference on Computer Vision, pp. 747–754 (2011)
8. Choi, W., Savarese, S.: Understanding collective activities of people from videos. IEEE Trans. Circuits Syst. Video Technol. **36**(6), 1242–1257 (2014)
9. Choi, W., Savarese, S.: A unified framework for multi-target tracking and collective activity recognition. In: Fitzgibbon, A., Lazebnik, S., Perona, P., Sato, Y., Schmid, C. (eds.) ECCV 2012. LNCS, vol. 7575, pp. 215–230. Springer, Heidelberg (2012). doi:10.1007/978-3-642-33765-9_16
10. Zhang, Y., Ge, W., Chang, M.C., Liu, X.: Group context learning for event recognition. In: Proceedings of IEEE Workshop on Application of Computer Vision, pp. 249–255 (2012)
11. Cheng, Z., Qin, L., Huang, Q., Yan, S., Tian, Q.: Recognizing human group action by layered model with multiple cues. Neurocomputing **136**, 124–135 (2014)
12. Mehran, R., Moore, B.E., Shah, M.: A streakline representation of flow in crowded scenes. In: Daniilidis, K., Maragos, P., Paragios, N. (eds.) ECCV 2010. LNCS, vol. 6313, pp. 439–452. Springer, Heidelberg (2010). doi:10.1007/978-3-642-15558-1_32
13. Bors, A.G., Pitas, I.: Prediction and tracking of moving objects in image sequences. IEEE Trans. Image Process. **9**(8), 1441–1445 (2000)
14. Doshi, A., Bors, A.G.: Robust processing of optical flow of fluids. IEEE Trans. Image Process. **19**(9), 2332–2344 (2010)
15. Bors, A.G., Nasios, N.: Kernel bandwidth estimation for nonparametric modelling. IEEE Trans. Syst. Man Cybern. Part B Cybern. **39**(6), 1543–1555 (2009)
16. Botev, Z., Grotowski, J., Kroese, D.: Kernel density estimation via diffusion. Ann. Stat. **38**(5), 2916–2957 (2010)
17. Cho, N.-G., Kim, Y.-J., Park, U., Park, J.-S., Lee, S.-W.: Group activity recognition with group interaction zone based on relative distance between human objects. Int. J. Pattern Recog. Artif. Intel. **29**(5), 1–15 (2015). #1555007
18. Amer, M.R., Todorovic, S., Fern, A., Zhu, S.C.: Monte Carlo tree search for scheduling activity recognition. In: Proceedings of IEEE International Conference on Computer Vision, pp. 1353–1360 (2013)

Segmentation

Deep Projective 3D Semantic Segmentation

Felix Järemo Lawin[✉], Martin Danelljan, Patrik Tosteberg, Goutam Bhat, Fahad Shahbaz Khan, and Michael Felsberg

Computer Vision Lab, Department of Electrical Engineering,
Linköping University, Linköping, Sweden
`felix.jaremo-lawin@liu.se`

Abstract. Semantic segmentation of 3D point clouds is a challenging problem with numerous real-world applications. While deep learning has revolutionized the field of image semantic segmentation, its impact on point cloud data has been limited so far. Recent attempts, based on 3D deep learning approaches (3D-CNNs), have achieved below-expected results. Such methods require voxelizations of the underlying point cloud data, leading to decreased spatial resolution and increased memory consumption. Additionally, 3D-CNNs greatly suffer from the limited availability of annotated datasets.

In this paper, we propose an alternative framework that avoids the limitations of 3D-CNNs. Instead of directly solving the problem in 3D, we first project the point cloud onto a set of synthetic 2D-images. These images are then used as input to a 2D-CNN, designed for semantic segmentation. Finally, the obtained prediction scores are re-projected to the point cloud to obtain the segmentation results. We further investigate the impact of multiple modalities, such as color, depth and surface normals, in a multi-stream network architecture. Experiments are performed on the recent Semantic3D dataset. Our approach sets a new state-of-the-art by achieving a relative gain of 7.9%, compared to the previous best approach.

Keywords: Point clouds · Semantic segmentation · Deep learning · Multi-stream deep networks

1 Introduction

The rapid development of 3D acquisition sensors, such as LIDARs and RGB-D cameras, has lead to an increased demand for automatic analysis of 3D point clouds. In particular, the ability to automatically categorize each point into a set of semantic labels, known as semantic point cloud segmentation, has numerous applications such as scene understanding and robotics. While the problem of semantic segmentation of 2D-images has gained a considerable amount of attention in recent years, semantic segmentation of point clouds has received little interest despite its significance. In this paper, we propose a framework for semantic segmentation of point clouds that greatly benefits from the recent developments in semantic image segmentation.

M. Felsberg et al. (Eds.): CAIP 2017, Part I, LNCS 10424, pp. 95–107, 2017.
DOI: 10.1007/978-3-319-64689-3_8

With the advent of deep learning, many tasks within computer vision have seen a rapid progress, including semantic segmentation of images. The key factors for this development are the introductions of large labeled datasets [2] and GPU implementations of Convolutional Neural Networks (CNNs). However, CNNs have not yet been successfully applied for semantic segmentation of 3D point clouds due to several challenges. In contrast to the regular grid-structure of image data, point clouds are in general sparse and unstructured. A common strategy is to resort to voxelization in order to directly apply CNNs in 3D. This introduces a radical increase in memory consumption and leads to a decrease in resolution. Additionally, labeled 3D data, which is crucial for training CNNs, is scarce due to difficulties in data annotation.

In this work, we investigate an alternative approach that avoids the afore-mentioned difficulties induced by 3D CNNs. As our first contribution, we propose a framework for 3D semantic segmentation that exploits the advantages of deep image segmentation approaches. The point cloud is first projected onto a set of synthetic images, which are then used as input to the deep network. The resulting pixel-wise segmentation scores are re-projected into the point cloud. The semantic label for each point is then obtained by fusing scores over the different views. As our second contribution, we investigate the impact of different input modalities, such as color, depth and surface normals, extracted from the point cloud. These modalities are fused in a multi-stream network architecture to obtain the final prediction scores.

Compared to semantic segmentation methods based on 3D CNNs [17], our approach has two major advantages. Firstly, our method benefits from the abundance of the already existing data sets for image segmentation and classification, such as ImageNet [2] and ADE20K [28]. This significantly reduces, or even eliminates the need of 3D data for training purposes. Secondly, by avoiding the large memory complexity induced by voxelization, our method achieves a higher spatial resolution which enables better segmentation quality.

We perform qualitative and quantitative experiments on the recently introduced Semantic3D dataset [6]. We show that different modalities contain complementary information and their fusion significantly improves the final segmentation performance. Further, our approach sets a new state-of-the-art performance on the Semantic3D dataset, outperforming both classical machine learning methods and 3D-CNN based approaches. Figure 4 shows an example segmentation result using our method.

2   Related Work

The task of semantic point cloud segmentation has received an increasing amount of attention due to the rapid development of sensors capable of capturing high-quality 3D data. RGB-D cameras, such as the Microsoft Kinect, have become popular for robotics and computer vision tasks. While RGB-D cameras are more suitable for indoors environments, terrestrial laser scanners capture large-scale point clouds for both indoors and outdoors applications. Both RGB-D cameras and modern laser scanners are capable of capturing color in association with the 3D information using calibrated RGB cameras. Besides visualization, this additional information is highly useful for automated analysis and processing of point clouds. While color is not a necessity for our approach, it alleviates the task of semantic segmentation and enables the use of large-scale image datasets.

Most previous works [1,7,11,13,16] in 3D semantic segmentation apply a combination of (i) hand-crafted features, (ii) discriminative classifiers and (iii) spatial smoothness models. In this setting, the construction of discriminative 3D-features (i) is arguably the most important task. Popular alternatives include features based on the 3D structure tensor [1,7,11,26], histogram-based descriptors [7,11,16] such as Spin Images [10] and SHOT [21], and simple color features [11,16,26]. The classifiers (ii) are often based on maximum margin methods [1,13] or employ random forests [7,11,16]. To utilize spatial correlation between semantic labels (iii), many methods apply graphical models, such as the Conditional Random Field (CRF) [1,13,26].

Recently, deep convolutional neural networks (CNNs) have been successfully applied for semantic segmentation of 2D images [15]. Their main strength is the ability to learn high-level discriminative features, which eliminates the need of hand-designed representations. The rapid progress of deep CNNs for a variety of computer vision problems is generally attributed to the introduction of large-scale datasets, such as ImageNet [2], and improved performance for GPU computing.

Despite its success for image data, the application of CNNs to 3D point cloud data [9,20,27] have been severely hindered due to several important factors. Firstly, a point cloud does not have the neighborhood structure of an image. The data is instead sparse and scattered. As a consequence, CNN-based methods resort to voxelization strategies of the underlying point cloud data to enable 3D-convolutions to be performed (3D-CNNs). Secondly, voxelization have several disadvantages, including loss of spatial resolution and large memory requirements. 3D-CNNs are therefore restricted to small volumetric models or processing data in many smaller chunks, which limits the use of context. Thirdly, annotated 3D data is extremely limited, especially for the 3D semantic segmentation task. This greatly limits the power of CNNs for semantic segmentation of generic 3D point clouds.

In contrast, our approach avoids these short comings by projecting the point cloud into dense 2D image representations, thus removing the need for voxelizations. The 2D images can then be efficiently processed using 2D convolutions. Also, performing segmentation in image space allows us to leverage well developed 2D segmentation techniques as well as large amount of annotated data.

3   Method

In this section we present our method for point cloud segmentation. The input is an unstructured point cloud and the objective is to assign a semantic label to each point. In our method we render the point cloud from different views by projecting the points into synthetic images. We render color, depth and other attributes extracted from the point cloud. The images are then processed by a CNN for image-based semantic segmentation, providing a prediction scores for the predefined classes in every pixel. We make the final class selection from the aggregated prediction scores, using all images where the particular points are visible. An overview of the method is illustrated in Fig. 1. A more detailed description is provided in the following sections.

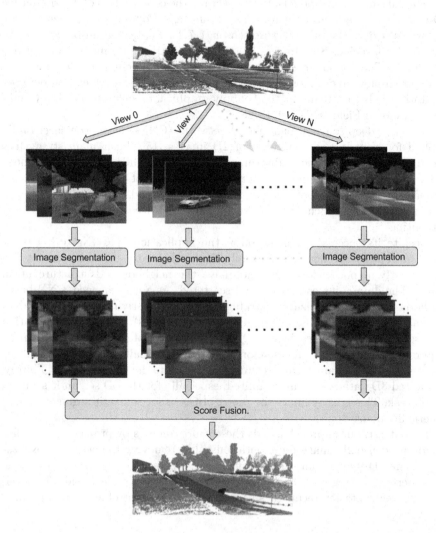

Fig. 1. An overview of the proposed method. The input point cloud is projected into multiple virtual camera views, generating 2D color, depth and surface normal images. The images for each view are processed by a multi-stream CNN for semantic segmentation. The output prediction scores from all views are fused into a single prediction for each point, resulting in a 3D semantic segmentation of the point cloud.

3.1   Render Views

The objective of the point cloud rendering is to produce structured 2D-images that are used as input to a CNN-based semantic segmentation algorithm. A variety of information stemming from the point cloud can be projected onto the synthetic images. In this work we particularly investigate the use of depth, color, and normals. However, the approach can be trivially extended to other features such as HHA [5] and other local information extracted from the point cloud. In order to map the semantic information back to the 3D points, we also need to keep track of the visibility of the projected points.

Our choice of rendering technique is a variant of point splatting [24,29], where the points are projected with a spread function into the image plane. While other rendering techniques, such as surface reconstruction as in [12], require demanding preprocessing steps of the point cloud in 3D space, splatting could be completely processed in image space. This further enables efficient and easily parallelizable implementations, which is essential for large-scale or dense point clouds.

Splatting-based rendering is performed by first projecting each 3D-point x_i of the point cloud into the image coordinates y_i of a virtual camera. The projected points are stored along with their corresponding depth values z_i and feature vectors c_i. The latter can include, e.g., the RGB-color and normal vector of the point x_i. The projection of a 3D-point is distributed by a Gaussian point spread function in the image plane,

$$w_{i,j} = G(y_i - p_j, \sigma^2). \tag{1}$$

Here, $w_{i,j}$ is the contributed weight of point x_i to pixel j in the projected image. It is obtained by evaluating an isotropic Gaussian kernel G with scale σ^2 at the pixel location p_j. In order to reduce computational complexity, the kernel

Fig. 2. Example of rendering output. Left: color image. Right: label image.

is truncated at a distance r. However, point spread functions, which originate from different surfaces, may still intersect in the image plane. Thus, the visibility of the projected points needs to be determined to avoid contributions of occluded surfaces. Moreover, the sensor data may contain significant foreground noise, such as scanning artifacts, which complicates this task. The challenge is to exclude the contribution from the noise and the occluded surfaces in the rendering process.

In traditional splatting [29], the resulting pixel value is obtained from the weighted average of the point spread functions in an accumulated fashion, using the weights $w_{i,j}$. If the depth of a new point significantly differs from the current weighted average, the pixel depth is either re-initialized with the new value if the point is closer than a specific threshold, or discarded if it is further away [29]. However, this implies that the resulting pixel value depends on both the threshold value and the order in which the points are projected. Furthermore, noise in the foreground will have significant impact on the resulting images, as it is always rendered.

Similar to the method proposed in [19], we perform mean-shift clustering [24] of the projected points in each pixel with respect to the depth z_i weighted with $w_{i,j}$ using a Gaussian kernel density estimator $G(d, s^2)$, where s^2 denotes the kernel width. Starting from the depth value $d_i^0 = z_i$ for each point $i \in I_j$ that contributes to the current pixel j, $I_j = \{i : \|p_j - y_i\| < r\}$, the following expression is iterated until convergence

$$d_i^{n+1} = \frac{\sum_{i \in I_j} w_{i,j} G(d_i^n - z_i, s^2) z_i}{\sum_{i \in I_j} w_{i,j} G(d^n - z_i, s^2)}. \tag{2}$$

The iterative process determines a set of unique cluster centers $\{d_k\}_1^K$ from the converged iterates $\{d_i^N\}_{i \in I_j}$. The kernel density of cluster center d_k is given by,

$$v_k = \frac{\sum_{i \in I_j} w_{i,j} G(d_k - z_i, s^2)}{\sum_{i \in I_j} w_{i,j}}. \tag{3}$$

We rank the clusters with respect to the kernel density estimates and the cluster centers,

$$s_k = v_k + \frac{D}{d_k}. \tag{4}$$

Here, the weight D rewards clusters that are near the camera. It is set such that foreground noise and occluded points are not rendered. We chose the optimal cluster as $\tilde{k} = \arg\max_k s_k$ and set the depth value of pixel j to the corresponding cluster center $d_{\tilde{k}}$. The feature value is calculated as the weighted average, where the weight is determined by the proximity to the chosen cluster,

$$\mathbf{c}_{\tilde{k}} = \frac{\sum_{i \in I_j} w_{i,j} G(d_{\tilde{k}} - z_i, s^2) \mathbf{c}_i}{\sum_{i \in I_j} w_{i,j} G(d_{\tilde{k}} - z_i, s^2)}. \tag{5}$$

Since the indices $i \in I_j$ of the contributing points i are stored, it is trivial to map the semantic segmentation scores produced by the CNN back to the point cloud itself.

An example of the rendering output is shown in Fig. 2.

3.2 Deep Multi Stream Image Segmentation

Following the current success of deep learning algorithms we deploy a CNN-based algorithm for performing semantic segmentation on the rendered images. We consider using multiple input modalities, which are combined using a multi-stream architecture [23]. The predictions from the streams are fused in a sum layer, as proposed in [4]. The full multi stream network can thus be trained end-to-end. However, note that our pipeline is agnostic to the applied image semantic segmentation approach.

In our method, each stream is processed using a Fully Convolutional Network (FCN) [15]. However, as previously mentioned, any CNN architecture can be employed. The FCN is based on the popular VGG16 network [22]. The weights in each stream are initialized by pre-training on the ImageNet dataset [2]. In this work, we investigate different combinations of input streams, namely color, depth, and surface normals. While the RGB-stream naturally benefits from pre-training on ImageNet, this is also the case for the depth stream. Previous work [3] has shown that a 3-channel jet colormap representation of the depth image better benefits from pre-training on RGB datasets, such as ImageNet. Finally, we also consider surface normals as input to a separate network stream. For this

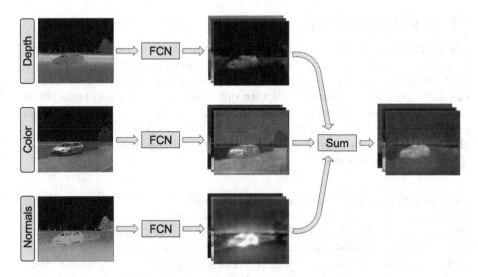

Fig. 3. Illustration of the proposed multi-stream architecture for 2D semantic segmentation. Each input stream is processed by a Fully Convolutional Network [15]. The prediction scores from each stream are summed to get the final prediction.

purpose, we deploy an efficient algorithm for approximate normals computation, which is based on direct differentiation of the depth map. An illustration of the multi-stream architecture is shown in Fig. 3.

3.3 Score Fusion

The deep network outputs a prediction score for each class for every pixel in the image. The scores from each rendered view are mapped to the corresponding 3D points using the indices $i \in I_j$ as described in Sect. 3.1. We fuse the scores by computing the sum over all projections. Finally, the points are assigned the labels corresponding to the largest sum of scores.

4   Experiments

4.1   Dataset

We conduct our experiments on the dataset Semantic3D [6], which provides a set of large scale 3D point clouds of outdoor environments. The point clouds were acquired by a laser scanner and include both urban and rural scenes. Colorization was performed using a cube map generated from a set of high-resolution camera images. In total, the dataset contains 30 separate scans and over 4 billion 3D-points. The points are labeled with 8 different semantic classes: man-made terrain, natural terrain, high vegetation, low vegetation, buildings, hard scape, scanning artifacts, and cars.

4.2   Experimental Setup

View Selection. In order to fully cover the point clouds in the rendered views, we collect images by rotating the camera 360° around a fix vertical axes. For each 360° rotation, we use 30 camera views at equally spaced angles. For each point cloud, we generate four such scans with different pitch angles and translations of the camera, resulting in a total of 120 camera views. To maintain a certain amount of contextual information, we remove images where more than 10% of the pixels have a depth less than five meters. Furthermore, images with less than 5% coverage were discarded.

Network Setup and Training. For the training we generated ground truth label images by selecting the most commonly occurring label in the optimal cluster from Sect. 3.1. An example is shown in Fig. 2. In addition to the 8 provided classes, we also included a 9th background class to label empty pixels, i.e. pixels without any intersecting point spread functions. We generated training data from the training set provided by Semantic3D [6], consisting of 15 point clouds from different scenes. Our training data set consists of 3132 labeled images including color, jet visualization of the depth, and surface normals.

We investigate the proposed multi stream approach using color, depth and surface normals streams as input. In order to determine the contribution of each input stream we also evaluate network configurations with a single stream. Since some point clouds may not have color information we also investigate a multi stream approach without the color stream. All network configurations are listed in Table 1.

Table 1. Network configurations with input streams in the left column

	RGB	D	N	RGB+D+N	D+N
Color	X			X	
Depth jet		X		X	X
Surface normals			X	X	X

All network configurations were trained using the same training parameters. We trained for 45 epochs with a batch size of 16. The initial learning rate was set to 0.0001 and divided by two every tenth epoch. Following the recommendations from [14], we used a momentum of 0.99. The networks were trained using MatConvNet [25].

Fig. 4. Qualitative results. Top: input point clouds. Bottom: Segmentation output using our proposed **RGB+D+N** network.

4.3   Results and Discussions

We evaluated our method for the different network configurations on the reduced test set provided by Semantic3D. The test set consists of four point clouds, containing 80 million points in total. All points are assigned a class label j, which is compared to the ground truth label i. A confusion matrix C is constructed, were each entry c_{ij} denotes the number of points with the ground truth label i that are assigned the label j. The quantitative measure provided by the benchmark [6] is the intersection over union for each class i, given by

$$\text{IoU}_i = \frac{c_{ii}}{c_{ii} + \sum_{j \neq i} c_{ij} + \sum_{k \neq i} c_{kj}}. \tag{6}$$

The over all accuracy is also provided and is given by

$$\text{IoU} = \frac{\sum_i c_{ii}}{\sum_j \sum_{jk} c_{jk}}. \tag{7}$$

The evaluation results are shown in Table 2. The single-stream network with RGB and surface normals as input performs significantly better than the single-stream depth network. However, the three streams seem to provide complementary information, and give a significant gain in performance when used together. Our best multi-stream approach significantly improves over the previous state-of-the art method [8]. Also our multi-stream approach without the color stream obtains results comparable to the previous state-of-the-art, showing that our method is applicable even if color information is absent. Interestingly, even our single-stream approaches with only RGB or surface normals as input achieves a remarkable gain compared to the 3D-CNN based VoxNet [6]. Figure 4 shows some qualitative results on the test set using our multi-stream **RBG+D+N** network.

Note that we are using a simple heuristic for generating camera views, and a basic segmentation network trained on limited data. Yet, we obtain very promising results. Replacing these blocks with better alternatives should improve the results even further. However, this is outside the scope of this paper.

Table 2. Benchmark results on the reduced test set in Semantic3D [6]. IoU for categories (1) man-made terrain, (2) natural terrain, (3) high vegetation, (4) low vegetation, (5) buildings, (6) hard scape, (7) scanning artefacts, (8) cars.

	Avg IoU	OA	IoU1	IoU2	IoU3	IoU4	IoU5	IoU6	IoU7	IoU8
TML-PCR [18]	0.384	0.740	0.726	0.730	0.485	0.224	0.707	0.050	0.000	0.150
DeepNet [6]	0.437	0.772	0.838	0.385	0.548	0.085	0.841	0.151	0.223	0.423
TLMC-MSR [8]	0.542	0.862	**0.898**	0.745	0.537	0.268	0.888	0.189	**0.364**	0.447
Ours RGB	0.515	0.854	0.759	0.791	0.720	**0.335**	0.857	0.209	0.123	0.326
Ours D	0.262	0.662	0.281	0.468	0.395	0.179	0.763	0.006	0.001	0.000
Ours N	0.511	0.846	0.815	0.622	0.679	0.164	0.903	**0.251**	0.186	0.470
Ours RGB+D+N	**0.585**	**0.889**	0.856	**0.832**	**0.742**	0.324	0.897	0.185	0.251	**0.592**
Ours D+N	0.543	0.872	0.839	0.736	0.717	0.210	**0.909**	0.153	0.204	0.574

5   Conclusion

We propose an approach for semantic segmentation of 3D point clouds that avoids the limitations of 3D-CNNs. Our approach first projects the point cloud onto a set of synthetic 2D-images. The corresponding images are then used as input to a 2D-CNN for semantic segmentation. Consequently, the segmentation results are obtained by re-projecting the prediction scores to the point cloud. We further investigate the impact of multiple modalities in a multi-stream deep network architecture. Experiments are performed on the Semantic3D dataset. Our approach outperforms existing methods and sets a new state-of-the-art on this dataset.

Acknowledgements. This work has been supported by the EU's Horizon 2020 Programme grant No 644839 (CENTAURO) and the Swedish Research Council in projects 2014-6227 (EMC2), the Swedish Foundation for Strategic Research (Smart Systems: RIT 15-0097) and the VR starting grant 2016-05543.

References

1. Anguelov, D., Taskar, B., Chatalbashev, V., Koller, D., Gupta, D., Heitz, G., Ng, A.Y.: Discriminative learning of markov random fields for segmentation of 3d scan data. In: 2005 IEEE Computer Society Conference on Computer Vision and Pattern Recognition (CVPR 2005), 20–26 June 2005, San Diego, CA, USA, pp. 169–176 (2005)
2. Deng, J., Dong, W., Socher, R., Li, L.J., Li, K., Fei-Fei, L.: ImageNet: a large-scale hierarchical image database. In: CVPR 2009 (2009)
3. Eitel, A., Springenberg, J.T., Spinello, L., Riedmiller, M., Burgard, W.: Multimodal deep learning for robust RGB-d object recognition. In: 2015 IEEE/RSJ International Conference on Intelligent Robots and Systems (IROS), pp. 681–687. IEEE (2015)
4. Feichtenhofer, C., Pinz, A., Zisserman, A.: Convolutional two-stream network fusion for video action recognition. In: 2016 IEEE Conference on Computer Vision and Pattern Recognition (CVPR 2016), Las Vegas, NV, USA, 27–30 June 2016, pp. 1933–1941 (2016). http://dx.doi.org/10.1109/CVPR.2016.213
5. Gupta, S., Girshick, R., Arbeláez, P., Malik, J.: Learning rich features from RGB-D images for object detection and segmentation. In: Fleet, D., Pajdla, T., Schiele, B., Tuytelaars, T. (eds.) ECCV 2014. LNCS, vol. 8695, pp. 345–360. Springer, Cham (2014). doi:10.1007/978-3-319-10584-0_23
6. Hackel, T., Savinov, N., Ladicky, L., Wegner, J.D., Schindler, K., Pollefeys, M.: Semantic3d.net: a new large-scale point cloud classification benchmark. arXiv preprint (2017). arXiv:1704.03847
7. Hackel, T., Wegner, J.D., Schindle, K.: Fast semantic segmentation of 3d point clouds with strongly varying density. In: ISPRS Annals - ISPRS Congress, Prague (2016)
8. Hackel, T., Wegner, J.D., Schindler, K.: Fast semantic segmentation of 3d point clouds with strongly varying density. ISPRS Ann. Photogram. Remote Sensing Spatial Inf. Sci. **3**, 177–184 (2016). Prague, Czech Republic
9. Huang, J., You, S.: Point cloud labeling using 3d convolutional neural network. In: International Conference on Pattern Recognition (ICPR) (2016)
10. Johnson, A.E., Hebert, M.: Using spin images for efficient object recognition in cluttered 3d scenes. IEEE Trans. Pattern Anal. Mach. Intell. **21**(5), 433–449 (1999)
11. Kähler, O., Reid, I.D.: Efficient 3d scene labeling using fields of trees. In: IEEE International Conference on Computer Vision (ICCV 2013), Sydney, Australia, 1–8 December 2013, pp. 3064–3071 (2013)

12. Kazhdan, M., Hoppe, H.: Screened poisson surface reconstruction. ACM Trans. Graph. (TOG) **32**(3), 29 (2013)
13. Kim, B., Kohli, P., Savarese, S.: 3d scene understanding by voxel-CRF. In: IEEE International Conference on Computer Vision (ICCV 2013), Sydney, Australia, 1–8 December 2013, pp. 1425–1432 (2013)
14. Liu, W., Rabinovich, A., Berg, A.C.: Parsenet: looking wider to see better. arXiv preprint (2015). arXiv:1506.04579
15. Long, J., Shelhamer, E., Darrell, T.: Fully convolutional networks for semantic segmentation. In: Proceedings of the IEEE Conference on Computer Vision and Pattern Recognition, pp. 3431–3440 (2015)
16. Martinovic, A., Knopp, J., Riemenschneider, H., Gool, L.J.V.: 3d all the way: semantic segmentation of urban scenes from start to end in 3d. In: IEEE Conference on Computer Vision and Pattern Recognition (CVPR 2015), Boston, MA, USA, 7–12 June 2015, pp. 4456–4465 (2015)
17. Maturana, D., Scherer, S.: Voxnet: a 3d convolutional neural network for real-time object recognition. In: 2015 IEEE/RSJ International Conference on Intelligent Robots and Systems (IROS), pp. 922–928. IEEE (2015)
18. Montoya-Zegarra, J.A., Wegner, J.D., Ladický, Ľ., Schindler, K.: Mind the gap: modeling local and global context in (road) networks. In: Jiang, X., Hornegger, J., Koch, R. (eds.) GCPR 2014. LNCS, vol. 8753, pp. 212–223. Springer, Cham (2014). doi:10.1007/978-3-319-11752-2_17
19. Ogniewski, J., Forssén, P.E.: Pushing the limits for view prediction in video coding. In: 12th International Conference on Computer Vision Theory and Applications (VISAPP 2017). Scitepress Digital Library, Porto, Portugal (2017)
20. Qi, C.R., Su, H., Nießner, M., Dai, A., Yan, M., Guibas, L.J.: Volumetric and multi-view CNNS for object classification on 3d data. In: 2016 IEEE Conference on Computer Vision and Pattern Recognition (CVPR 2016), Las Vegas, NV, USA, 27–30 June, pp. 5648–5656 (2016)
21. Salti, S., Tombari, F., di Stefano, L.: SHOT: unique signatures of histograms for surface and texture description. Comput. Vis. Image Underst. **125**, 251–264 (2014)
22. Simonyan, K., Zisserman, A.: Very deep convolutional networks for large-scale image recognition. CoRR abs/1409.1556 (2014)
23. Simonyan, K., Zisserman, A.: Two-stream convolutional networks for action recognition in videos. In: Advances in Neural Information Processing Systems 27: Annual Conference on Neural Information Processing Systems 8–13 2014, Montreal, Quebec, Canada, pp. 568–576 (2014). http://papers.nips.cc/paper/5353-two-stream-convolutional-networks-for-action-recognition-in-videos
24. Szeliski, R.: Computer Vision: Algorithms and Applications. Springer, New York (2010)
25. Vedaldi, A., Lenc, K.: Matconvnet - convolutional neural networks for matlab. In: Proceeding of the ACM International Conference on Multimedia (2015)
26. Wolf, D., Prankl, J., Vincze, M.: Fast semantic segmentation of 3d point clouds using a dense CRF with learned parameters. In: IEEE International Conference on Robotics and Automation (ICRA 2015), Seattle, WA, USA, 26–30 May 2015, pp. 4867–4873 (2015)
27. Wu, Z., Song, S., Khosla, A., Yu, F., Zhang, L., Tang, X., Xiao, J.: 3d shapenets: a deep representation for volumetric shapes. In: IEEE Conference on Computer Vision and Pattern Recognition, (CVPR 2015), Boston, MA, USA, 7–12 June 2015, pp. 1912–1920 (2015). http://dx.doi.org/10.1109/CVPR.2015.7298801

28. Zhou, B., Zhao, H., Puig, X., Fidler, S., Barriuso, A., Torralba, A.: Scene parsing through ADE20K dataset. In: Proceedings of the IEEE Conference on Computer Vision and Pattern Recognition (2017)
29. Zwicker, M., Pfister, H., Van Baar, J., Gross, M.: Surface splatting. In: Proceedings of the 28th annual conference on Computer graphics and interactive techniques, pp. 371–378. ACM (2001)

Detection of Curved Lines with *B*-COSFIRE Filters: A Case Study on Crack Delineation

Nicola Strisciuglio[1]([✉]), George Azzopardi[1,2], and Nicolai Petkov[1]

[1] Johann Bernoulli Institute for Mathematics and Computer Science,
University of Groningen, Groningen, The Netherlands
`n.strisciuglio@rug.nl`
[2] Intelligent Computer Systems, University of Malta, Msida, Malta

Abstract. The detection of curvilinear structures is an important step for various computer vision applications, ranging from medical image analysis for segmentation of blood vessels, to remote sensing for the identification of roads and rivers, and to biometrics and robotics, among others. This is a nontrivial task especially for the detection of thin or incomplete curvilinear structures surrounded with noise. We propose a general purpose curvilinear structure detector that uses the brain-inspired trainable *B*-COSFIRE filters. It consists of four main steps, namely non-linear filtering with *B*-COSFIRE, thinning with non-maximum suppression, hysteresis thresholding and morphological closing. We demonstrate its effectiveness on a data set of noisy images with cracked pavements, where we achieve state-of-the-art results (F-measure = 0.865). The proposed method can be employed in any computer vision methodology that requires the delineation of curvilinear and elongated structures.

Keywords: Line detection · Curved lines · Non-linear filtering · COSFIRE · Crack delineation

1 Introduction

The detection of curvilinear and elongated structures is of great importance in image processing due to its application to numerous problems. The delineation of blood vessels in medical images, the detection and measure of cracks in walls and roads for damage estimation, the segmentation of river and roads in aerial and satellite images to prevent disasters or accidents are few applications of algorithms for the detection of curvilinear patterns.

In the literature, various approaches for curvilinear structure detection were proposed, for which a survey was recently published in [5]. Existing methodologies range from parametric methods to approaches based on filtering techniques or region growing, point processes and machine learning. For instance, the Hough transform is a parametric method, which converts an input image to a parameter space where line or circle segments can be detected. It requires a mathematical model of the patterns of interest. Different elongated structures, such as lines,

M. Felsberg et al. (Eds.): CAIP 2017, Part I, LNCS 10424, pp. 108–120, 2017.
DOI: 10.1007/978-3-319-64689-3_9

circles and Y-junctions, require different mathematical models to transform an image into the particular parameter space where the patterns of interest can be distinguished.

Approaches based on filtering techniques employ local derivatives in a multi-scale analysis [7] or model the profile of elongated structures by means of a two-dimensional Gaussian kernel [10]. Region-growing that considers multi-scale information about width, length and orientation of lines was proposed in [14]. In [15], mathematical morphology and tracking techniques were combined with a-priori information about the line network of interest. These methods are intuitive but require a-priori structural information about the patterns of interest.

Despite their high computational complexity, point and object processes were proposed to detect line networks in images, especially in applications of road and river detection in aerial images. Methods in this group are based on tracking elongated structures by simulation of complex mathematical models. In [11], a line network is considered as a set of interacting line segments which are reconstructed by object processes. Point processes based on the Gibbs model and Monte Carlo simulations were introduced in [12,25], respectively. In [6,24], point processes were combined with a graph-based representation and classification to improve the accuracy of segmentation. Point processes and graph-based approaches require high computational resources, consequently reducing the applicability to high resolution images.

In the last group, there are methods that employ machine learning techniques. They are based on the construction of pixel-wise feature vectors and the use of classifiers to decide whether a pixel is part of an elongated structure or not. In [16], the responses of multi-scale Gaussian filters were used in combination with a k-NN classifier, while in [19] a feature vector was constructed with the responses of a bank of ridge detectors. In [18], the coefficients of multi-scale Gabor wavelets were used to form a feature vector and to train a Bayesian classifier. Recently, a convolutional neural network trained with image patches of lines was proposed in [13]. These methods are more complex than filtering-based approaches and require long training time. Furthermore, the classifiers can be trained only when ground truth is available, which is not always possible or prohibitively expensive to obtain.

In this work, we present a method for the detection of curvilinear structures, composed of four steps: 1. B-COSFIRE filtering, 2. thinning with non-maximum suppression, 3. hysteresis thresholding and 4. morphological closing. The basic idea of the B-COSFIRE filters, that we originally proposed for retinal vessel segmentation in [4,21], is inspired by the functions of simple cells in area V1 of visual cortex selective to elongated patterns of certain widths and orientations.

The B-COSFIRE filter is trainable as its structure is not fixed in the implementation, but it is learned in an automatic configuration process performed on a prototype pattern. The concept of trainable filters was introduced in [3] and employed in image analysis [2], object recognition [8] and adapted to audio analysis [23]. The trainability of the COSFIRE approach concerns the learning of the structure of the filters directly from prototype patterns. This aspect

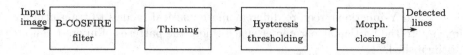

Fig. 1. High-level architecture of the proposed curved line operator. A rotation-tolerant *B*-COSFIRE filter responds to lines of preferred width and length. Thinning, hysteresis thresholding and morphological closing are performed to obtain the final elongated and curvilinear binary structures.

can be considered a kind of *representation learning*. Similarly to deep learning but considering a single training sample at time, it aims at avoiding a feature engineering process and building adaptive pattern recognition systems.

We perform experiments on a publicly available data set, namely Crack_PV14 data set [26] and compare the resulting performance to those of existing methods. We show the effectiveness of the proposed method for the detection of curvilinear and elongated structures, the robustness of *B*-COSFIRE filters to incomplete lines, noise and tortuosity, and their application in a pipeline for crack detection in pavement and road images.

The paper is organized as follows. In Sect. 2 we present the proposed curved line operator. In Sect. 3 we describe the Crack_PV14 data set and the experimental protocol that we followed, compare the results that we achieved with the ones reported in the literature and discuss certain aspects of the proposed method. Finally, we draw conclusions in Sect. 4.

2 Method

2.1 Overview

In Fig. 1, we show the main steps of the proposed curved line detector. The architecture for the detection of curvilinear and elongated patterns, which we apply to the detection of cracks in pavement and road images, is composed of four steps: 1. *B*-COSFIRE filtering, 2. thinning with non-maximum suppression, 3. hysteresis thresholding and 4. morphological closing.

Below we present the *B*-COSFIRE filter and show how it is incorporated into the proposed curved line operator that is suitable for the detection of cracks in roads and walls.

2.2 Configuration of a *B*-COSFIRE Filter

A *B*-COSFIRE filter takes input from of a group of Difference-of-Gaussians functions DoG_σ, with the outer Gaussian function having standard deviation σ:

$$DoG_\sigma(x,y) = \frac{\exp\left(-\frac{x^2+y^2}{2(0.5\sigma)^2}\right)}{2\pi(0.5\sigma)^2} - \frac{\exp\left(-\frac{x^2+y^2}{2\sigma^2}\right)}{2\pi\sigma^2} \qquad (1)$$

In [17], it was shown that for the above function the maximum response is elicited for a spot with a radius of 0.96σ or a line with a width of 1.92σ. Based on this finding we set the outer standard deviation $\sigma = w/1.92$ where w is the preferred width of the line of interest.

The structure of a B-COSFIRE filter, i.e. the positions at which we consider the DoG responses, is determined in an automatic configuration process on a given prototype pattern. For details about the configuration we refer the reader to [4]. We configure a B-COSFIRE filter on a prototype line structure of width w, length l and orientation ϕ. The result of the configuration is a set $B_{w,l,\phi}$:

$$B_{w,l,\phi} = \{(0,0), (\lambda, \phi), (\lambda, 2\pi - \phi)\} \cup \{(\rho_i, \phi)\} \cup \{(\rho_i, 2\pi - \phi)\} \tag{2}$$

where $\lambda = \lfloor (l-1)/2 \rfloor$ and $\rho_i = \eta i$ with $i = 1, \ldots, \lfloor (\lambda-1)/\eta \rfloor - 1$. ϕ is the preferred orientation of the line. The two-tuples in the set $B_{w,l,\phi}$ indicate the positions (distances and polar angles) with respect to the B-COSFIRE filter support center at which we take the responses of a center-on difference-of-Gaussians (DoG) filter. The parameter η (with $1 \leq \eta \leq \lambda$) represents the pixel spacing between the considered DoG responses. When $\eta = 1$ we configure a tuple for every location along a line with preferred width w, length l, and orientation ϕ, and when $\eta = \lfloor (l-1)/2 \rfloor$ (i.e. the maximum possible value) the resulting filter consists of only three tuples: the tuple $(0,0)$ that refers to the DoG response at the center, and the two tuples $(\lfloor (l-1)/2 \rfloor, \phi)$ and $(\lfloor (l-1)/2 \rfloor, 2\pi - \phi)$ that refer to the farthest distances on both sides of the support. The selectivity of a B-COSFIRE filter increases with decreasing η value.

2.3   Response of a B-COSFIRE Filter

The response of a B-COSFIRE filter $B_{w,l,\phi}(x,y)$ is computed in four steps, namely *filter-blur-shift-combine*.

In the first step we *filter* an input image I with the DoG kernel $DoG_{\sigma=w/1.92}$ and denote the resulting image by C:

$$C(x,y) = \left| \sum_{x'=-3\sigma}^{3\sigma} \sum_{y'=-3\sigma}^{3\sigma} I(x,y) DoG_\sigma(x - x', y - y') \right|^+ \tag{3}$$

where the operation $|.|^+$ denotes half-wave rectification, recently also known as rectifying linear unit (ReLU).

Then, in order to allow for some tolerance with respect to the preferred positions we *blur* the DoG responses by a nonlinear blurring operation that consists in a weighted maximum. The weighting is given by a Gaussian function whose standard deviation σ_i' increases linearly with an increasing distance from the support center of the B-COSFIRE filter: $\sigma_i' = \sigma_0' + \alpha\rho_i$. The values σ_0' and α are parameters and regulate the tolerance to deformations of the prototype pattern.

In the third step, we *shift* the i-th blurred DoG response by a vector $(\rho_i, 2\pi - \phi_i)$. In this way, all involved DoG responses meet at the same location, that is the support center of the concerned B-COSFIRE filter. We denote by $s_{\rho_i,\phi_i}(x,y)$ the blurred and shifted DoG response at the location (x,y) of the i-th tuple:

$$s_{\rho_i,\phi_i}(x,y) = \max_{x',y'}\{C(x - x' - \Delta x, y - y' - \Delta y)\, G_{\sigma_i'}(x',y')\} \qquad (4)$$

where $\Delta x = -\rho_i \cos\phi_i$ and $\Delta y = -\rho_i \sin\phi_i$.

Finally, we denote by $r_{B_{w,l,\phi}}(x,y)$ the response of a B-COSFIRE filter, which we compute by geometric mean:

$$r_{B_{w,l}}(x,y) = \left(\prod_{i=1}^{|B_{w,l}|} s_{\rho_i,\phi_i}(x,y) \right)^{1/|B_{w,l}|}. \qquad (5)$$

2.4   Orientation Bandwidth and Tolerance to Rotation

The orientation bandwidth of a B-COSFIRE filter is controlled by the parameters σ_0' and α. In the example of Fig. 2, the B-COSFIRE filter that is selective for lines of length 59 pixels achieves an orientation bandwidth (full width at 75% of the maximum) of circa $\pi/8$ radians, for $\sigma_0' = 5$ and $\alpha = 1$.

We configure a set $\beta = \{B_{w,l,\phi=\theta} \mid \theta = 0, \pi/8, \ldots, 7\pi/8\}$ of B-COSFIRE filters with eight orientation preferences. With a bandwidth of circa $\pi/8$ radians, a set of eight B-COSFIRE filters with equidistant orientation preference is sufficient to respond to lines in any orientation. We denote by $\hat{r}_\beta(x,y)$ and $\Phi_\beta(x,y)$ the rotation-tolerant response and the orientation map of the rotation-tolerant B-COSFIRE filter:

$$\hat{r}_\beta(x,y) = \max\{r_{B_{w,l,\phi=0}}(x,y), r_{B_{w,l,\phi=\pi/4}}(x,y), \ldots, r_{B_{w,l,\phi=7\pi/8}}(x,y)\} \qquad (6)$$

$$\Phi_\beta(x,y) = \underset{\phi}{\operatorname{argmax}}\{r_{B_{w,l,\phi=0}}(x,y), r_{B_{w,l,\phi=\pi/4}}(x,y), \ldots, r_{B_{w,l,\phi=7\pi/8}}(x,y)\} \qquad (7)$$

2.5   Binary Map with Thinning and Hysteresis Thresholding

In order to obtain a binary map of curvilinear structures, we apply a thinning and hysteresis thresholding operations to the response map of the rotation-tolerant B-COSFIRE filter. We use the thinning algorithm described in [9] that takes as input the response map $\hat{r}_\beta(x,y)$ and the orientation map $\Phi_\beta(x,y)$ and applies non-maximum suppression to thin areas in the response map, where the responses are non-zero, to one pixel wide candidate points belonging to curvilinear structures.

The hysteresis thresholding requires a low and a high threshold parameter values, denoted by t_l and t_h, respectively. We set $t_l = 0.5t_h$ according to [9]. The resulting binary image depends on the given high threshold t_h: the lower that value the more line pixels in the binary image as less responses are suppressed. In Fig. 2d we show the thinned and binarized response map of the proposed curved line operator applied to the image in Fig. 2a.

We finally perform a morphological closing operation, with a 3×3 square structuring element, so as to fill eventual small gaps in the detected lines.

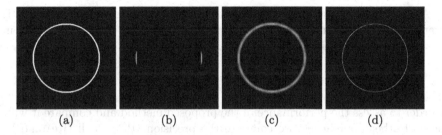

(a)   (b)   (c)   (d)

Fig. 2. (a) An input image (of size 300×300 pixels) containing a circle with a circumference of radius 100 pixels and a line width of 5 pixels. (b) The response map obtained by the B-COSFIRE filter $B_{w=5,l=59,\phi=0}$. For $\sigma_0' = 5$ and $\alpha = 1$ the orientation bandwidth at 75% of the maximum is circa $\pi/8$ radians. (c) The rotation-tolerant response map with eight orientations and (d) the thinned and binary image.

3   Experiments

3.1   Data Set

We carried out experiments on a publicly available data set of pavement images called Crack_PV14 [26]. This data set is composed of 14 images taken with a laser range imaging appliance, mounted on the back of a car. We show an example image in Fig. 3a. The images are distributed in BMP format and have resolution of 200×300 pixels. Each image is provided together with a manually annotated image that serves as ground truth for performance evaluation, Fig. 3b. The ground truth annotation for each image is a one-pixel wide line-network that delineates the center-line of the cracks.

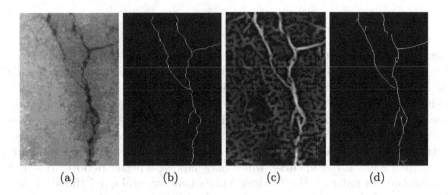

(a)   (b)   (c)   (d)

Fig. 3. (a) Example image of a pavement crack with (b) Its corresponding manually annotated ground truth. (c) The response of a rotation-tolerant B-COSFIRE filter and (d) The final thinned and binarized output image.

As an example, we show the response of a rotation-tolerant B-COSFIRE filter applied to the image in Fig. 3a and its thinned an binarized version in Figs. 3c and d, respectively.

3.2 Evaluation

In order to assess the performance of the proposed method and compare it with those of other approaches, we compute the precision (Pr), recall (Re) and F-measure (F) for each image in the Crack_PV14 data set, as follows:

$$Pr = \frac{TP}{TP + FP}, Re = \frac{TP}{TP + FN}, F = \frac{2 \cdot Pr \cdot Re}{Pr + Re}, \tag{8}$$

where TP are true positive pixels, FP are false positives and FN are false negatives. For each image we compute these three measurements for values of the threshold t_h from 0 to 1 in steps of 0.01. Then, we compute the average F-measure $\bar{F}$ for each threshold value on the whole data set and choose the value of t_h that contributes to the highest $\bar{F}$ value.

According to [26], we consider some tolerance when computing the performance measures to compensate for some imprecision in the ground truth. If the Euclidean distance d of a detected crack point to the nearest crack point in the ground truth is lower than a value d^*, we consider that point as a true positive, otherwise it is a false positive. The points of the ground truth that are not detected within a distance d^* in the output image are considered false negatives. As suggested in [26] we set $d^* = 2$. Furthermore, we evaluate the overall performance of the proposed approach by plotting the Precision-Recall curve. This curve shows the trade-off between the Precision and Recall metrics as the value t_h for the hysteresis thresholding varies.

3.3 Results and Discussion

Using this approach, we obtained an average F-measure $\bar{F}$ on the Crack_PV14 data set equals to 0.865 (with a standard deviation of 0.0975). In Fig. 4 we plot the Precision-Recall curve obtained by the proposed method. On the same plot we indicate the points that correspond to the results achieved by other methods. The points corresponding to the CrackTree [27] and FoSA [1] approaches are considerably below our curve, and hence they are much less effective than our method. The point that represents the average results reported in [26] is slightly above the Precision-Recall curve, which may indicate a better performance than our method. In order to clarify these indications, we evaluated the statistical significance of the results that we obtained with respect to the ones achieved by other methods by means of a paired t-test statistic. It turns out that there is significant statistical difference between the results of our method and those obtained by CrackTree and FoSA, but no statistical difference with respect to the ones obtained by Zou *et al.* [26]. Although we achieved comparable results with the ones obtained by the method of Zou *et al.*, the proposed approach is

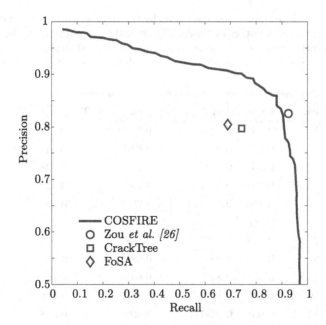

Fig. 4. Precision-Recall curve achieved by the proposed approach on the Crack_PV14 data set. The point ◊ corresponds to the results of the FoSA method ($Pr = 0.8045$, $Re = 0.6896$), the point □ to the results of the CrackTree method ($Pr = 0.7972$, $Re = 0.7441$), while the point o to the ones of Zou *et al.* [26] ($Pr = 0.8254$, $Re = 0.9253$).

based on a general algorithm for delineation of curvilinear structures in images, while other approaches are designed to solve a specific problem.

In Table 1 we report the F-measure that we obtained for each image in the Crack_PV14 data set and compare them with the ones obtained by other methods. The results that we report are obtained by setting the hysteresis threshold t_h equal to 0.49, the one that contributed to the best overall results.

The curved line operator based on *B*-COSFIRE filters that we propose can be employed in image processing pipelines that require the delineation of elongated structures. In this work, we demonstrated the effectiveness of the proposed operator in the application of crack detection in images with noisy pavements.

The configuration parameters of a *B*-COSFIRE filter determine its selectivity for lines of given width, length and orientation. For our experiments, we chose these values in a way that the configured filter is selective for average characteristics of the patterns of interest (i.e. the cracks) in the application at hand. We configured a single *B*-COSFIRE filter with the following parameters: $w = 6.34$, $l = 29$, $\eta = 2$, $\sigma_0 = 2$ and $\alpha = 1$, which we determined by a grid search on 50% of the images in the Crack_PV14 data set.

In contrast to existing approaches for line detection, the *B*-COSFIRE filters are not restricted to the detection of elongated structures. They can be configured to be selective for any pattern of interest in an automatic configuration

Table 1. Detailed F-measure values achieved by the proposed approach in comparison with the ones obtained by other methods. The statistical significance of the F-measure difference with other methods is evaluated with a paired t-test statistic ($h = 0$ indicates that the difference is not statistically significant while $h = 1$ statistical significance).

F-measure per image in the Crack_PV14 data set				
Image	Ours	Zou *et al.* [26]	CrackTree [27]	FoSA [1]
1	0.899	**0.916**	0.751	0.721
2	0.867	**0.872**	0.614	0.64
3	**0.97**	0.874	0.79	0.728
4	0.822	**0.845**	0.764	0.722
5	0.886	**0.944**	0.70	0.691
6	**0.95**	0.747	0.708	0.729
7	0.818	**0.937**	0.648	0.623
8	0.836	**0.886**	0.682	0.698
9	0.69	**0.867**	0.695	0.70
10	**0.99**	0.892	0.892	0.808
11	0.668	**0.906**	0.898	0.858
12	0.854	**0.893**	0.883	0.835
13	**0.872**	0.639	0.739	0.679
14	**0.982**	0.913	0.966	0.901
$\overline{F}$	0.865	**0.866**	0.766	0.738
σ_F	0.0975	**0.0811**	0.1057	0.0821
Results of paired t-test statistic				
h	–	0	1	1
p-value	–	0.9608	0.0131	0.0015

step that is performed on a given prototype pattern. This possibility is a kind of *representation learning*, which involves the construction of a data representation learned directly from training data. Similarly to deep learning and in contrast with traditional pattern recognition approaches, the COSFIRE approach avoids engineering of hand-crafted features and allows for the construction of flexible and adaptive pattern recognition systems.

One can configure filters that are selective for curvilinear structures of different sizes or shapes and combine their responses in order to improve the performance of the concerned method. As an example, in [4] we demonstrated how the responses of a B-COSFIRE filter selective for vessels and one for vessel-endings were combined to improve the quality of the vessel delineation. Alternatively, machine learning techniques proposed in [20, 22], based on genetic algorithms and learning vector quantization, can be utilized to select and combine the responses of the most discriminant filters from a large set of pre-configured ones.

One of the strengths of the *B*-COSFIRE filter approach is the tolerance it uses in its application phase, which is controlled by the parameters σ_0 and α. This accounts for generalization capabilities and robustness with respect to variations of the patterns of interest.

The characteristics of the *B*-COSFIRE filters, namely the possibility of combining the responses of filters selective for structures with different characteristics, the automatic configuration step, which is explained in detail in [4], and the tolerance introduced in their application phase, make them a flexible tool for image processing and pattern recognition. The *B*-COSFIRE filters can be used for the design and implementation of systems which can be easily adapted and applied to various problems.

For the processing of a single image in the Crack_PV14 data set (of 200×300 pixels), our straightforward Matlab implementation takes an average time of 0.3575 s (with a standard deviation of 0.0228) on a machine with a 2.7 GHz processor. The processing of a *B*-COSFIRE filter is, however, parallelizable and hence the efficiency of the proposed approach can be further improved.

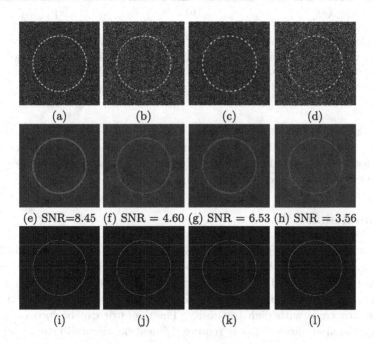

Fig. 5. (First row) Four stimuli (of size 300×300 pixels) of dashed circles with radii of 100 pixels and line width of 5 pixels, together with added Gaussian noise. The small arcs along the circumference are separated by (a, b) $3°$ and (c, d) $5°$ and the Gaussian noise in each image has zero mean and a variance of 0.2 in (a, c) and 0.5 in (b, d). (Second row) The corresponding rotation-tolerant response maps of the *B*-COSFIRE filter. (Third row) The final thinned and binarized output maps with $t_h = 0.75$.

3.4   Robustness to Noise, Incomplete Lines and Tortuosity

Figure 5 provides insights about the robustness of the proposed method to noise and incomplete lines. In the first row we use four stimuli with dashed circles added with Gaussian noise of different variance. For each response map of the rotation-tolerant B-COSFIRE filter (second row) we compute the signal-to-noise ratio $SNR = 20\log_{10}(A_s/A_n)$, where A_s is the average of all responses along the circumference of the circle in the input image, and A_n is the average of all the other responses. The thinned and binarized results presented in the last row demonstrate the robustness of the proposed method.

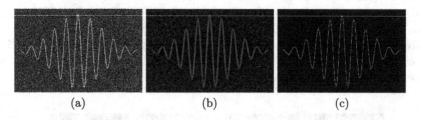

(a)   (b)   (c)

Fig. 6. (a) A synthetic image (of size 500×300 pixels) with Gaussian noise (zero mean and variance of 0.2) superimposing a curvilinear structure that follows a one-dimensional Gabor function. (b) The response map of a B-COSFIRE filter and (c) its thinned and binarized output ($t_h = 0.75$).

We also demonstrate the robustness of the proposed method with respect to tortuosity. In Fig. 6a we illustrate a stimulus with different degrees of tortuosity surrounded with Gaussian noise and in Fig. 6(b–c) we show the response map of the B-COSFIRE filter and the final binary output.

4   Conclusions

The detector of curvilinear and elongated structures that we propose is highly effective in noisy images. We achieve state-of-the-art results (F-measure equals to 0.865) on the Crack_PV14 benchmark data set of images with noisy and cracked pavements. The proposed method is also very robust to incomplete lines and to linear structures with high tortuosity. The operator can be incorporated in computer vision applications that require delineation of curvilinear structures.

References

1. Li, Q., Zou, Q., Zhang, D., Mao, Q.: Fosa: F* seed-growing approach for crack-line detection from pavement images. Image Vis. Comput. **29**(12), 861–872 (2011)
2. Azzopardi, G., Petkov, N.: A CORF computational model of a simple cell that relies on lgn input outperforms the gabor function model. Biol. Cybern. **106**, 177–189 (2012)

3. Azzopardi, G., Petkov, N.: Trainable COSFIRE filters for keypoint detection and pattern recognition. IEEE Trans. Pattern Anal. Mach. Intell. **35**, 490–503 (2013)
4. Azzopardi, G., Strisciuglio, N., Vento, M., Petkov, N.: Trainable COSFIRE filters for vessel delineation with application to retinal images. Med. Image Anal. **19**(1), 46–57 (2015)
5. Bibiloni, P., González-Hidalgo, M., Massanet, S.: A survey on curvilinear object segmentation in multiple applications. Pattern Recognit. **60**, 949–970 (2016)
6. Chai, D., Forstner, W., Lafarge, F.: Recovering line-networks in images by junction-point processes. In: CVPR (2013)
7. Frangi, A.F., Niessen, W.J., Vincken, K.L., Viergever, M.A.: Multiscale vessel enhancement filtering. In: Wells, W.M., Colchester, A., Delp, S. (eds.) MICCAI 1998. LNCS, vol. 1496, pp. 130–137. Springer, Heidelberg (1998). doi:10.1007/BFb0056195
8. Gecer, B., Azzopardi, G., Petkov, N.: Color-blob-based COSFIRE filters for object recognition. Image Vis. Comput. **57**, 165–174 (2017)
9. Grigorescu, C., Petkov, N., Westenberg, M.A.: Contour and boundary detection improved by surround suppression of texture edges. Image Vis. Comput. **22**(8), 609–622 (2004)
10. Hoover, A., Kouznetsova, V., Goldbaum, M.: Locating blood vessels in retinal images by piecewise threshold probing of a matched filter response. IEEE Trans. Med. Imag. **19**(3), 203–210 (2000)
11. Lacoste, C., Descombes, X., Zerubia, J.: Point processes for unsupervised line network extraction in remote sensing. IEEE Trans. Pattern Anal. Mach. Intell. **27**(10), 1568–1579 (2005)
12. Lafarge, F., Gimel'farb, G., Descombes, X.: Geometric feature extraction by a multimarked point process. IEEE Trans. Pattern Anal. Mach. Intell. **32**(9), 1597–1609 (2010)
13. Liskowski, P., Krawiec, K.: Segmenting retinal blood vessels with deep neural networks. IEEE Trans. Med. Imag. **35**(11), 2369–2380 (2016)
14. Martinez-Pérez, M.E., Hughes, A.D., Thom, S.A., Bharath, A.A., Parker, K.H.: Segmentation of blood vessels from red-free and fluorescein retinal images. Med. Image Anal. **11**(1), 47–61 (2007)
15. Mendonca, A.M., Campilho, A.: Segmentation of retinal blood vessels by combining the detection of centerlines and morphological reconstruction. IEEE Trans. Med. Imag. **25**(9), 1200–1213 (2006)
16. Niemeijer, M., Staal, J., van Ginneken, B., Loog, M., Abramoff, M.: Comparative study of retinal vessel segmentation methods on a new publicly available database. In: SPIE Medical Imaging (2004)
17. Petkov, N., Visser, W.: Modifications of center-surround, spot detection and dot-pattern selective operators. University of Groningen, Johann Bernoulli Institute for Mathematics and Computer Science (2005)
18. Soares, J.V.B., Leandro, J.J.G., Cesar Jr., R.M., Jelinek, H.F., Cree, M.J.: Retinal vessel segmentation using the 2-D Gabor wavelet and supervised classification. IEEE Trans. Med. Imag. **25**(9), 1214–1222 (2006)
19. Staal, J., Abramoff, M., Niemeijer, M., Viergever, M., van Ginneken, B.: Ridge-based vessel segmentation in color images of the retina. IEEE Trans. Med. Imag. **23**(4), 501–509 (2004)
20. Strisciuglio, N., Azzopardi, G., Vento, M., Petkov, N.: Multiscale blood vessel delineation using B-COSFIRE filters. In: Azzopardi, G., Petkov, N. (eds.) CAIP 2015. LNCS, vol. 9257, pp. 300–312. Springer, Cham (2015). doi:10.1007/978-3-319-23117-4_26

21. Strisciuglio, N., Azzopardi, G., Vento, M., Petkov, N.: Unsupervised delineation of the vessel tree in retinal fundus images. In: VIPIMAGE, pp. 149–155 (2015)
22. Strisciuglio, N., Azzopardi, G., Vento, M., Petkov, N.: Supervised vessel delineation in retinal fundus images with the automatic selection of B-COSFIRE filters. Mach. Vis. Appl. 1–13 (2016)
23. Strisciuglio, N., Vento, M., Petkov, N.: Bio-inspired filters for audio analysis. In: Amunts, K., Grandinetti, L., Lippert, T., Petkov, N. (eds.) Brain-Comp 2015. LNCS, vol. 10087, pp. 101–115. Springer, Cham (2016). doi:10.1007/978-3-319-50862-7_8
24. Türetken, E., Benmansour, F., Andres, B., Głowacki, P., Pfister, H., Fua, P.: Reconstructing curvilinear networks using path classifiers and integer programming. IEEE Trans. Pattern Anal. Mach. Intell. **38**(12), 2515–2530 (2016)
25. Verdié, Y., Lafarge, F.: Efficient Monte Carlo Sampler for detecting parametric objects in large scenes. In: Fitzgibbon, A., Lazebnik, S., Perona, P., Sato, Y., Schmid, C. (eds.) ECCV 2012. LNCS, vol. 7574, pp. 539–552. Springer, Heidelberg (2012). doi:10.1007/978-3-642-33712-3_39
26. Zou, Q., Li, Q., Zhang, F., Wang, Z.X.Q., Wang, Q.: Path voting based pavement crack detection from laser range images. In: IEEE ICDSP, pp. 432–436 (2016)
27. Zou, Q., Cao, Y., Li, Q., Mao, Q., Wang, S.: Cracktree: automatic crack detection from pavement images. Pattern Recognit. Lett. **33**(3), 227–238 (2012)

Motion-Coherent Affinities for Hypergraph Based Motion Segmentation

Kai Cordes[(⊠)], Christopherus Ray'onaldo, and Hellward Broszio

VISCODA GmbH, Hannover, Germany
{cordes,rayonaldo,broszio}@viscoda.com
http://www.viscoda.com

Abstract. Motion segmentation is the task of classifying the feature trajectories in an image sequence to different motions. Hypergraph based approaches use a specific graph to incorporate higher order similarities for the estimation of motion clusters. They follow the concept of hypothesis generation and validation. For the sampling of hypotheses, a high probability of selecting clean samples, i.e. samples consisting of points from the same cluster, is desired. Many approaches use spatial proximity to build an auxiliary graph for the sampling. But, spatial proximity is often not sufficient to capture the main affinities for motion segmentation. Thus, we introduce a simple but effective model for incorporating motion-coherent affinities into the auxiliary graph. The evaluation on two state of the art benchmarks shows that the hypotheses generated from the resulting hypergraphs lead to a significant decrease of the segmentation error. Additionally, less computation time is required due to a reduced hypergraph complexity.

1 Introduction

Motion segmentation algorithms classify feature trajectories in an image sequence to a number of motions. Most approaches are multi-frame methods [1,4,6,22]. They take trajectories from many frames (e.g. 30) as input. On the other hand, there are two-view methods [8,13,15,20,22] which use correspondences of only two frames as input. In practice, it is often impossible to establish a sufficient number of trajectories on moving objects with large trajectory length. Often, a short response time is desired. Thus, our aim is to solve the motion segmentation task using small trajectory lengths, such as 3–10 frames. The clustering of trajectories can be used estimate the three-dimensional geometry of the scene including moving objects [5].

Recently, hypergraph based methods have been proposed for clustering tasks, such as motion segmentation [9,16,22], geometric model fitting [21,22] or face clustering [16]. A hypergraph contains higher order similarities instead of pairwise similarities. This approach results in more accurate results with the same number of samples. In contrast to previous approaches [2,10] where only small degrees for hyperedges are used, recent approaches employ large hyperedge

© Springer International Publishing AG 2017
M. Felsberg et al. (Eds.): CAIP 2017, Part I, LNCS 10424, pp. 121–132, 2017.
DOI: 10.1007/978-3-319-64689-3_10

Fig. 1. Feature distribution for the example *cars5* from the Hopkins155 benchmark [19] (left) and for the example *van* from the MPTV benchmark [8] (right). For *cars5*, feature points on different objects mostly have large spatial distances while for *van*, features all features show small spatial distances, but different motion vectors.

degrees [16,22]. Using large degrees of hyperedges yields better clustering accuracy because more information of the relationship between vertices is included. To incorporate large hyperedges, the approach [16] makes use of a special sampling technique [14,18]. Therefor, a neighborhood structure is required which guides the hypothesis sampling. The neighborhood structure is encoded in the auxiliary graph. In [16], the auxiliary graph is based on spatial proximity. The idea is that adjacent trajectories are likely to belong to the same motion segment. Thus, it is advantageous that they share the same hyperedge. Sampling from these hyperedges guarantee a relatively high probability of selecting candidates from the same motion segment, so-called pure hyperedges. Experiments [16] show superior performance of large hyperedges for the applications face clustering (Yale face database [7]) and motion segmentation (Hopkins155 data set [19]).

For the evaluation of motion segmentation approaches the well known Hopkins155 benchmark [19] is widely used. It consists of image sequences and trajectory data together with ground truth information providing the motion segmentation result (cf. Figure 1, left: red, green, and yellow colors indicate three motion segments). All trajectories in the Hopkins benchmark consist of points which are visible in every frame of the sequence. Trajectories which discontinue within the sequence were deleted from the data set. Thus, only large trajectory lengths are provided in the benchmark. Since its publication in 2007, many approaches result in low segmentation errors on the Hopkins data set.

The MPTV benchmark [8] includes eight challenging sequences. Like in Hopkins155, a ground truth labeling of the trajectories is provided. In contrast to Hopkins155, feature tracks may start and end in any frame. Thus, trajectories of arbitrary length are included in MPTV. For applications such as motion estimation from a camera mounted on a car, short trajectories are of special importance since long trajectories are rarely available on moving objects.

We propose a method for building the auxiliary graph which incorporates the motion of a trajectory instead of using spatial proximity only [16]. The new auxiliary graph leads to improved hypotheses which lead to a decrease of the segmentation error on both benchmarks. Furthermore, less computation time is required since the complexity of the graph is reduced. The extension is evaluated on Hopkins155 and MTPV. For the evaluation, the sequences are divided into subsequences of length f_s to provide results for limited trajectory length. Trajectories which are visible in each image of the subsequence are included in the evaluation. This leads to a large number of experiments for each sequence. For MPTV, this step is necessary since the trajectories have arbitrary length.

To summarize, the contributions of this paper are as follows:

- an improved method for the generation of the auxiliary using motion cues from the trajectory data
- the evaluation on two reference benchmarks [8,19] based on limited trajectory lengths
- the evaluation of the applicability of the benchmarks [8,19] since Hopkins155 only includes long trajectories while MTPV provides arbitrary trajectory lengths

In the following Sect. 2, the hypergraph based approach of motion segmentation is introduced. In Sect. 3, the new auxiliary graph is presented. Section 4 shows experimental results. In Sect. 5, the paper is concluded.

2   Motion Segmentation Based on Hypergraphs

We briefly review the reference method proposed in [16]. An overview diagram is shown in Fig. 2. Based on the input data matrix X and a neighborhood measure, the *auxiliary graph* is generated. This graph guides the generation of hyperedge of the hypergraph $G^{(t)}$ at iteration t. The hypotheses are generated using the RCM (Random Cluster Model) [14] and Swendsen-Wang sampling [18]. The sampling procedure is designed for larger than minimal subsets and a large hyperedge degree [14]. After several iterations, the motion segmentation result is achieved based on a minimal subspace approximation error resulting from the subspace estimation.

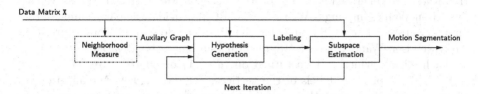

Fig. 2. Motion segmentation using hypergraphs (cf. Sect. 2). The hypotheses are generated using the auxiliary graph as neighborhood structure. The proposed neighborhood measure which incorporates motion affinities exchanges the spatial proximity proposed in [16] in the box with the dotted border. It leads to improved hypotheses.

Hypergraphs. A hypergraph $H = (V, E)$ consists of a set of vertices V and a set of hyperedges E. In a weighted hypergraph H, a weight $w(e)$ is associated with each hyperedge e. The degree of a hyperedge $\delta(e)$ is defined as the number of vertices in edge e. The degree of a vertex $d(v)$ is the sum of all weights of the hyperedges incident with vertex v. If all hyperedges have the same degree r, the hypergraph is a r-uniform. A two-uniform hypergraph describes a common graph in which an edge connects two vertices.

2.1 Auxiliary Graph

For the RCM sampling, a spatial neighborhood structure is required. This structure is represented as a graph G. In [16], the spatial proximity, i.e. the Euclidean distance of trajectory vectors, is encoded in G.

Initially, mean subtracted coordinates of the trajectory vectors are computed from the data matrix X. In the case of motion segmentation, X consists of all point coordinates of a trajectory in the images. The graph is built by firstly using a PCA on the trajectory vectors to reduce the data dimension to dim (default: $dim = 5$) followed by a k-nearest-neighbor algorithm (default: $k = 3$). The k-nearest-neighbor employs the Euclidean distance for each vector in the reduced data matrix $Y = \{y_i\}_{i=i}^{N}$. Two vertices v_i, v_j are connected if v_j is a k-nearest neighbor of v_i and vice versa. The distance between two data points determines the weight p_e of the edge e connecting these points as follows [16]:

$$p_e = exp\left(-\frac{||y_i - y_j||^2}{2\sigma_e^2}\right) \qquad (1)$$

The weight p_e indicates the probability of regarding two data points of the same structure. σ_e is computed as the standard deviation of the nearest neighbor distances in Y. The auxiliary graph shares the same vertices as the hypergraph $G^0 = (V, E^0)$.

2.2 Hypothesis Generation Using RCM

Hypothese are generated using the Random Cluster Model (RCM) [14] as well as Swendsen-Wang sampling (SWS) [18]. The RCM provides the partitioning of the auxiliary graph while SWS is an enhanced Monte Carlo sampling procedure [12]. The Swendsen-Wang approach [18] introduces a binary bond variable $d \in \{0, 1\}$ for each edge e which can be turned on ($d = 1$) or off ($d = 0$). The vector $f = \{f_i\}_{i=1}^{N}$ represents labels of the vertices, $f \in \{1, \ldots, K\}$. A realization of (f, d) effectively partitions the vertices into a set of connected components. Each connected component is a subset of V such that all the bond variables between vertices in the component are turned on. Additionally, vertices in a connected component must have the same label. This leads to the graph $G^{(t)} = (V, E^{(t)})$ with $E^{(t)} = \{e = <i, j> | f_i^{(t)} = f_j^{(t)}\}$ at iteration t.

In [16], the vector f and the samples d are updated alternatingly. Given f, a number of samples d provides a set of hyperedges. Given the hyperedges,

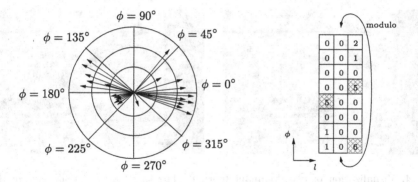

Fig. 3. Calculation of the dominant motions from a set of feature correspondences. The three dominant motions in this example are marked with red color. (Color figure online)

f is updated using NCut [17,23]. Each subcluster is a set of vertices connected by bond variables that have been turned on. Subclusters with a size less than a given hyperedge size D ($D = 10$ by default) are removed. The newly generated subset is added to the set of all generated hyperedges. The iterative process is terminated when either the labels f do not change significantly, or a maximum number of iterations is reached. Initially, the bond variables $d \in \{0,1\}$ are determined probabilistically by comparing a random number $r \in [0;1]$ with the edge weight p_e (cf. Eq. (1)). If $r \leq$ the edge probability p_e, the bond variable is turned on.

2.3 Subspace Estimation

In the last step of each iteration, the subspace error of the current labeling f is determined. The weight $w(e)$ of a hyperedge e is calculated as:

$$w(e) = \begin{cases} \exp(\frac{-r^2(v,\phi_s)}{2\sigma^2}) & \text{if } v \notin s \\ 0 & \text{otherwise} \end{cases} \tag{2}$$

Here, s is a sampled $D-1$ tuple, and v is an arbitrary vertex in V. ϕ_s is the model fitted in a least squares manner on s, and $r(v, \phi_s)$ is the residual of v with respect to ϕ_s. The parameter σ is problem dependent and needs to be tuned [16]. It is determined as the sigma which gives the lowest subspace approximation error [3].

3 Auxiliary Graph Based on Dominant Motions

The input data for the auxiliary graph is given by the trajectories of all feature points in the current image I_k to the corresponding features in image I_{k-s}. The auxiliary graph serves as initialization for the hypergraph $G^{(0)}$. Edges in the graph between differently moving objects should be avoided. The reference algorithm [16] employs spatial proximity to generate the auxiliary graph. We argue that this is not an appropriate measure as shown in the example input

Fig. 4. Visualization of two dominant motions. The trajectories classified as in the histogram bins as dominant motions are marked with yellow and red color, respectively. The image subsequences is taken from the *van* sequence [8], cf. Fig. 1, right. (Color figure online)

sequence *van* in Fig. 4. Many data point pairs have a low spatial distance from each other but belong to different motions, e.g. points on the trees in the background are near to points on the upper region of the van. But, their motion vectors are very different from each other (one motion pointing to the left, the other pointing to the right). Thus, the motion of the trajectories should be incorporated in the computation of the auxiliary graph.

The proposed approach determines the dominant motions in the scene. The main idea is borrowed from the task of computing the main orientation for scale invariant features, such as SIFT [11]. In SIFT, the main orientation of a texture patch surrounding a feature is estimated using orientation histograms build from the angles of gradients. The angles are grouped into bins and the maximum is determined. The maximum gives the dominant orientation of a feature.

For our task of computing the dominant motions in the scene, we extend this idea to two dimensions, angle ϕ and length l of a trajectory. The two dimensional histogram array of size $d_\phi \times d_l$ is shown in Fig. 3. The angle ϕ is determined from corresponding points $\mathbf{x}_{i-s} = (x_{i-s}, y_{i-s})^T$, $\mathbf{x}_i = (x_i, y_i)^T$ as $\phi_i = \tan^{-1}\left(\frac{y_i - y_{i-s}}{x_i - x_{i-s}}\right)$ while the length l_i is $l_i = |\mathbf{x}_i - \mathbf{x}_{i-s}|$. The lengths l_i are normalized using the maximum length of all trajectories. All correspondences are grouped into a two dimensional histogram array as shown in Fig. 3. Then, local maxima are determined (marked with red color in Fig. 3). The feature correspondences included in the maximum bin determine the dominant motion of the object. A threshold value M_{thres} ensures that a dominant motion has at least M_{thres} entries. Otherwise, it is discarded. The dimensions of the 2D histogram are given by parameters d_ϕ and d_l. In Fig. 3, $d_\phi = 8$ and $d_l = 3$ are illustrated while in all experiments in this paper $d_\phi = 10$ and $d_l = 4$ are used. The number of minimum bin entries for a dominant motion is set to $M_{thres} = 8$.

The dimensions of the 2D histogram are given by parameters d_ϕ and d_l. In Fig. 3, the configuration $d_\phi = 8$, $d_l = 3$ is shown while in all experiments in this paper $d_\phi = 16$ and $d_l = 8$ are used. The threshold M_{thres} for the minimal number of trajectories to build a dominant motion cluster is set to $M_{thres} = 5$. An example for the determination of a dominant motion is shown in Fig. 4. The trajectories of the dominant motion induced by the moving van is marked with red color while the dominant motion of the background is shown with yellow trajectories.

The information of trajectories located on the extracted dominant motions are incorporated in the generation of the auxiliary graph such that vertices in the auxiliary graph which belong to different dominant motions are not allowed to share an edge in the hypergraph (cf. Sect. 2). Thus, no hypotheses are generated from points on different dominant motions. This leads to a decreased number of samples required for accurate segmentation results. Furthermore, the hypergraph has reduced complexity which leads to smaller computation times. Both improvements are evaluated in the following section.

4   Experimental Results

The performance of the proposed approach is evaluated on the *Hopkins155* [19] and the MTPV benchmark [8]. Both benchmarks provide image sequences together with annotated feature tracks as ground truth information. The benchmarks do not contain outliers. While the trajectories in the *Hopkins155* benchmark consist of points visible in every frame of the sequence (up to 53 frames), feature tracks may start and end in arbitrary frames in MTPV [8].

For the evaluation, the matlab code provided by the authors of [16] is used. The proposed approach using motion-coherent affinities is implemented as an extension of this code. We focus on the performance with limited trajectory lengths. For all evaluations, subsequences with a fixed number $f_s, f_s = 3, \dots, 10$ of consecutive images are generated. Then, all trajectories of length f_s in this subsequence are used as input data. To get a sufficient number of experiments per sequence, the next subsequence starts with a frame distance of f_d ($f_d = 2$ used

Fig. 5. Visualization of the auxiliary graphs for Pulak et al. [16] (left) and the proposed method (right). Edges between the bus (blue) and the background (red) are not established by the proposed method since their motions differ. Even spatially long edges are possible for small subgraphs. The image example is taken from the *bus* sequence [8]. (Color figure online)

Table 1. Segmentation error [%] on benchmark **Hopkins155** [19] for Pulak et al. [16] and the proposed method. Bold numbers depict the lower error.

f_s	Mean		Median	
	reference	proposed	reference	proposed
3	7.96	**7.03**	0.84	**0.67**
4	5.76	**5.32**	0.0	**0.0**
5	4.94	**4.71**	0.0	**0.0**
6	4.64	**4.36**	0.0	**0.0**
7	3.79	**3.61**	0.0	**0.0**
8	4.03	**3.58**	0.0	**0.0**
9	3.76	**3.75**	0.0	**0.0**
10	3.87	**3.83**	0.0	**0.0**

for all experiments). As the results are based on random samples, 50 runs with random initializations are employed. The segmentation error is defined as [19]:

$$\text{segmentation error} = \frac{\#\text{ of misclassified points}}{\text{total }\#\text{ of points}} \qquad (3)$$

Additionally to the segmentation error mean, we report the median values. The segmentation results for Hopkins155 and MTPV are subsumed in Tables 1 and 2, respectively. Generally, the mean segmentation error decreases with increasing f_s. While the improvement of the proposed method is small for Hopkins155, its performance is significantly better for the MPTV benchmark. The decrease in the mean segmentation error is larger than 30% for $f_s \in \{4, \ldots, 10\}$. For $f_s = 7$ the largest gain of 41% is achieved.

Table 2. Segmentation error [%] on benchmark **MTPV** [8] for Pulak et al. [16] and the proposed method. Bold numbers depict the lower error.

f_s	Mean		Median	
	reference	proposed	reference	proposed
3	26.96	**21.06**	32.27	**22.60**
4	21.30	**14.20**	22.83	**0.51**
5	19.91	**12.69**	17.87	**0.29**
6	18.27	**12.60**	9.74	**0.30**
7	19.01	**11.09**	12.30	**0.23**
8	18.66	**12.09**	10.75	**0.25**
9	17.46	**11.18**	1.21	**0.24**
10	16.83	**11.16**	1.87	**0.24**

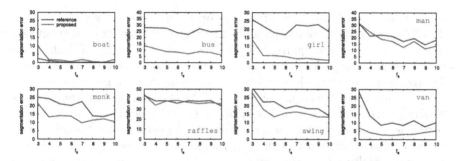

Fig. 6. Mean segmentation error [%] on the **MTPV** benchmark for each individual sequence with varying length of input trajectories f_s. In blue (solid line) the reference method [16] is shown, while in red (dashed line) the proposed auxiliary graph is employed. Note that the y axes have different scalings. (Color figure online)

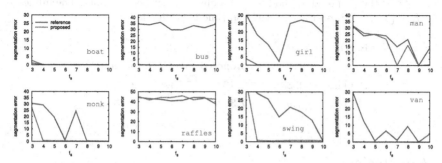

Fig. 7. Median of segmentation error [%] on the **MTPV** benchmark for each individual sequence with varying length of input trajectories f_s. In blue (solid line) the reference method [16] is shown, while in red (dashed line) the proposed auxiliary graph is employed. Note that the y axes have different scalings. (Color figure online)

Obviously, the reference auxiliary graph leads to a much lower probability of sampling pure hyperedges from the input data MTPV. Since only trajectories with a large lengths are included in Hopkins155, the spatial proximity appears to provide a suitable metric for the auxiliary graph. This is not the case for MTPV since trajectories with short lengths are not excluded from this data set. Thus, trajectories on differently moving objects may have a small spatial distance in the images (cf. Fig. 5). This is often the case in applications using trajectory data extracted from natural sequences since trajectories with small length are required to capture moving objects. Thus, we can conclude that:

(1) our approach provides significantly improved results for the MTPV benchmark
(2) Hopkins155 may not be a suitable benchmark for the evaluation of motion segmentation approaches if short trajectory lengths are of interest.

The MTPV results for different trajectory lengths f_s are shown in Figs. 6 and 7 in detail. For 6 of the 8 sequences, the proposed method leads to a significant decrease in the segmentation error. For the other 2 sequences (boat and raffles), the results are comparable. The sequence boat leads to small segmentation errors for both approaches. The sequence raffles fails for both methods. The proposed approach can not recover the motion because too few trajectories are located on the object and no dominant motion can be established.

The computation times are shown in Table 3. They are measured on a 3.20 GHz AMD Processor with Matlab2016. On the one hand, the computation time is heavily dependent on the number of trajectories. On the other, it is dependent on the complexity of the hypergraphs built at each iteration. As the hypergraphs resulting from the proposed auxiliary graph are less complex, the computation times are much smaller compared to the reference.

Table 3. Mean **Computation time** in seconds for a segmentation process for Pulak et al. [16] and the proposed method. Bold numbers depict better values. The influence of f_s, $f_s \in \{3, \dots, 10\}$, on the computation time is negligible.

Benchmark	Mean		Median	
	reference	proposed	reference	proposed
Hopkins	8.83 s	**2.49 s**	5.53 s	**2.12 s**
MPTV	7.98 s	**4.30 s**	4.26 s	**2.75 s**

5   Conclusions

Hypergraph based approaches for motion segmentation reduce the number of required samples by exploiting higher order similarities for the hypothesis generation. The reference method proposed by Pulak et al. [16] uses an auxiliary graph which guides the sampling process. The auxiliary graph is based on spatial proximity of the trajectories.

The approach proposed in this paper uses a simple but effective scheme to combine motion direction and length of the trajectories for the determination of an improved auxiliary graph. Thus, less samples are required for accurate segmentation results.

The evaluation of the approaches employs two benchmarks, the well-known *Hopkins155* and the *MTPV* benchmark. While for Hopkins155, the proposed approach provides similar results regarding classification accuracy, it decreases the segmentation error significantly for MTPV. Since only long trajectories are included in Hopkins155, spatial proximity is an adequate measure to discriminate differently moving objects. It is shown that this measure is suboptimal for the MTPV benchmark. The proposed auxiliary provides a significantly lower segmentation error of up to 41%. Additionally, the computation time decreases since the resulting hypergraphs have reduced complexity.

References

1. Brox, T., Malik, J.: Large displacement optical flow: descriptor matching in variational motion estimation. IEEE Trans. Pattern Anal. Mach. Intell. (PAMI) **33**(3), 500–513 (2011)
2. Brox, T.: Higher order motion models and spectral clustering. In: IEEE Conference on Computer Vision and Pattern Recognition (CVPR), pp. 614–621. IEEE Computer Society (2012)
3. Chen, G., Lerman, G.: Spectral curvature clustering (SCC). Int. J. Comput. Vision (IJCV) **81**(3), 317–330 (2008)
4. Chin, T.J., Yu, J., Suter, D.: Accelerated hypothesis generation for multistructure data via preference analysis. IEEE Trans. Pattern Anal. Mach. Intell. (PAMI) **34**(4), 625–638 (2012)
5. Cordes, K., Hockner, M., Ackermann, H., Rosenhahn, B., Ostermann, J.: WM-SBA: weighted multibody sparse bundle adjustment. In: 14th IAPR International Conference on Machine Vision Applications (MVA), pp. 162–165 (2015)
6. Delong, A., Osokin, A., Isack, H.N., Boykov, Y.: Fast approximate energy minimization with label costs. Int. J. Comput. Vision (IJCV) **96**(1), 1–27 (2012)
7. Georghiades, A.S., Belhumeur, P.N., Kriegman, D.J.: From few to many: illumination cone models for face recognition under variable lighting and pose. IEEE Trans. Pattern Anal. Mach. Intell. (PAMI) **23**(6), 643–660 (2001)
8. Li, Z., Guo, J., Cheong, L.F., Zhou, S.: Perspective motion segmentation via collaborative clustering. In: IEEE International Conference on Computer Vision (ICCV), pp. 1369–1376 (2013)
9. Li, Z., Cheong, L.-F., Yang, S., Toh, K.C.: Simultaneous clustering and model selection for tensor affinities. In: IEEE Conference on Computer Vision and Pattern Recognition (CVPR) (2016)
10. Liu, H., Yan, S.: Efficient structure detection via random consensus graph. In: IEEE Conference on Computer Vision and Pattern Recognition (CVPR), pp. 574–581 (2012)
11. Lowe, D.G.: Distinctive image features from scale-invariant keypoints. Int. J. Comput. Vision (IJCV) **60**(2), 91–110 (2004)
12. MacKay, D.J.: The Swendson-Wang method. In: Information Theory, Inference and Learning Algorithms. Cambridge University Press (2003)
13. Ozden, K., Schindler, K., Van Gool, L.: Simultaneous segmentation and 3D reconstruction of monocular image sequences. In: IEEE International Conference on Computer Vision (ICCV), pp. 1–8 (2007)
14. Pham, T.T., Chin, T.J., Yu, J., Suter, D.: The random cluster model for robust geometric fitting. IEEE Trans. Pattern Anal. Mach. Intell. (PAMI) **36**(8), 1658–1671 (2014)
15. Poling, B., Lerman, G.: A new approach to two-view motion segmentation using global dimension minimization. Int. J. Comput. Vision (IJCV) **108**(3), 165–185 (2014)
16. Purkait, P., Chin, T.-J., Ackermann, H., Suter, D.: Clustering with hypergraphs: the case for large hyperedges. In: Fleet, D., Pajdla, T., Schiele, B., Tuytelaars, T. (eds.) ECCV 2014. LNCS, vol. 8692, pp. 672–687. Springer, Cham (2014). doi:10. 1007/978-3-319-10593-2_44
17. Shi, J., Malik, J.: Normalized cuts and image segmentation. IEEE Trans. Pattern Anal. Mach. Intell. (PAMI) **22**(8), 888–905 (2000)

18. Swendsen, R.H., Wang, J.S.: Nonuniversal critical dynamics in Monte Carlo simulations. Phys. Rev. Lett. **58**, 86–88 (1987)
19. Tron, R., Vidal, R.: A benchmark for the comparison of 3D motion segmentation algorithms. In: IEEE Conference on Computer Vision and Pattern Recognition (CVPR) (2007)
20. Vidal, R., Ma, Y., Soatto, S., Sastry, S.: Two-view multibody structure from motion. Int. J. Comput. Vision (IJCV) **68**(1), 7–25 (2006)
21. Wang, H., Xiao, G., Yan, Y., Suter, D.: Mode-seeking on hypergraphs for robust geometric model fitting. In: The IEEE International Conference on Computer Vision (ICCV), pp. 2902–2910 (2015)
22. Xiao, G., Wang, H., Lai, T., Suter, D.: Hypergraph modelling for geometric model fitting. Pattern Recogn. **60**, 748–760 (2016)
23. Zhou, D., Huang, J., Schölkopf, B.: Learning with hypergraphs: clustering, classification, and embedding. In: Advances in Neural Information Processing Systems (NIPS), vol. 19. MIT Press (2006)

Poster Session 1

Robust Long-Term Aerial Video Mosaicking by Weighted Feature-Based Global Motion Estimation

Holger Meuel$^{(\boxtimes)}$, Stephan Ferenz, Florian Kluger, and Jörn Ostermann

Institut für Informationsverarbeitung,
Leibniz Universität Hannover, Hannover, Germany
{meuel,ferenz,kluger,office}@tnt.uni-hannover.de

Abstract. Aerial video images can be stitched together into a common panoramic image. For that, the global motion between images can be estimated by detecting Harris corner features which are linked to correspondences by a feature tracker. Assuming a planar ground, a homography can be estimated after an appropriate outlier removal. Since Harris features tend to occur clustered at highly structured 3D objects, these features are located in various different planes leading to an inaccurate global motion estimation (GME). Moreover, if only a small number of features is detected or features are located at moving objects, the accuracy of the GME is also negatively affected, leading to severe stitching errors in the panorama.

To overcome these issues, we propose: Firstly, the feature correspondences are weighted to approximate a uniform distribution over the image. Secondly, we enforce a fixed number of correspondences of highest possible quality. Thirdly, we propose a temporally variable tracking distance approach to remove outliers located at slowly moving objects.

As a result we improve the GME accuracy by 10% for synthetic data and highly reduce the structural dissimilarity (DSSIM) caused by stitching errors from 0.12 to 0.035.

1 Introduction

For the visualization of aerial videos, *e. g.* captured from Unmanned Aerial Vehicles (UAVs) in a nadir view (orthorectified video), one common approach is to stitch the video images together to a panoramic image by mosaicking. For the generation of this panorama, each video image is registered into a common coordinate system. Since GPS/IMS systems can not provide a satisfactory accuracy, the global motion has to be estimated from the video images. One common approach is the detection of features, *e. g.* Harris Corner features [4] in one video image and its correspondence in the preceding image (feature correspondence) by a KLT feature tracker [16]. Assuming a planar ground and thus a uniform motion of detected feature points, RANSAC [2] can be used to remove feature correspondences not matching the global motion (outliers). From the remaining feature correspondences (inliers), a homography can be estimated. However, for a

© Springer International Publishing AG 2017
M. Felsberg et al. (Eds.): CAIP 2017, Part I, LNCS 10424, pp. 135–147, 2017.
DOI: 10.1007/978-3-319-64689-3_11

small number of detected features – *e. g.* due to unstructured, blurry or low quality content – and small local displacements of moving objects between images
(*e. g.* for pedestrians), RANSAC is not able to remove wrong correspondences
anymore. Thus, a reliable estimation of a projective transform representing the
global motion of the surface of the earth in the video is not possible. Moreover,
features are often detected on non-planar structures, *e. g.* houses or trees whose
motion does not match the motion of the ground plane of the scene. Furthermore, those features tend to be spatially clustered, which is known to negatively
influence the quality of the global motion estimation [3]. Figure 1 shows an example of a wrong stitching based on the global motion estimation (GME) from [8]
and using a standard mosaicking approach like [7,10].

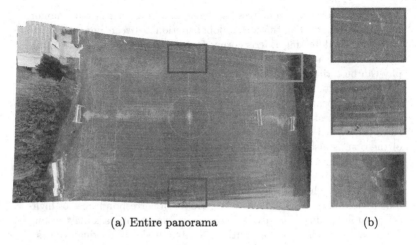

(a) Entire panorama   (b)

Fig. 1. Panoramic image from 3000 images of the self-recorded *Soccer* sequence and
magnifications in (b).

In this paper we propose different methods to increase the quality of the
global motion estimation, which are mainly based on the usage of weighted
features. To prevent an over-proportional weighting of feature clusters at highly
structured areas in the image (like 3D objects), we propose to approximate a
uniform distribution of the features in the entire image, considering the detected
feature positions (Subsect. 3.1). In order to provide enough features for a reliable
motion estimation, we propose to use a high, fixed number of features of highest
possible quality (Subsect. 3.2). To further improve the quality of the resulting
estimation, we rely on tracking over long temporal distances in order to remove
features positioned at (slowly) moving objects which are not detected as outliers
by a common RANSAC in case of small motion (Subsect. 3.3).

The remaining paper is organized as follows: Sect. 2 gives a short overview of
global motion estimation for aerial videos. In Sect. 3 we describe our proposed
robust long-term mosaicking approach. Our weighting algorithm for RANSAC

which approximates a uniform distribution of the features in the image is introduced in Subsect. 3.1. Furthermore, we introduce a straight forward approach for detecting sufficient high quality features in the image in Subsect. 3.2. The tracking over long temporal distances is explained in Subsect. 3.3. In Sect. 4 we present experimental results for synthetic as well as real-world data, using the structural dissimilarity DSSIM [12] as quality metric. Finally, Sect. 5 concludes the paper.

2   Related Work: Global Motion Estimation for Aerial Videos

A lot of research has been done for the reliable estimation of the global motion in video sequences. Typical approaches are based on defining discriminative features like SIFT/SURF [1], Harris corners [4], MSER [6] etc. in one video image [9,15,20,22], the generation of trajectories for these features (e. g. by feature relocation [16], dense [14] or sparse optical flow [11]), and finally the estimation of the global motion according to an assumed scene model, e. g. using RANSAC [2].

In this work we extend the global motion estimation framework from [9] which is designed for the usage onboard of UAVs with limited energy and processing power. We also rely on KLT tracking of Harris corners, which are highly efficient to be computed compared to other features like SIFT or SURF. Whereas the common approach consisting of feature detection, RANSAC and least-square-minimization works well for a lot of applications, it fails for certain conditions as outlined above based on the example from Fig. 1. Thus, we aim at the improvement of the global motion estimation using RANSAC for videos captured from UAVs with low translational movement and slowly moving objects in the scene, e. g. in an aerial police surveillance scenario for soccer games.

3   Robust Long-Term Global Motion Estimation for Aerial Videos

Assuming the surface of the earth to be planar – which is valid for flight altitudes of several hundred meters – we can project one camera image I_k into the previous image I_{k-1} using a homography $\mathbf{H}_k^{k-1}$ which is described by a projective transform with 8 parameters $\vec{a}k = (a_{1,k}, a_{2,k}, \ldots, a_{8,k})^\top$:

$$\mathbf{H}_k^{k-1} = \begin{pmatrix} a_{1,k} & a_{2,k} & a_{3,k} \\ a_{4,k} & a_{5,k} & a_{6,k} \\ a_{7,k} & a_{8,k} & 1 \end{pmatrix}. \tag{1}$$

We can calculate the transformed pixel coordinates (x_{k-1}, y_{k-1}) in image $k-1$ from the image coordinates (x_k, y_k) in image k:

$$x_{k-1} = \frac{a_{1,k}x_k + a_{2,k}y_k + a_{3,k}}{a_{7,k}x_k + a_{8,k}y_k + 1}, \quad y_{k-1} = \frac{a_{4,k}x_k + a_{5,k}y_k + a_{6,k}}{a_{7,k}x_k + a_{8,k}y_k + 1}. \tag{2}$$

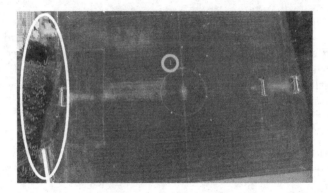

Fig. 2. Video image from the *Soccer* sequence with inliers and their trajectories (yellow lines) after KLT & RANSAC. The inliers are highly clustered at 3D structures (trees/houses) on the left (white ellipse). Moreover, a correspondence located at a player was errouneously considered as inlier (red circle). (Color figure online)

However, for a reliable homography estimation, the detected feature correspondences have to be located in one plane which becomes even more important for the projection of several video images into one common panoramic image. This plane optimally should be the ground plane, i. e. the feature correspondences have to be located on the surface of the earth. Whereas RANSAC is often capable of removing correspondences not matching the global motion, it may fail in removing correspondences not matching the global motion of the ground plane, if from the set of all correspondences C the amount of correspondences located on the ground $J \in C$ (inliers) is small compared to the amount of correspondences located on various different planes $O \in C$ (outliers). As a consequence, the estimated plane does not reflect the real ground plane which leads to an estimated global motion not reflecting the true motion of the surface of the earth. If $O \gg J$ (Fig. 2, white ellipse), the ground plane estimation becomes instable, resulting in stitching errors (Fig. 1).

3.1   Weighted Feature-Based Global Motion Estimation

Since only a few high quality features are typically located in unstructured areas (e. g. on the lawn in our example) compared to the number of features located at 3D structures (e. g. trees or houses), the former features have to be considered stronger within the least-square optimization in order to retain a homography representing the real global motion. Based on this idea, we formulate the least-squared minimization problem for the set of inliers J as:

$$\min \sum_{j \in J} \left((\tilde{x}_{j,k-1} - x_{j,k-1})^2 + (\tilde{y}_{j,k-1} - y_{j,k-1})^2 \right) \cdot (W_{j,k})^2, \qquad (3)$$

where $(\tilde{x}_{j,k-1}, \tilde{y}_{j,k-1})$ are the estimated coordinates and $W_{j,k}$ is a weighting function in dependence of $x_{j,k}$ and $y_{j,k}$. Based on Eqs. (3) and (2) we build a linear equation system which can be solved with a least-squares approach.

The weighting function $W_{j,k}$ is modeled with an instance reweighting approach, such that a uniform distribution $p_e(x, y)$ of the feature correspondences is approximated over the entire image.

The real feature distribution $p_{\text{feat}}(x, y)$ in the image for the (discrete) feature positions with the kernel function K is given as:

$$p_{\text{feat},k}(x,y) = \frac{1}{J} \sum_{i=1}^{J} K(x - x_{i,k}, y - y_{i,k}). \tag{4}$$

We approximate K by a Gaussian probability density function (pdf) $p_g(x, y)$ to model the neighborhood of each feature [18]:

$$p_g(x,y) = \frac{1}{2\pi\sigma_x\sigma_y} \exp\left[-\frac{1}{2}\left(\frac{x^2}{\sigma_x^2} + \frac{y^2}{\sigma_y^2}\right)\right] \tag{5}$$

As suggested in [13], we define the variances σ_x and σ_y being the mean value of the pairwise distances of all feature correspondences and κ being a scaling factor:

$$\sigma_x = \sigma_y = \kappa \cdot \frac{2}{J^2} \sum_{j=1}^{J} \sum_{i=1}^{j-1} \sqrt{(x_i - x_j)^2 + (y_i - y_j)^2} \tag{6}$$

The weighting function $W_{j,k}$ finally is calculated by dividing p_t by p_{feat} [17,19], i. e. the weighting for each feature is the reciprocal of the real feature distribution:

$$W_{j,k} = \frac{p_e}{p_{\text{feat},k}(x_{j,k}, y_{j,k})} = J \cdot \frac{2\pi\sigma_x\sigma_y}{\sum_{i=1}^{J} \exp\left[-\frac{1}{2}\left(\frac{(x_{j,k}-x_{i,k})^2}{\sigma_x^2} + \frac{(y_{j,k}-y_{i,k})^2}{\sigma_y^2}\right)\right]} \tag{7}$$

3.2 Increase of the Number of Features with Highest Possible Quality ("More Features")

The approximation of a uniform distribution of the feature correspondences over the entire image as described in the last subsection leads to highly improved global motion results. However, if only a small number of features can be detected e. g. due to bad input image quality or unstructured areas, an accurate solution for the global motion can not be determined.

Therefore, we propose to include a predefined minimum number of Harris features in the global motion estimation, always using the best available detected features. First, we calculate the Jacobian matrix and its lowest eigenvalue for each image pixel and sort them in a list. As a second step, we select the n-best features from the sorted list, with n being a predefined number of features. These n features are fed into subsequent motion estimation steps (RANSAC and homography estimation).

3.3   Variable Tracking Distance

Whereas we focused on the improvement of feature correspondences based on their spatial position in the image in Subsect. 3.1 and on the number of detected features in Subsect. 3.2, feature correspondences located at slow moving objects may not be recognized as wrong correspondences and thus not be removed as outliers by RANSAC (Fig. 2, red circle). As a consequence, these correspondences negatively influence the accuracy of the homography estimation. To overcome this issue, we propose to increase the temporal distance d between the images used for the homography estimation. Thereby, local motion tends to be larger and RANSAC is more likely able to remove features located on moving objects as outliers. Furthermore, to reduce drift as it may occur in image-to-image-based approaches, we aim at tracking against one specific image (reference image) as long as possible. Whereas in general it is beneficial to have a larger temporal tracking distance d, it may be disadvantageous, if the temporal distance between the images becomes too large. In such a case, KLT may not be able to reliably find correspondences due to shape changes or rotations which impairs the feature correspondence accuracy. Thus, we propose to use a constraint variable tracking distance d between the images. Summarizing, we aim at using one specific reference image for the estimation of homographies of several consecutive video images, whereas we limit the temporal distance to a predefined maximum value $d_{\max}$ and try to prefer large tracking distances. For each image k, we first calculate the distance d:

$$d = (k \quad \mathrm{mod} \ \frac{d_{\mathrm{ref}}^{\mathrm{curr}}}{2}) + 1 + \frac{d_{\mathrm{ref}}^{\mathrm{curr}}}{2}, \tag{8}$$

with $d_{\mathrm{ref}}^{\mathrm{curr}}$ being an intermediate tracking distance (initialized to $d_{\max}$ for each image). The first term of Eq. (8) selects the same reference image as long as possible, whereas the last term enforces high tracking distances. Assuming a linear global motion, we approximate an estimated homography $\tilde{\mathbf{H}}_k^{k-d} = \mathbf{H}_{k-1}^{k-d} \cdot \mathbf{H}_{k-1}^{k-2}$ from already known homographies and transform all features from the current image using this $\tilde{\mathbf{H}}_k^{k-d}$. Then we check, if the following conditions are fulfilled:

1. Are enough transformed features located within the area of image I_{k-d}?
2. Is the intersection area of images I_k and I_{k-d} large enough?

If at least one of these conditions is violated, we halve $d_{\mathrm{ref}}^{\mathrm{curr}}$ and restart again with the computation of d. If all conditions are fulfilled, we use a guided tracking for the generation of accurate feature correspondences. For that, we apply the extrapolated homography $\tilde{\mathbf{H}}_k^{k-d}$ to all features in image I_k and use the result as seed position for the KLT search, resulting in accurate correspondences. The latter are used for the subsequent outlier removal and for the estimation of the improved, final homography $\mathbf{H}_k^{k-d}$.

4   Experiments

We present results for synthetic data in the Subsect. 4.1 before we evaluate our approach in detail for camera captured (real world) data in Subsect. 4.2.

4.1   Synthetic Data

In order to show that our method reliably improves the homography estimation, we generated a synthetic scene. We defined an array containing 30×17 blocks, each of size 64×64 pixels, which is approximately the size of one HDTV resolution image. For each block we randomly defined if it is supposed to be a block containing 3D structure ("house block") or not, and limited the amount of house blocks to 25%. In order to simulate a unequal feature distribution, we randomly draw a predefined mean number of feature positions $n_h = [0 \ldots 50]$ for the house blocks (green) and for the non-house blocks (blue) $n_n = 4$ (Fig. 3).

Fig. 3. Visualization of a synthetic image with "house blocks" (green), non-house blocks (blue) and randomly drawn features (white dots) and their simulated movement (white arrows). (Color figure online)

Furthermore, we manually generated homography parameters a_{syn_k} similar to those which we observed in real multicopter videos (Table 1).

Table 1. Example synthetic homography parameters a_{syn_k}.

k	$a_{k,1}$	$a_{k,2}$	$a_{k,3}$	$a_{k,4}$	$a_{k,5}$	$a_{k,6}$	$a_{k,7}$	$a_{k,8}$	$a_{k,9}$
1	1	0	0.6	0.0001	1	−0.5	0	0	1
2	1	0	−0.8	0	1	−2.8	0	0	1
3	1	0	−0.1	0	0.9999	0.1	0	0	1
4	1	0	0.6	0	1	0.7	0	0.0001	1
⋮					⋮				
30	1	0	−0.6	0	1	0	0	0	1

The feature points from the current image I_k were transformed according to the synthetic homographies. We simulated motion parallax effects by moving all features on house blocks after the global motion compensation in the direction of the image center by m pixels. Since m should correspond to the motion parallax which can be observed in real scenes, we linearly increase m dependent on its distance to the image center up to a maximum of $m = 50$ pixels (which is a realistic motion parallax to be observed for high 3D structures and relatively low flight altitudes). Afterwards we applied zero-mean Gaussian noise with a variance of $\sigma^2 = 2$ pel to all feature positions.

Finally we used the synthetic scene as input for the motion estimation system, one time without and one time with our proposals, and compared the accuracy of the estimated homographies. For the improvement measure we applied each estimated homography to the corner pixels of the image and calculate the errors compared to the projected point position using the real homography parameters a_{syn}. We varied the mean number of features n_h located in each house block between $10 \ldots 50$. The average error at the corner points was decreased from 10.1 to 9.0 pel which corresponds to 10.6% for $n_h = 10$ and from 18.1 to 16.4 pel for $n_h = 50$ (9.4%).

4.2   Camera Captured Videos

In this subsection we present results for real world data. Since the amount of test sequences providing a nadir view of the camera and containing 3D structured areas as well as plain areas is limited (although it may be the predominant view for aerial surveillance missions from UAVs), we recorded a test sequence of a soccer game (*Soccer* sequence) and present detailed results for this sequence. To underline the versatility of our proposals, we also provide results for the *1500* m sequence from the *TNT Aerial Video Testset* (TAVT) [5,9]. We will show that we can improve the homography estimation leading to subjectively highly improved results in panoramic images, especially in terms of line consistency.

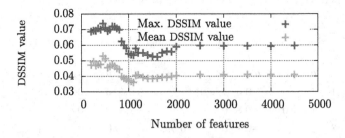

Fig. 4. Structural dissimilarity (DSSIM) [12] values (smaller is better) of reconstructed video images from panoramic image for different numbers of features for the *Soccer* sequence.

We generate a mosaic from the videos based on the estimated homographies. From this, we reconstruct video images again as described in [7,10]. For the quality measure we reconstruct video images from the mosaic and compare them image-wise with the input sequence. Due to the image reconstruction from the mosaic, no motion parallax is contained in the reconstructed video images. Thus, we cannot rely on a PSNR-based quality evaluation but use the structural dissimilarity (DSSIM) [12] instead. The structural dissimilarity is based on the well-known structural similarity (SSIM) [21] and lies between 0 (identical images) and ∞ (no similarity). It reflects the subjective impression in terms of cross-correlation between both images (structure), luminance similarity as well as contrast similarity.

Quality measures for the self-recorded *Soccer* sequence and the *1500 m* sequence from the data set TAVT [5,9] are presented in Table 2 and in Fig. 5 for each proposed method alone and all combinations.

Table 2. Results of different methods for the *Soccer* sequence, 3000 images ((*): manual reference only for 100 images) and the *1500* m sequence from TAVT [5,9].

Sequence	*Soccer* seq. DSSIM		*1500* m seq. DSSIM	
Method	Mean	Max	Mean	Max
Manual reference	$0.036^{(*)}$	$0.060^{(*)}$	—	—
Baseline (w/o proposed methods)	**0.120**	**0.146**	**0.067**	**0.156**
Weighting of correspondences	0.123	0.151	0.066	0.155
More features	0.094	0.129	0.065	0.133
Weighting & more features	0.094	0.128	0.064	0.133
Variable tracking	0.054	0.079	0.062	0.094
Weighting & variable tracking	0.045	0.071	0.063	0.099
Weighting & more feat. & var. track.	**0.035**	**0.051**	**0.061**	**0.088**

From the detailed results it is obvious, that our proposed weighting algorithm can improve the quality of the global motion estimation, if *enough* features are in the image (*Weighting & more features* in the tables: 0.120 to 0.094 for the *Soccer* sequence, Fig. 5c, 0.067 to 0.064 for the *1500* m sequence). Simulations for the HDTV resolution *Soccer* sequence lead to an optimal value of about $n = 1050$ features (Fig. 4), which is in the range of $n = [900 \ldots 1200]$ we found as optimal number of features also for other sequences we tested. If the number of features is too small, we only can observe small average gains (0.067 to 0.066 for the *1500* m sequence) or even small (average) losses (0.120 to 0.123 for the *Soccer* sequence, Fig. 5a) if – like in the latter case – not enough features of high quality are contained due to a low image quality. Thus, the combination of weighting and more features is always beneficial for low as well as for high quality videos. The usage of a variable tracking distance is recommendable in any case, since it improves the line accuracy by enforcing tracking against one reference image for several video images. Thus, drift is highly reduced and the objective and subjective results are improved on average (0.120 to 0.054 for the *Soccer* seq., 0.067 to 0.062 for the *1500* m sequence) as well as for the maximum DSSIM

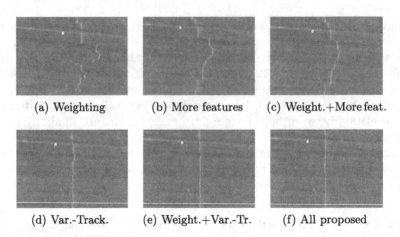

(a) Weighting   (b) More features   (c) Weight.+More feat.

(d) Var.-Track.   (e) Weight.+Var.-Tr.   (f) All proposed

Fig. 5. Subjective comparison of different proposed methods and combinations for the self-recorded *Soccer* sequence.

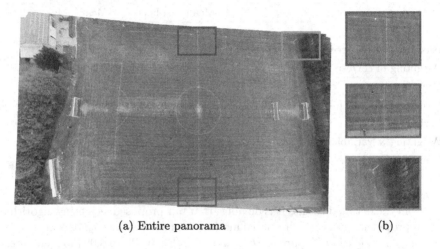

(a) Entire panorama   (b)

Fig. 6. Final panorama using all proposed improvements for global motion estimation with uniform distribution and weight of $\kappa = 0.575$. (b) magnifications.

values (0.146 to 0.079 for the *Soccer* sequence, Fig. 5d, 0.156 to 0.094 for the *1500* m sequence). This holds also true for the combined approaches with the variable tracking (Figs. 5e and f).

Combining our approaches, we observe that we highly improve the DSSIM from 0.12 to 0.035 for the *Soccer* sequence. Our combined methods even slightly outperform a manually generated reference, which matches the subjective impression. For the *1500* m sequence we achieve an improvement from 0.067 to 0.061 in terms of mean DSSIM. Although the average gain for the latter sequence is smaller than for the *Soccer* sequence, the maximal structural dissimilarity was drastically reduced (*Soccer* seq.: 0.146 to 0.051; *1500* m seq.: 0.156 to

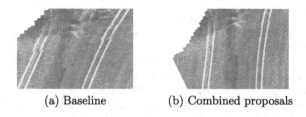

(a) Baseline   (b) Combined proposals

Fig. 7. Subjective results for the *1500* m sequence from the TAVT data set [5,9].

0.088) which results in smaller maximal distortions leading to subjectively much more pleasing results, especially in terms of line accuracy (Figs. 5f and 7b). In Fig. 6 we present the final long-term panoramic image after the fully automatic processing of 3000 images. A subjective impression for the *1500* m sequence is shown in the magnifications from the panoramic image in Fig. 7.

5 Conclusions

In this paper, we aim at a robust global motion estimation for UAV captured ortho-videos which contain distinct 3D structures (*e. g.* houses, trees) as well as real ground.

We propose to tackle the problem of a unequal feature correspondence distribution over the image by introducing a weighting function which approximates a uniform distribution over the image. In order to provide enough features also in scenarios with only a small number of high-quality features, we additionally propose to use a high but fixed number of features based on the feature quality. Finally, our third contribution is to track over long temporal distances with a variable tracking distance. The benefits of this approach are twofold: firstly, we use the same reference image for several images which reduces drift. Secondly, the motion of small and slow moving objects can more likely be removed by an outlier removal (RANSAC).

We show, using synthetic data, that our feature correspondence weighting proposal improve the estimation accuracy by up to 10% for realistic assumptions. For camera captured data, the resulting panoramic images which were generated based on the estimated global motions were improved and provide much better and virtually drift free reconstruction of linear structures (*e. g.* lines at a Soccer play ground). The structural dissimilarity (DSSIM) for reconstructed images from the panoramic image was highly reduced, *e. g.* from 0.120 to 0.035 on average for the self-recorded *Soccer* sequence.

References

1. Bay, H., Ess, A., Tuytelaars, T., Van Gool, L.: Speeded-up robust features (surf). Comp. Vis. Image Underst. **110**(3), 346–359 (2008). http://dx.doi.org/10.1016/j.cviu.2007.09.014

2. Fischler, M.A., Bolles, R.C.: Random sample consensus: a paradigm for model fitting with applications to image analysis and automated cartography. Commun. ACM **24**(6), 381–395 (1981)
3. Han, Y., Choi, J., Byun, Y., Kim, Y.: Parameter optimization for the extraction of matching points between high-resolution multisensor images in urban areas. IEEE Trans. Geosci. Remote Sens. **52**(9), 5612–5621 (2014)
4. Harris, C., Stephens, M.: A combined corner and edge detection. In: Proceeding of the Fourth Alvey Vision Conference, pp. 147–151 (1988)
5. Institut für Informationsverarbeitung (TNT), Leibniz Universität Hannover: TNT Aerial Video Testset (TAVT) (2010–2014). https://www.tnt.uni-hannover.de/project/TNT_Aerial_Video_Testset/
6. Matas, J., Chum, O., Urban, M., Pajdla, T.: Robust wide baseline stereo from maximally stable extremal regions. In: Proceedings of the British Machine Vision Conference, pp. 36.1–36.10. BMVA Press (2002)
7. Meuel, H., Kluger, F., Ostermann, J.: Illumination change robust, codec independent low bit rate coding of stereo from singleview aerial video. In: 10th IEEE International 3DTV Conference, pp. 1–4, July 2014. http://ieeexplore.ieee.org/document/7548961/
8. Meuel, H., Munderloh, M., Ostermann, J.: Low bit rate ROI based video coding for HDTV aerial surveillance video sequences. In: Proceeding of the IEEE Conference on Computer Vision and Pattern Recognition - Workshops (CVPRW), pp. 13–20, June 2011
9. Meuel, H., Munderloh, M., Reso, M., Ostermann, J.: Mesh-based piecewise planar motion compensation and optical flow clustering for ROI coding. In: APSIPA Transactions on Signal and Information Processing, vol. 4 (2015). http://journals.cambridge.org/article_S2048770315000128
10. Meuel, H., Schmidt, J., Munderloh, M., Ostermann, J.: Region of interest coding for aerial video sequences using landscape models. In: Advanced Video Coding for Next-Generation Multimedia Services. Intech, January 2013. http://tinyurl.com/ntx7u29
11. Munderloh, M., Meuel, H., Ostermann, J.: Mesh-based global motion compensation for robust mosaicking and detection of moving objects in aerial surveillance. In: Proceeding of IEEE Conference on Computer Vision and Pattern Recognition Workshop (CVPRW), pp. 1–6 (2011)
12. Pornel: RGBA Structural Similarity (2016). https://kornel.ski/dssim/
13. Reddi, S.J., Ramdas, A., Póczos, B., Singh, A., Wasserman, L.: On the decreasing power of kernel and distance based nonparametric hypothesis tests in high dimensions. In: Proceeding of the AAAI Conference on Artificial Intelligence, pp. 3571–3577 (2015)
14. Reso, M., Jachalsky, J., Rosenhahn, B., Ostermann, J.: Temporally consistent superpixels. In: Proceeding of the IEEE International Conference on Computer Vision (ICCV), pp. 385–392, December 2013
15. Shi, G., Xu, X., Dai, Y.: SIFT feature point matching based on improved RANSAC algorithm. In: International Conference on Intelligent Human-Machine Systems and Cybernetics, vol. 1, pp. 474–477, August 2013
16. Shi, J., Tomasi, C.: Good features to track. In: Proceeding of the IEEE Conference on Computer Vision and Pattern Recognition (CVPR), Seattle, June 1994
17. Shimodaira, H.: Improving predictive inference under covariate shift by weighting the log-likelihood function. J. Stat. Planning Infer. **90**(2), 227–244 (2000). http://www.sciencedirect.com/science/article/pii/S0378375800001154

18. Silverman, B.W.: Density Estimation for Statistics and Data Analysis, vol. 26. CRC Press (1986)
19. Sugiyama, M., Nakajima, S., Kashima, H., von Bünau, P., Kawanabe, M.: Direct importance estimation with model selection and its application to covariate shift adaptation. In: Proceeding of the Conference on Neural Information Processing Systems (NIPS), pp. 1433–1440 (2007). http://tinyurl.com/jt5rdz8
20. Wang, Y., Fevig, R., Schultz, R.R.: Super-resolution mosaicking of UAV surveillance video. In: IEEE International Conference on Image Processing, pp. 345–348, October 2008
21. Wang, Z., Bovik, A.C., Sheikh, H.R., Simoncelli, E.P.: Image quality assessment: from error visibility to structural similarity. IEEE Trans. Image Process. 13(4), 600–612 (2004)
22. Xu, Y., Li, X., Tian, Y.: Automatic panorama mosaicing with high distorted fisheye images. In: International Conference on National Computation, vol. 6, pp. 3286–3290 (2010)

Multilinear Methods for Spatio-Temporal Image Recognition

Hayato Itoh[1,2(✉)], Atsushi Imiya[3], and Tomoya Sakai[4]

[1] Graduate School of Informatics, Nagoya University, Nagoya, Japan
hitoh@mori.m.is.nagoya-u.ac.jp
[2] School of Advanced Integration Science, Chiba University, Chiba, Japan
[3] Institute of Management and Information Technologies,
Chiba University, Chiba, Japan
[4] Graduate School of Engineering, Nagasaki University, Nagasaki, Japan

Abstract. We introduce multilinear dimension-reduction and classification methods for video image sequences. Tensor-to-tensor projection methods for spatio-temporal data are derived as dimension-reduction methods using the three-mode tensor representation. The tensor-to-tensor projection methods transform a tensor to a product of smaller tensors. Furthermore, we construct efficient and robust multiclass classifiers for multilinear forms by using tensorial expressions of spatio-temporal video sequences.

1 Introduction

We introduce multilinear pattern recognition methods for video image sequences. Sequences of video images, such as gait sequences and images of temporal MRI, are spatio-temporal sequences. These sequential data are expressed as three-mode tensors in spatio-temporal spaces. Multilinear forms allow us to deal with three-mode tensors without embedding a vector space, which is a traditional data space for pattern recognition. The tensor-to-tensor projection (TTP) methods for multilinear forms were derived as dimension-reduction methods. The TTP methods transform a tensor to a product of smaller tensors. For the construction of a TTP, tensor decompositions give the bases of tensor subspaces. Using these tensor subspaces, we construct classifiers for tensorial data.

For the dimension reduction of tensors, tensor-based PCA methods have been proposed [1–4]. As an extension of principal component analysis to higher-order tensors, tensor principal component analysis (TPCA) has been proposed [1]. By adding an uncorrelation constraint [2], a sparsity constraint [3] and a nonnegativity constraint [5] to TPCA, TPCA has been further extended. However, closed-form formulations do not exist for the decompositions. Therefore, decompositions are generally based on iterative procedures. For the construction of classifiers, supervised tensor learning frameworks have been proposed [6–8]. As an extension of linear discriminant analysis, multilinear discriminant analysis has been proposed [6]. Also, as an extension of linear support vector machines

© Springer International Publishing AG 2017
M. Felsberg et al. (Eds.): CAIP 2017, Part I, LNCS 10424, pp. 148–159, 2017.
DOI: 10.1007/978-3-319-64689-3_12

to tensors, a support tensor machine has been proposed [7,8]. These methods are basically two-class classifiers.

By reorganising TPCA methods in the manner of ref. [9], we introduce multilinear projections for video image sequences. The linear projection for an n-mode unfolded matrix is an n-mode product of a tensor. We obtain a multilinear dimension-reduction method by applying a linear dimension-reduction method for vector data to each unfolded tensor as an n-mode tensor projection. Since an n-mode tensor projection is commutative, we have a unique representation of a TTP as an extension of a linear dimension-reduction method for vector data. In this paper, we introduce the three-dimensional discrete cosine transform (3DDCT) as an approximation of three-mode TPCA. Furthermore, we introduce multiclass linear classifiers, a tensor subspace method (TSM) [9] and a mutual tensor subspace method (MTSM) [10] for video image sequences. In numerical examples, by combining tensor-based dimension reduction and multilinear classifiers, we demonstrate the efficient and robust recognition of image sequences. In these examples, we adopt image sequences of gait patterns in the OU-ISIR dataset [11].

2 Tensor Expression and Processing

We briefly summarise the multilinear projection for multidimensional arrays from ref. [12]. A tensor $\mathcal{M} \in \mathbb{R}^{m \times n}$, which is a matrix, is expressed as $((x_{ij}))$ for $1 \leq i \leq I_1$, $1 \leq j \leq I_2$. Therefore, as an extension of the matrix, a third-order tensor is defined in $\mathbb{R}^{I_1 \times I_2 \times I_3}$. A third-order tensor is expressed as $\mathcal{X} = ((x_{ijk}))$ with three indices $1 \leq i \leq I_1$, $1 \leq j \leq I_2$, $1 \leq k \leq I_3$. i, j, k denote the mode of the tensor $\mathcal{X}$. For $\mathcal{X}$, the n-mode vectors, $n = 1, 2, 3$, are defined as the I_n-dimensional vectors obtained from unfolding of $\mathcal{X}$ by varying index i_n while fixing all the other indices. For $n = 1, 2, 3$, the matricising of $\mathcal{X}$ along the n-mode vectors of $\mathcal{X}$ is defined as

$$\mathcal{X}_{(1)} \in \mathbb{R}^{I_1 \times I_{23}}, \ \mathcal{X}_{(2)} \in \mathbb{R}^{I_2 \times I_{13}}, \ \mathcal{X}_{(3)} \in \mathbb{R}^{I_3 \times I_{12}}, \tag{1}$$

where $I_{23} = I_2 \times I_3$, $I_{13} = I_1 \times I_3$, $I_{12} = I_1 \times I_2$ and the column vectors of $\mathcal{X}_{(n)}$ are the n-mode vectors of $\mathcal{X}$. For example, 1- and 2-mode vectors are column and row vectors of $\mathcal{X}$, respectively. Therefore, the column vectors of $\mathcal{X}_{(1)}$ and $\mathcal{X}_{(2)}$ are the column and row vectors of $\mathcal{X}$, respectively. Figure 1 shows an example of n-mode matricising for a third-order tensor. The 1-, 2- and 3-mode products of matrices $U^{(1)} \in \mathbb{R}^{P_1 \times I_1}$, $U^{(2)} \in \mathbb{R}^{P_2 \times I_2}$ and $U^{(3)} \in \mathbb{R}^{P_3 \times I_3}$ and a tensor $\mathcal{X}$ are given by

$$\mathcal{X} \times_1 U^{(1)} = \hat{\mathcal{X}}^{(1)}, \ \hat{\mathcal{X}}_{(1)}^{(1)} = U^{(1)} \mathcal{X}_{(1)}, \tag{2}$$

$$\mathcal{X} \times_2 U^{(2)} = \hat{\mathcal{X}}^{(2)}, \ \hat{\mathcal{X}}_{(2)}^{(2)} = U^{(2)} \mathcal{X}_{(2)}, \tag{3}$$

$$\mathcal{X} \times_3 U^{(3)} = \hat{\mathcal{X}}^{(3)}, \ \hat{\mathcal{X}}_{(3)}^{(3)} = U^{(3)} \mathcal{X}_{(3)}, \tag{4}$$

where $\hat{\mathcal{X}}_{(1)}^{(1)}$, $\hat{\mathcal{X}}_{(2)}^{(2)}$ and $\hat{\mathcal{X}}_{(3)}^{(3)}$ are matricised tensors of $\hat{\mathcal{X}}^{(1)}$, $\hat{\mathcal{X}}^{(2)}$ and $\hat{\mathcal{X}}^{(3)}$, respectively. Therefore, n-mode products of $\mathcal{X}$ are achieved by matricising tensor, computing the product of the matricised tensor with a matrix and tensorising the results of the product. For two matrices U and V, n-mode and m-mode tensor products are commutative [13], that is,

$$\mathcal{X} \times_n U \times_m V = \mathcal{X} \times_m V \times_n U. \tag{5}$$

We define the inner product of two tensors $\mathcal{X}_1, \mathcal{X}_2 \in \mathbb{R}^{I_1 \times I_2 \times I_3}$ by

$$\langle \mathcal{X}_1, \mathcal{X}_2 \rangle = \sum_{i_1}^{I_1} \sum_{i_2}^{I_2} \sum_{i_3}^{I_3} \mathcal{X}_1(i_1, i_2, i_3) \cdot \mathcal{X}_2(i_1, i_2, i_3). \tag{6}$$

Using this inner product, the Frobenius norm of a tensor $\mathcal{X}$ is

$$\|\mathcal{X}\|_F = \sqrt{\langle \mathcal{X}, \mathcal{X} \rangle} = \|\text{vec}\,\mathcal{X}\|_2, \tag{7}$$

where vec and $\|\cdot\|_2$ are the vectorisation operator and Euclidean norm of a tensor, respectively. For the two tensors $\mathcal{X}_1$ and $\mathcal{X}_2$, we define the distance between them as

$$d(\mathcal{X}_1, \mathcal{X}_2) = \|\mathcal{X}_1 - \mathcal{X}_2\|_F. \tag{8}$$

Although this definition is a tensor-based measure, this distance is equivalent to the Euclidean distance between the vectorised tensors $\mathcal{X}_1$ and $\mathcal{X}_2$.

As the tensor $\mathcal{X}$ is in the tensor space $\mathbb{R}^{I_1} \otimes \mathbb{R}^{I_2} \otimes \mathbb{R}^{I_3}$, the tensor space can be interpreted as the Kronecker product of three vector spaces $\mathbb{R}^{I_1}, \mathbb{R}^{I_2}, \mathbb{R}^{I_3}$. To project $\mathcal{X} \in \mathbb{R}^{I_1} \otimes \mathbb{R}^{I_2} \otimes \mathbb{R}^{I_3}$ to another tensor $\mathcal{Y}$ in a lower-dimensional tensor space $\mathbb{R}^{P_1} \otimes \mathbb{R}^{P_2} \otimes \mathbb{R}^{P_3}$, where $P_n \leq I_n$ for $n = 1, 2, 3$, we need three matrices $\{U^{(n)} \in \mathbb{R}^{I_n \times P_n}\}_{n=1}^3$. For a tensor, a multilinear projection maps the input tensor data from one space to another space. Using the three matrices, the TTP is given by

$$\mathcal{Y} = \mathcal{X} \times_1 U^{(1)} \times_2 U^{(2)} \times_3 U^{(3)}. \tag{9}$$

This projection is established in three steps, where at the nth step, each n-mode vector is projected to a P_n-dimensional space by $U^{(n)}$. Figure 1(b) shows an example of a 1-mode linear projection for a third-order tensor. Figure 1(c) shows the procedure used to project third-order tensors.

3   Decompositions and Dimension Reductions

A third-order tensor $\mathcal{X} \in \mathbb{R}^{I_1 \times I_2 \times I_3}$, which is the array $X \in \mathbb{R}^{I_1 \times I_2 \times I_3}$, is denoted as a triplet of indices (i_1, i_2, i_3). We set the identity matrices I_j, $j = 1, 2, 3$ in $\mathbb{R}^{I_j \times I_j}$. Here we summarise higher-order singular value decomposition (HOSVD) [14] for third-order tensors. For a collection of tensors $\{\mathcal{X}_i\}_{i=1}^N \in \mathbb{R}^{I_1 \times I_2 \times I_3}$ satisfying the zero expectation condition $\mathrm{E}(\mathcal{X}_i) = 0$, we compute

$$\hat{\mathcal{X}}_i = \mathcal{X}_i \times_1 U^{(1)\top} \times_2 U^{(2)\top} \times_3 U^{(3)\top}, \tag{10}$$

where $U^{(j)} = [u_1^{(j)}, \ldots, u_{I_j}^{(j)}]$ that minimises the criterion

$$J_- = \mathrm{E}\left(\|\mathcal{X}_i - \hat{\mathcal{X}}_i \times_1 U^{(1)} \times_2 U^{(2)} \times_3 U^{(3)}\|_F^2\right) \tag{11}$$

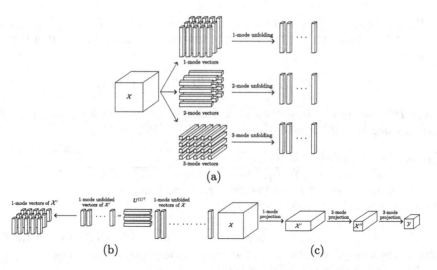

Fig. 1. (a) Matricising of a third-order tensor showing 1-, 2- and 3-mode unfoldings of the third-order tensor $\mathcal{X} \in \mathbb{R}^{4 \times 5 \times 3}$. (b) 1-mode projection that projects $\mathcal{X} \in \mathbb{R}^{4 \times 5 \times 3}$ to a lower-dimensional tensor $\mathcal{Y} \in \mathbb{R}^{3 \times 5 \times 3}$. (c) Multilinear map that consists of three linear projections.

with respect to the condition $U^{(j)\top} U^{(j)} = I_j$.

Eigendecomposition problems are derived by computing the extremes of

$$E_j = J_j + tr((I_j - U^{(j)\top} U^{(j)}) \Sigma^{(j)}), j = 1, 2, 3, \tag{12}$$

where we set

$$J_j = \mathrm{E} \left(\| U^{(j)\top} \mathcal{X}_{i,(j)} \mathcal{X}_{i,(j)}^\top U^{(j)} \|_{\mathrm{F}}^2 \right). \tag{13}$$

For matrices $M^{(j)} = \frac{1}{N} \sum_{i=1}^{N} \mathcal{X}_{i,(j)} \mathcal{X}_{i,(j)}^\top$, $j = 1, 2, 3$, the optimisation of J_j derives the eigenvalue decomposition

$$M^{(j)} U^{(j)} = U^{(j)} \Sigma^{(j)}, \tag{14}$$

where $\Sigma^{(j)} \in \mathbb{R}^{I_j \times I_j}$, $j = 1, 2, 3$, are diagonal matrices satisfying the relationships $\lambda_k^{(j)} = \lambda_k^{(j')}$, $k \in \{1, 2, \ldots, K\}$ for

$$\Sigma^{(j)} = \mathrm{diag}(\lambda_1^{(j)}, \lambda_2^{(j)} \cdots, \lambda_K^{(j)}, 0 \cdots, 0). \tag{15}$$

For the optimisation of $\{J_j\}_{j=1}^3$, there is no closed-form solution to this maximisation problem [14]. Algorithm 1 is the iterative procedure of multilinear principal component analysis (MPCA) [1]. For Algorithm 1, we have the following property.

Property 1. [9] *MPCA without iteration is equivalent to HOSVD if the dimensions of a projected tensor are coincident with those of each mode of the original tensor.*

For $p_k \in \{e_k\}_{k=1}^K$, we set orthogonal projection matrices $P^{(j)} = \sum_{k=1}^{k_j} p_k p_k^\top$ for $j = 1, 2, 3$. Using these $\{P^{(j)}\}_{j=1}^3$, the low-rank tensor approximation [14] is given by

$$\mathcal{Y} = \mathcal{X} \times_1 (P^{(1)} U^{(1)})^\top \times_2 (P^{(2)} U^{(2)})^\top \times_3 (P^{(3)} U^{(3)})^\top, \qquad (16)$$

where $P^{(j)}$ selects k_j bases of projection matrices $U^{(j)}$. The low-rank approximation using Eq. (16) is used for compression in TPCA.

Algorithm 1. Iterative method for third-order tensors

Input: A set of tensors $\{\mathcal{X}_i\}_{i=1}^N$. The dimension of projected tensors $\{k_j\}_{j=1}^3$.
   A maximum number of iterations K. A small number η.
Output: A set of projection matrices $\{U^{(j)}\}_{j=1}^3$.
1: Compute the eigendecomposition of a covariant matrix.
   $M^{(j)} = \frac{1}{N} \sum_{i=1}^N \mathcal{X}_{i,(j)} \mathcal{X}_{i,(j)}^\top$, where $\mathcal{X}_{i,(j)}$ is a j-mode unfolded $\mathcal{X}_i$, for $j = 1, 2, 3$.
2: Construct projection matrices by selecting eigenvectors corresponding to
   the k_j largest eigenvalues for $j = 1, 2, 3$.
3: Compute $\Psi_0 = \sum_{i=1}^N \|\mathcal{X}_i \times_1 U^{(1)\top} \times_2 U^{(2)\top} \times_3 U^{(3)\top}\|_F$
4: Iteratively compute the following procedure.
   for $k = 1, 2, \ldots, K$.
     for $j = 1, 2, 3$.
       Update $U^{(j)}$ by decomposing the matrix.
       $\sum_{i=1}^N (\mathcal{X} \times_{\xi_\alpha} U^{(\xi_\alpha)\top} \times_{\xi_\beta} U^{(\xi_\beta)\top})(\mathcal{X} \times_{\xi_\alpha} U^{(\xi_\alpha)\top} \times_{\xi_\beta} U^{(\xi_\beta)\top})^\top$,
       where $\xi_\alpha, \xi_\beta \in \{1, 2, 3\} \setminus \{j\}$, $\xi_\alpha \neq \xi_\beta$
     end
     Compute $\Psi_k = \sum_{i=1}^N \|\mathcal{X}_i \times_1 U^{(1)\top} \times_2 U^{(2)\top} \times_3 U^{(3)\top}\|_F$
     if $|\Psi_k - \Psi_{k-1}| < \eta$.
       break
   end

For HOSVD for third-order tensors, we have the following theorem.

Theorem 1. *The compression computed by HOSVD is equivalent to the compression computed by TPCA.*

(*Proof*) The projection that selects $K = k_1 k_2 k_3$ bases of the tensor space spanned by $u_{i_1}^{(1)} \circ u_{i_2}^{(2)} \circ u_{i_3}^{(3)}$, $i_j = 1, 2, \ldots, k_j$ for $j = 1, 2, 3$, is

$$\begin{aligned} (P^{(3)} U^{(3)} \otimes P^{(2)} U^{(2)} \otimes P^{(1)} U^{(1)}) \\ = (P^{(3)} \otimes P^{(2)} \otimes P^{(1)})(U^{(3)} \otimes U^{(2)} \otimes U^{(1)}) = PW, \end{aligned} \qquad (17)$$

where W and P are the projection matrix and a unitary matrix, respectively. Therefore, HOSVD is equivalent to TPCA for third-order tensors. $\square$

Furthermore, we have the following theorem.

Theorem 2. *The HOSVD method is equivalent to the vector PCA method.*

(*Proof*). The equation

$$\mathcal{Y} = \mathcal{X} \times_1 (\boldsymbol{P}^{(1)} \boldsymbol{U}^{(1)})^\top \times_2 (\boldsymbol{P}^{(2)} \boldsymbol{U}^{(2)})^\top \times_3 (\boldsymbol{P}^{(3)} \boldsymbol{U}^{(3)})^\top \qquad (18)$$

is equivalent to

$$\mathrm{vec}\,\mathcal{Y} = (\boldsymbol{P}^{(3)} \boldsymbol{U}^{(3)} \otimes \boldsymbol{P}^{(2)} \boldsymbol{U}^{(2)} \otimes \boldsymbol{P}^{(1)} \boldsymbol{U}^{(1)})^\top \mathrm{vec}\,\mathcal{X} = (\boldsymbol{PW})^\top \mathrm{vec}\,\mathcal{X}. \qquad (19)$$

(Q.E.D.)

This theorem implies that the 3DDCT is an acceptable approximation of HOSVD for third-order tensors [9] since this is an analogy of the approximation of PCA for two-dimensional images by the two-dimensional discrete cosine transform [15].

In our application, an $n \times n \times n$ digital array is directly compressed by the 3DDCT-II [16,17] with order $\mathcal{O}(n^3)$. If we apply the fast Fourier transform to the computation of the 3DDCT-II, the computational complexity is $\mathcal{O}(n \log n)$.

4   Classifiers

4.1   Tensor Subspace Method

We introduce the linear TSM for third-order tensors [9]. Setting $\boldsymbol{U}^{(j)}$, $j = 1, 2, 3$, to be orthogonal projections, for a collection of matrices $\{\mathcal{X}_i\}_{i=1}^M$, such that $\mathcal{X}_i \in \mathbb{R}^{I_1 \times I_2 \times I_3}$ and $\mathrm{E}(\mathcal{X}_i) = 0$, the solutions of

$$\{\boldsymbol{U}^{(j)}\}_{j=1}^3 = \arg \max \mathrm{E} \left(\frac{\|\mathcal{X} \times_1 \boldsymbol{U}^{(1)\top} \times_2 \boldsymbol{U}^{(2)\top} \times_3 \boldsymbol{U}^{(3)\top}\|_F}{\|\mathcal{X}_i\|_F} \right) \qquad (20)$$

with respect to $\boldsymbol{U}^{(j)\top} \boldsymbol{U}^{(j)} = \boldsymbol{I}$ for $j = 1, 2, 3$ define a trilinear subspace that approximates $\{\mathcal{X}_i\}_{i=1}^M$. Therefore, using projection matrices $\{\boldsymbol{U}_k^{(j)}\}_{j=1}^3$ obtained as the solutions of Eq. (20) for the kth category, if a query tensor $\mathcal{G}$ satisfies the condition

$$\arg \left(\max_l \frac{\|\mathcal{G} \times_1 \boldsymbol{U}_l^{(1)\top} \times_2 \boldsymbol{U}_l^{(2)\top} \times_3 \boldsymbol{U}_l^{(3)\top}\|_F}{\|\mathcal{G}\|_F} \right) = \{\boldsymbol{U}_k^{(j)}\}_{j=1}^3, \qquad (21)$$

we conclude that $\mathcal{G} \in \mathcal{C}_k$, $k, l = 1, 2, \ldots, N_C$, where $\mathcal{C}_k$ and N_C are the tensor subspace of kth category and the number of categories, respectively. For the practical computation of projection matrices $\{\boldsymbol{U}_k^{(j)}\}_{j=1}^3$, we adopt the iterative method of MPCA described in Algorithm 1.

4.2   Mutual Tensor Subspace Method

We define a classifier for two tensor subspaces. For each of N_C categories of volume data, we set a collection of third-order tensors $\{\mathcal{X}_i\}_{i=1}^M$, such that $\mathcal{X}_i \in \mathbb{R}^{I_1 \times I_2 \times I_3}$ and $\mathrm{E}(\mathcal{X}_i) = 0$. For the kth category, we have an orthogonal projection by $\{\boldsymbol{U}_{k,j}\}_{j=1,k=1}^{3,N_c}$, which satisfies Eq. (20). For the practical computation of bases of tensor subspaces, we use MPCA. We have a collection of query tensors $\{\mathcal{G}_{i'}\}_{i'=1}^{M'}$. Using projection matrices $\{U_k^{(j)}\}_{j=1,k=1}^{3,N_c}$, we have the projected tensor

$$\mathcal{A}_{i'} = \mathcal{G}_{i'} \times_1 U_k^{(1)\top} \times_2 U_k^{(2)\top} \times_3 U_k^{(3)\top}. \tag{22}$$

Furthermore, assuming that queries belong to one of the N_{C} categories, we have orthogonal projection matrices $\{V_j\}_{j=1}^{3}$, which are given by Eq. (20), for a tensor subspace of queries. This orthogonal projection gives the projected tensor

$$\mathcal{B}_{i'} = \mathcal{G}_{i'} \times_1 V^{(1)\top} \times_2 V^{(2)\top} \times_3 V^{(3)\top}. \tag{23}$$

For a tensor subspace $\mathcal{C}_k$ of a category and a tensor subspace $\mathcal{C}_q$ of queries, we define the dissimilarity of subspaces $d(\mathcal{C}_k, \mathcal{C}_q)$ by

$$\mathrm{E}\left(\|\mathcal{A}_{i'} \times_1 PU_k^{(1)} \times_2 PU_k^{(2)} \times_3 PU_k^{(3)} - \mathcal{B}_{i'} \times_1 PV^{(1)} \times_2 PV^{(2)} \times_3 PV^{(3)}\|_{\mathrm{F}}^2\right), \tag{24}$$

where a unitary matrix P selects bases for each mode of tensors. Therefore, using the dissimilarity given by Eq. (24), if queries $\{\mathcal{G}_{i'}\}_{l=1}^{M'}$ satisfy the condition

$$\arg\left(\min_l d(\mathcal{C}_l, \mathcal{C}_q)\right) = \mathcal{C}_k, \tag{25}$$

we conclude that $\{\mathcal{G}_{i'}\}_{i'=1}^{M'} \in \mathcal{C}_k(\delta)$ for $k, l = 1, 2, \ldots, N_{\mathrm{C}}$.

5   Numerical Examples

To evaluate the numerical relationship between HOSVD and the 3DDCT, we compute recognition rates using OU-ISIR treadmill dataset A [11]. Figure 2 shows examples of sequences of silhouette images from two categories in the OU-ISIR dataset. Table 1 summarises the sizes of the tensors of the OU-ISIR dataset. For the compression of the silhouette-image sequences, we use HOSVD and the 3DDCT. For the practical computation of HOSVD, we use the iterative method described in Algorithm 1 [1]. If we set the number of iterations to 0 in Algorithm 1, we have the three-dimensional version of HOSVD. If we set the number of bases to the size of the original tensors in Algorithm 1, we call the method the full projection (FP). If we set the number of bases to less than the size of the original tensors in Algorithm 1, we call the method the full-projection truncation (FPT).

Table 1. Sizes and number of frames of the resampled OU-ISIR dataset. ♮class and ♮data/class represent the number of classes and the number of data in each class, respectively.

	♮class	♮data/class	Tensor size	Reduced tensor size
OU-ISIR	34	9	$128 \times 88 \times 90$	$32 \times 32 \times 32$

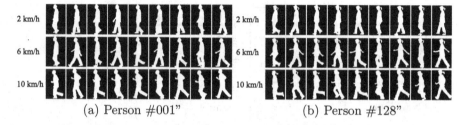

(a) Person #001"   (b) Person #128"

Fig. 2. Examples of sequences of silhouette images, which are binary images whose pixel values are 0 or 255. The figures illustrate the 1st, 11th, 21st, ..., 81st silhouette images of sequences of two people walking at different speeds. Each sequence consists of 90 silhouette images of four steps. The resolution of these silhouette images is 128×88 pixels. For each sequence, we manually selected the first and last frames of the sequence.

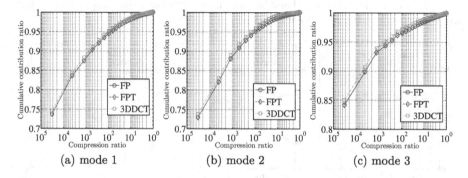

(a) mode 1   (b) mode 2   (c) mode 3

Fig. 3. Comparison of CCRs for three types of compressed tensor. For the compression of tensors, we use Algorithm 1 and the 3DDCT. In Algorithm 1, we respectively adopt sizes of $128 \times 88 \times 90$ and $32 \times 32 \times 32$ for the computation by the FP and FPT. For the three types of compressed tensor of $32 \times 32 \times 32$, we apply 10 iterations of Algorithm 1. In (a)-(c), the horizontal and vertical axes represent the compression ratio and CCR, respectively.

First, we compute the cumulative contribution ratios (CCRs) of the eigenvalues obtained by 10 iterations of Algorithm 1 for compressed tensors. For the compression of the tensors from $128 \times 88 \times 90$ to $32 \times 32 \times 32$, we adopt the FP, the FPT and the 3DDCT. Figure 3 shows the CCRs of each mode for the three types of compressed tensor. Figure 4 summarises the computational time required for dimension reduction. The tensors compressed by the 3DDCT give larger eigenvalues than those compressed by the FP and FPT with a smaller number of bases. The FP and FPT give the same CCRs for each mode. The computational time of the 3DDCT is smaller than those of the FP and FPT.

Then, we compute the recognition rates of sequences of silhouette images using the TSM and MTSM. In this validation, we use the original sizes of the tensors and compressed tensors for comparison. For the compression, we adopt the HOSVD, the FP, the FPT and the 3DDCT. Using these four methods, we

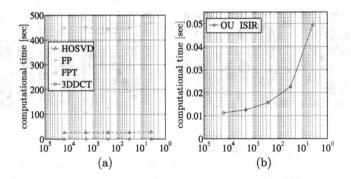

Fig. 4. Computational time for dimension reduction for third-order tensors. (a) Computational time for construction of projection matrices for 306 sequences of silhouette images. (b) Mean computational times for projecting images to low-dimensional tensor space. In (a) and (b), the vertical and horizontal axes represent the computational time and compression ratio, respectively.

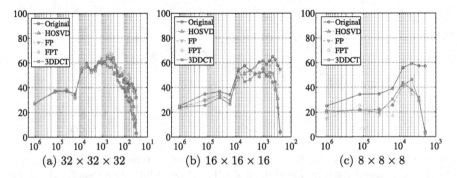

Fig. 5. Recognition rates of gait patterns for original and compressed tensors. We adopt the reduced sizes of $32 \times 32 \times 32$, $16 \times 16 \times 16$ and $8 \times 8 \times 8$. (a)–(c) Recognition rate obtained by the TSM for the three reduced sizes. For compression, we use HOSVD, the FP, the FPT and the 3DDCT. In (a)–(c), the horizontal and vertical axes represent the compression ratio and CCR, respectively. For the original size $D = 128 \times 88 \times 90$ and the reduced size $K = k \times k' \times k''$, the compression ratio is given as D/K.

compress the tensors to the sizes $32 \times 32 \times 32$, $16 \times 16 \times 16$ and $8 \times 8 \times 8$. The OU-ISIR dataset contains sequences of images of 34 people with nine different walking speeds. We use the sequences with walking speeds of 2, 4, 6, 8 and 10 km/h for learning data and the sequences with walking speeds of 3, 5, 7 and 9 km/h for test data. The recognition rate is defined as the successful label estimation ratio for 1000 label estimations. In each estimation of a label for a query, categories and queries are randomly chosen from the test dataset. For the 1-, 2- and 3-modes, we evaluate the results for multilinear subspaces with sizes from one to the dimension of the compressed tensors.

Figures 5 and 6 show the recognition rates obtained by the TSM and MTSM, respectively, for the four compression methods with three different sizes of the

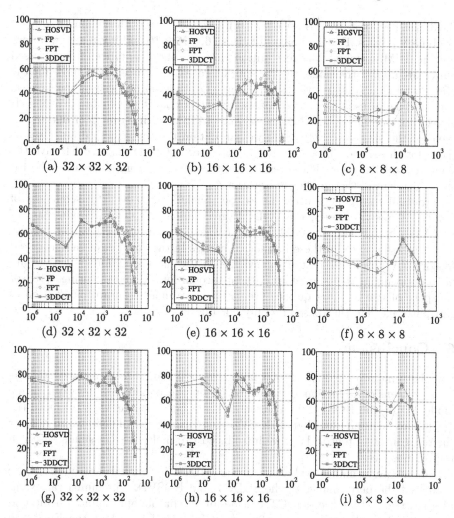

Fig. 6. Recognition rates of gait patterns for compressed tensors obtained by the MTSM. We adopt the reduced sizes of $32 \times 32 \times 32$, $16 \times 16 \times 16$ and $8 \times 8 \times 8$. For compression, we use HOSVD, the FP, the FPT and the 3DDCT. (a)–(i) Recognition rates for the three reduced sizes. (a)–(c), (d)–(f) and (g)–(i) show the recognition rates for the case of using one query, two queries and three queries, respectively. In (a)–(i), the horizontal and vertical axes represent the compression ratio and CCR, respectively. For the original size $D = 128 \times 88 \times 90$ and the reduced size $K = k \times k' \times k''$, the compression ratio is given as D/K.

compressed tensors. For the images of size $32 \times 32 \times 32$, the recognition rates for all four types of compressed tensor obtained by the TSM and MTSM are almost coincident when the compression ratio is higher than 10^3. When the compression ratio is less than 10^3, for the TSM and MTSM, the recognition ratio of the FPT is higher than those of HOSVD, the FP and the 3DDCT. The recognition ratio of the 3DDCT is lower than those of the other methods since the silhouette images are binary images. For the images of sizes $16 \times 16 \times 16$ and $8 \times 8 \times 8$, when we use the TSM, although the recognition rates for the four types of compressed tensor are almost the same, the recognition rates are smaller than those for the original tensors. This recognition property depends on the size of the images, and the images used for the comparison are too small to evaluate our methods of recognition. However, when we use the MTSM with more than one query, even for the size of $16 \times 16 \times 16$, we can obtain the same recognition rates as those for the size of $32 \times 32 \times 32$. In all cases, HOSVD and the FP give the same recognition rates. These results imply that the decomposition for the FP is independent of the number of iterations.

From all the numerical examples, we conclude that even for sequences of binary silhouette images, the 3DDCT gives an acceptable approximation of HOSVD, the FP and the FPT in the context of tensor-based pattern recognition. Furthermore, in these methods, changes in the energies of the projected tensors and the CCRs of the eigenvalues in the decomposition of tensors are not important in the context of pattern recognition. Moreover, using a tensor subspace of two or more queries, we can achieve more accurate recognition of gait patterns than that based on only one query.

6   Conclusions

For the dimension reduction of spatio-temporal data, we introduced tensor-based dimension-reduction methods based on the equivalence between three-mode TPCA and HOSVD for volumetric data. Furthermore, we introduced the 3DDCT as an approximation of HOSVD for spatio-temporal data. Moreover, for the recognition of tensor data, we introduced two multiclass classifiers based on tensor subspaces.

For the performance evaluation of the 3DDCT in tensor-based methods, we evaluated the performance of the 3DDCT and HOSVD. Our numerical examples illustrated that the 3DDCT can be an acceptable approximation method for HOSVD in the recognition of volumetric data if we adopt the Euclidean distance as the metric of the pattern space. Furthermore, observations of the properties of the iterative algorithm showed that the accuracy of the recognition of volume data is independent of the number of iterations in MPCA. Moreover, the numerical examples showed that our proposed classifiers can achieve the accurate recognition of multiclass volume data.

This research was supported by "Object oriented data-analysis for understanding and recognition of higher-dimensional multimodal data" by grant for Scientific Research from JSPS, Japan.

References

1. Lu, H., Plataniotis, K.N., Venetsanopoulos, A.N.: MPCA: Multilinear principal component analysis of tensor objects. IEEE Trans. Neural Networks **19**(1), 18–39 (2008)

2. Lu, H., Plataniotis, K.N., Venetsanopoulos, A.N.: Uncorrelated multilinear principal component analysis for unsupervised multilinear subspace learning. IEEE Trans. Neural Networks **20**(11), 1820–1836 (2009)
3. Shen, H., Huang, J.Z.: Sparse principal component analysis via regularized low rank matrix approximation. J. Multivariate Anal. **99**(6), 1015–1034 (2008)
4. Vasilescu, M.A.O., Terzopoulos, D.: Multilinear (tensor) ICA and dimensionality reduction. In: Proceedings of the 7th International Conference on Independent Component Analysis and Signal Separation, pp. 818–826 (2007)
5. Panagakis, Y., Kotropoulos, C., Arce, G.R.: Non-negative multilinear principal component analysis of auditory temporal modulations for music genre classification. IEEE Trans. Audio Speech Lang. Process. **8**(3), 576–588 (2010)
6. Yan, S., Xu, D., Yang, Q., Zhang, L., Tang, X., Zhang, H.J.: Multilinear discriminant analysis for face recognition. IEEE Trans. Image Process. **16**(1), 212–220 (2007)
7. Tao, D., Li, X., Wu, X., Hu, W., Maybank, S.J.: Supervised tensor learning. Knowl. Inf. Syst. **13**(1), 1–42 (2007)
8. Kotsia, I., Guo, W., Patras, I.: Higher rank support tensor machines for visual recognition. Pattern Recogn. **45**(12), 4192–4203 (2012)
9. Itoh, H., Imiya, A., Sakai, T.: Pattern recognition in multilinear space and its applications: mathematics, computational algorithms and numerical validations. Mach. Vis. Appl. **27**(8), 1259–1273 (2016)
10. Itoh, H., Imiya, A., Sakai, T.: Approximation of n-way principal component analysis for organ data. In: Proceedings of the ACCV Workshop on Mathematical and Computational Methods in Biomedical Imaging and Image Analysis, pp. 16–31 (2017)
11. Makihara, Y., Mannami, H., Tsuji, A., Hossain, M.A., Sugiura, K., Mori, A., Yagi, Y.: The OU-ISIR gait database comprising the treadmill dataset. IPSJ Trans. Comput. Vis. Appl. **4**, 53–62 (2012)
12. Lu, H., Plataniotis, K.N., Venetsanopoulos, A.N.: A survey of multilinear subspace learning for tensor data. Pattern Recogn. **44**, 1540–1551 (2011)
13. Cichoki, A., Zdunek, R., Phan, A.H., Amari, S.: Nonnegative Matrix and Tensor Factorizations. Wiley, Chichester (2009)
14. Lathauwer, L.D., Moor, B.D., Vandewalle, J.: On the best rank-1 and rank-(r_1, r_2, r_n) approximation of higher-order tensors. SIAM J. Matrix Anal. Appl. **21**(4), 1324–1342 (2000)
15. Oja, E.: Subspace Methods of Pattern Recognition. Research Studies Press, New York (1983)
16. Roese, J., Pratt, W., Robinson, G.: Interframe cosine transform image coding. IEEE Trans. Commun. **25**(11), 1329–1339 (1977)
17. Natarajan, T., Ahmed, N.: On interframe transform coding. IEEE Trans. Commun. **25**(11), 1323–1329 (1977)

On the Essence of Unsupervised Detection of Anomalous Motion in Surveillance Videos

Abdullah A. Abuolaim[1,2(✉)], Wee Kheng Leow[1], Jagannadan Varadarajan[2], and Narendra Ahuja[2,3]

[1] Department of Computer Science,
National University of Singapore, Singapore, Singapore
{abdullah,leowwk}@comp.nus.edu.sg
[2] Advanced Digital Sciences Center, Singapore, Singapore
vjagan@adsc.com.sg
[3] Department of Electrical and Computer Engineering,
University of Illinois at Urbana-Champaign, Champaign, IL, USA
n-ahuja@illinois.edu

Abstract. An important application in surveillance is to apply computerized methods to automatically detect anomalous activities and then notify the security officers. Many methods have been proposed for anomaly detection with varying degree of accuracy. They can be characterized according to the approach adopted, which is supervised or unsupervised, and the features used. Unfortunately, existing literature has not elucidated the essential ingredients that make the methods work as they do, despite the fact that tests have been conducted to compare the performance of various methods. This paper attempts to fill this knowledge gap by studying the videos tested by existing methods and identifying key components required by an effective unsupervised anomaly detection algorithm. Our comprehensive test results show that an unsupervised algorithm that captures the key components can be relatively simple and yet perform equally well or better compared to existing methods.

1 Introduction

In recent decades, surveillance cameras have been widely used in public places to monitor human activities and provide security measures. A security officer typically has to monitor a dozen or more surveillance videos at the same time. Most of the time, there is no significant anomalous activity, which tends to lower the guard of the officer. After monitoring for long hours, he can get tired and miss important events that happen suddenly. Therefore, automatic detection of anomalous activities by computerized methods has attracted much research effort. These methods can also be used for criminal investigation to sieve through video archives to detect anomalous activities that have happened in the past.

Many methods have been proposed for anomaly detection with varying degree of accuracy. They can be characterized according to the approach adopted, which is supervised [1–16] or unsupervised [17–21], and the features used, which range from

© Springer International Publishing AG 2017
M. Felsberg et al. (Eds.): CAIP 2017, Part I, LNCS 10424, pp. 160–171, 2017.
DOI: 10.1007/978-3-319-64689-3_13

Fig. 1. Sample frames from test videos. (a) UCSDped1, (b) UCSDped2, (c) Subway entrance, (d) Subway exit, (e) UMN, (f) PETS2009 scene 1, (g) PETS2009 scene 2.

low-level optical flow to high-level multiple object trajectories. Unfortunately, existing literature has not elucidated the essential ingredients that make the methods work as they do, despite the fact that tests have been conducted to compare the performance of various methods. For example, test results (Sect. 4) seem to suggest that there is no significant advantage in offline training performed by supervised methods compared to well-crafted unsupervised methods. It is also uncertain whether the time taken to process high-level features necessarily leads to better detection accuracy. This situation makes it difficult to optimize the methods for real-time online detection and efficient video archive analysis.

This paper attempts to fill this knowledge gap by studying the videos tested by existing methods and identifying key components required by an effective unsupervised anomaly detection algorithm. We have chosen to investigate unsupervised method instead of supervised method for the following reasons: (1) Unsupervised method does not require tedious and time-consuming manual labeling of training data. (2) It does not require an offline training phase. Therefore, it can be more easily extended to handle new normal and abnormal motion patterns that have not happened in the past. (3) Without the need of offline training, it can be more easily adapted to real-time online applications by implementing incremental algorithms. We focus on surveillance videos of pedestrians captured by stationary cameras because they are widely tested in the literature. Our comprehensive test results on these videos show that an unsupervised algorithm that captures the key components can be relatively simple and yet perform equally well or better compared to existing methods.

2   Existing Methods

Regardless of the approach, all existing methods begin by extracting features from the input videos and then making detection decisions based on the features.

The extracted features include optical flow [1,7,8,17,18], histogram of optical flow (HOF) [2,4,14–16,19,20], histogram of oriented gradient (HOG) [4,14], 3D SIFT [4,21], histogram of edge orientation [16], descriptors of intensity, gradient, object persistence, motion direction, optical flow orientation, speed, etc. [6,9,12], structural descriptors based on HOF [19], particle advection based on optical flow [3,13], tracked interest points or targets [3,7,19,20], dynamic texture [5,11], and pedestrian regions [19]. In addition, auto-encoder neural network has also been used to extract features from video images [1,10,14]. These features may be extracted for image pixels [3,7,13,17], 2D spatial regions [1,2,5,6,8,10–12,15,16,18–20] or 3D spatio-temporal regions [4,6,9,14,15,21] of the video. Simple features, such as optical flow and intensity gradient, take much less time to extract compared to features extracted by complex algorithms, such as pedestrian detection and multiple target tracking [19,20]. Auto-encoders can extract features efficiently but it takes a large amount of time to train them.

Existing methods for detecting anomalous motion in surveillance videos can be grouped into two categories: supervised and unsupervised. Supervised methods [1–16] typically work in two phases: training and testing. In the training phase, these methods use labeled training data to train a classifier or a probabilistic model. Various algorithms have been used for training, including SVM [1,7], conjugate Bayesian analysis [2], EM [3,5,11–13], Gaussian process regression [4], Bayesian network propagation [6], recurrent neural network [10], and sparse reconstruction [15]. Methods that use k-nn [9,16] do not need the training phase. Methods that model simple probability distributions such as Gaussian distributions [8,14] have a simple training phase that estimates the distribution parameters. In the testing phase, trained classifier or probabilistic model is used to classify features as normal or abnormal. Well-trained supervised methods can be accurate. Moreover, their testing phases are typically efficient enough for real-time applications, provided the features can be extracted efficiently. However, manual labeling of training data is tedious and time-consuming. Therefore, it is difficult to extend supervised methods to include new scenario.

Unsupervised methods [17–21] typically group extracted features into clusters without relying on labeled data. The clustering algorithms that have been used include hierarchical cluster merging [18], k-means [20], online weighted clustering [20], and fuzzy probabilistic clustering [21]. After clustering, these methods label dominant clusters (i.e., clusters with the most members) as normal and the other clusters as abnormal. The threshold for deciding which clusters are dominant is empirically set. The methods of [17,19], on the other hand, do not perform clustering. Instead, the method of [17] performs line intersection to detect the center of crowd dispersion, and the method of [19] measures dissimilarity between features to detect anomalies. Unsupervised methods do not require manually labeled training data and do not perform offline training. Therefore, they can be easily extended to handle new normal and abnormal motion. Moreover, unsupervised methods that use incremental algorithms are very suitable for real-time online applications.

The above methods have been tested on one or more of the following surveillance videos on pedestrians (Fig. 1):

- UCSDped1 [22]: 36 videos, tested in [1,4–6,8–13,15,16,19–21].
- UCSDped2 [22]: 12 videos, tested in [5,8,11–16,19–21].
- Subway [8]: 2 videos in 2 scenes, tested in [4–6,8,9,12,15,21].
- UMN [23]: 3 videos in 3 scenes, tested in [2,3,5,7,9,13–15,17–19].
- PETS2009 [24]: 8 videos in 2 scenes, tested in [2,3,13,18].

For UCSDped1, UCSDped2, UMN, and PETS2009, the walking pedestrians constitute the dominant motion and they are regarded as normal. Abnormal motion is elicited by carts, cyclists, skaters, escaping humans, etc., which move at faster speeds. That is, normal and abnormal motion in these videos differ primarily in motion speed. On the other hand, for Subway, the passengers entering and existing the subway gates in an orderly manner constitute the dominant motion and are regarded as normal. Passengers who move along directions other than entering or exiting the gates are regarded as abnormal. That is, normal and abnormal motion in these videos differ primarily in motion direction.

3 Unsupervised Anomaly Detection

Our research goal is to identify the essential ingredients for effective unsupervised detection of anomalies in pedestrian surveillance videos. To achieve this goal, we apply the principle of Occam's razor: given several equally effective alternatives, we choose the simplest alternative. Therefore, we call our method OCCAM. Similar to unsupervised methods based on clustering, OCCAM consists of three stages: (1) feature extraction, (2) features clustering, and (3) cluster labeling.

3.1 Feature Extraction

Analysis of common test videos used in existing work (Sect. 2) shows that normal and abnormal motion may be differentiated by either motion speed or motion direction alone, depending on the test videos. Therefore, OCCAM uses motion speed or motion direction as the feature. It applies the method of [25] to extract trajectories of distinctive image feature points. This method samples feature points at multiple spatial scales and tracks feature points using median filtering to obtain optical flow. Stationary feature points and those with large displacements between two consecutive frames are removed to reduce tracking error. Tracked feature trajectories have a fixed length l, and long trajectories are split into short trajectories of length l. Trajectories with length shorter than l are removed because they are insignificant.

Let $\{\mathbf{x}_i(t), \ldots, \mathbf{x}_i(t+l)\}$ denote the trajectory of feature point p_i, $i = 1, \ldots, n$, from frame t to $t + l$, where $\mathbf{x}_i(t)$ is the position of p_i in frame t. Then, the direction θ_i and speed s_i of feature point p_i are computed as the direction and magnitude divided by trajectory length of the vector $\mathbf{x}_i(t + l) - \mathbf{x}_i(t)$.

In UCSDped1 videos, humans and other objects move toward or away from the camera resulting in noticeable perspective distortion. As a result, objects nearer to the camera appears to move faster than those further from the camera even though they may move at the same actual speed. To overcome this distortion, the feature points are projected onto the ground plane using an estimated homography. Then, the speeds of the feature points are computed after projection.

3.2   Feature Clustering

Feature clustering is performed on either motion speed or motion direction. Let us denote the extracted feature values as f_i, $i = 1, \ldots, n$. Since the features are 1-D, the simplest way to cluster f_i is to divide the feature value range (minimum to maximum) into m equal intervals, and regard each interval as a cluster C_j, $j = 1, \ldots, m$. Then, features f_i can be clustered efficiently into their respective clusters in a fixed $O(n)$ time. Each cluster C_j is characterized by the cluster size $|C_j|$ and the cluster center, which is the average feature value $\bar{f}_j$ of the features in C_j. This simple and efficient clustering method ensures that the intra-cluster differences are much smaller than the inter-cluster differences.

After clustering, normalized cluster size S_j and normalized cluster center F_j are computed for each cluster C_j. Let us denote the dominant cluster, the cluster with the largest size, as C^+ and the largest feature value as f^*. Then, S_j and F_j are computed as follows:

$$S_j = |C_j|/|C^+|, \quad F_j = \bar{f}_j/f^*. \tag{1}$$

Therefore, these normalized values range between 0 and 1. Each cluster C_j is now characterized by a characteristic vector of two components, namely normalized cluster size S_j and normalized cluster center F_j.

3.3   Cluster Labeling

Unlike existing methods, OCCAM labels the clusters into three types: normal, abnormal, and ambiguous. The ambiguous clusters allow the normal and abnormal clusters to be separated as widely as possible. Since the characteristic vectors of the clusters are 2-D, 2-D k-means clustering is used to group the clusters C_j into three groups G_h, $h = 1, 2, 3$.

First, k-means clustering is initialized as follows: The center of group G_1 is initialized as the characteristic vector of the dominant cluster C^+. Similarly, the abnormal group G_2 is initialized with the cluster C^- whose cluster center is the furthest from that of C^+ because C^- is most likely to be abnormal. The ambiguous group G_3 is initialized with the cluster that is approximately equidistant to C^+ and C^-.

Next, k-means clustering is executed to group the remaining clusters C_j into the three groups G_h. The distance between a cluster and a group is measured in terms of the Euclidean distance between their characteristic vectors. After clustering, all the clusters in group G_1 are labeled as normal, those in G_2 abnormal, and those in G_3 ambiguous. In addition, the abnormal cluster that is nearest to G_1 is re-labeled as ambiguous so as to widen the separation between normal and abnormal clusters.

After cluster labeling, the features f_i in abnormal clusters are labeled as abnormal features. The corresponding trajectory positions $\mathbf{x}_i(t)$ of f_i are labeled as abnormal feature points. Finally, the video frames that contain abnormal feature points are labeled as abnormal frames.

4 Experiments and Discussions

4.1 Data Preparation and Procedure

Five sets of common test videos discussed in Sect. 2 were used in the experiments, namely UCSDped1, UCSDped2, Subway, UMN, and PETS2009. For OCCAM, motion directions were extracted from Subway video whereas motion speeds were extracted from the other videos. Next, feature clustering and cluster labeling were performed to detect abnormal feature points and abnormal frames. Then, true positive rate (TPR), false positive rate (FPR), and accuracy of detected abnormal frames were computed.

To determine a suitable value for the number of clusters m in the feature clustering stage, a test was performed on one video each from UCSDped1, UCS-Dped2, PETS2009 scene 1 and PETS2009 scene 2 test sets with varying values of m. The test shows that OCCAM achieves the overall highest accuracy with $m = 10$. Therefore, m is set to 10 for all the tests.

4.2 Benefit of Ambiguous Clusters

This test illustrates the benefit of having ambiguous clusters. A variant of OCCAM, denoted as OCCAM−, was tested such that its cluster labeling stage ran k-means clustering with $k = 2$ for normal and abnormal groups, without ambiguous group. Existing methods also label their clusters as either normal or abnormal, without ambiguous clusters. Both OCCAM and OCCAM− were tested on the common test videos discussed in Sect. 2. True positive rate (TPR) and false positive rate (FPR) were measured for the detected abnormal frames.

Table 1 compares the results of OCCAM and OCCAM−. For all test videos, OCCAM's TPR is slightly smaller than that of OCCAM−, but OCCAM's FPR is significantly smaller than that of OCCAM−. That is, by regarding some clusters as ambiguous, OCCAM makes significantly fewer false detections than does OCCAM− without significantly sacrificing its true detection rate.

4.3 Performance Comparison

OCCAM's results are compared with all of the existing methods discussed in Sect. 2. These methods belong to the following categories:

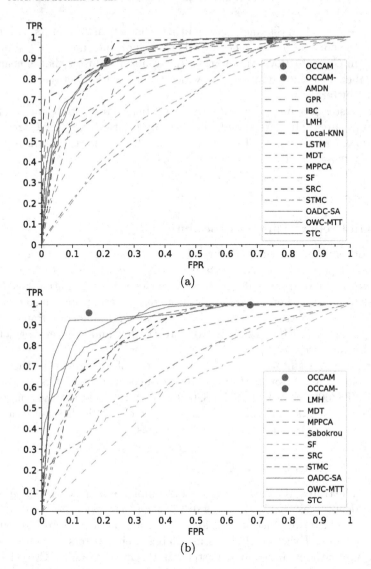

Fig. 2. Performance comparison. 14 methods are available for comparison on (a) UCS-Dped1 videos and 10 methods on (b) UCSDped2 videos. Supervised methods (dashed lines), unsupervised methods (solid lines).

- Supervised: AMDN [1], BM [2], CI [3], GPR [4], H-MDT-CRF [5], IBC [6], IEP [7], LMH [8], Local-KNN [9], LSTM [10], MDT [11], MPPCA [12], OF [13], SF [13], Sabokrou [14], SRC [15], and STMC [16]. [13] tested both OF and SF methods.
- Unsupervised: DC [17], FF [18], OADC-SA [19], OWC-MTT [20], and STC [21].

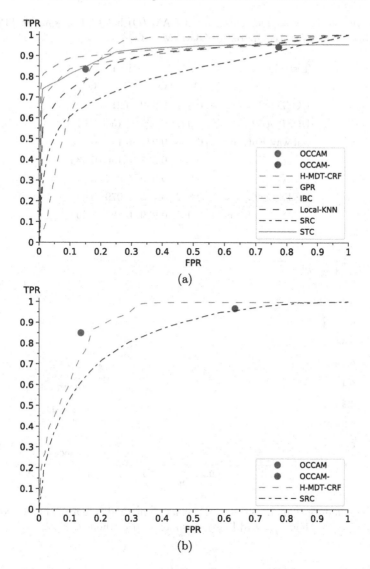

Fig. 3. Performance comparison. 6 methods are available for comparison on (a) Subway entrance video and 2 methods on (b) Subway exit video. Supervised methods (dashed lines), unsupervised methods (solid lines).

Most of these methods were tested only on some of the test videos. The test results on UCSDped1, UCSDped2, and Subway were reported as ROC curves. For the test results on UMN, some papers reported ROC curves whereas others reported only accuracy. For PETS2009, only accuracy was reported. ROC curves are not reported for H-MDT-CRF [5] on UCSDped1 and UCSDped2, LMH [8] and MPPCA [12] on Subway, and Sabokrou [14] on UMN. Therefore, they are

Table 1. Benefit of ambiguous clusters. OCCAM (O) has slightly smaller TPR, but significantly smaller FPR compared to OCCAM− (O−).

Test videos	TPR		FPR	
	O	O−	O	O−
UCSDped1	0.887	0.982	0.214	0.741
UCSDped2	0.957	0.994	0.154	0.677
Subway Entrance	0.835	0.942	0.152	0.773
Subway Exit	0.850	0.967	0.136	0.634
UMN	0.910	0.999	0.002	0.818
PETS2009 Scene 1	0.892	0.973	0.079	0.482
PETS2009 Scene 2	0.987	0.999	0.125	0.395

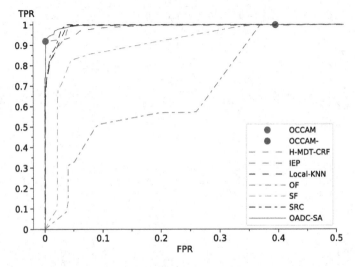

Fig. 4. Performance comparison on UMN video. 7 methods are available for comparison. Supervised methods (dashed lines), unsupervised methods (solid lines).

not included in our ROC graphs. The ROC curves reported in this paper are plotted using either the test results provided by the authors or a software that traces the curves' points presented in existing papers.

For UCSD (Fig. 2) and UMN videos (Fig. 4), OCCAM is among the best performers compared to existing methods. For the Subway videos (Fig. 3), OCCAM's performance is comparable to those of existing methods that are far more complex than OCCAM. For the same FPR, OCCAM achieves the highest TPR compared to existing methods for UCSDped2 (Fig. 2b), Subway exit (Fig. 3b), and UMN (Fig. 4), the 3rd highest TPR for UCSDped1 (Fig. 2a), and the 4th highest TPR for Subway entrance (Fig. 3a). In applications where high FPR is tolerable, OCCAM can run as OCCAM− without ambiguous clusters.

Table 2. Performance comparison on UMN and PETS2009 videos. OCCAM has the highest overall accuracy. (S) Supervised method, (U) unsupervised method.

Method	Type	UMN	PETS2009 scene 1	PETS2009 scene 2
OCCAM	U	**0.98**	**0.91**	**0.99**
BM [2]	S	0.96	0.89	0.94
CI [3]	S	0.88	0.60	0.93
SF [13]	S	0.85	0.59	0.85
SRC [15]	S	0.85	–	–
DC [17]	U	0.96	–	–
FF [18]	U	0.81	0.38	0.88

Then, OCCAM− achieves TPR of close to 1.0 for all test cases. Figures 2, 3 and 4 also show that existing unsupervised methods can perform as well as or better than supervised methods.

Some existing papers reported only accuracy on UMN and PETS2009 videos. Table 2 shows that OCCAM is more accurate than these methods for both UMN and PETS2009.

For UCSDped1 and UCSDped2 videos, Li and Mahadevan [5,11] also proposed a pixel-level criterion to measure the spatial accuracy of detected abnormal frames. This error measure depends on the number of detected abnormal pixels in an abnormal region. Since OCCAM detects only selected pixels in these regions instead of the whole regions, pixel-level criterion is not appropriate for OCCAM. Instead, this paper measures spatial accuracy in terms of precision, which is the percentage of detected abnormal pixels that are true positives. OCCAM achieves abnormal pixel detection precision of 0.72 for UCSDped1 and 0.78 for UCSDped2. Moreover, most of the false positive pixels are located around the abnormal regions. On the other hand, the spatial precision of OCCAM− on UCSDped1 and UCSDped2 is, respectively, 0.37 and 0.40, which is much lower than that of OCCAM. Therefore, ambiguous clusters are important for OCCAM to achieve high spatial accuracy in detecting abnormal pixels.

5   Conclusions

This paper investigated the essential components required for effective unsupervised detection of anomalies in surveillance videos of pedestrians. It shows that relatively simple but well-designed unsupervised algorithm like OCCAM can perform as well as or better than existing supervised and unsupervised methods. In particular, simple but informative features such as motion direction and motion speed are sufficient for achieving high TPR with low FPR. Moreover, inclusion of ambiguous clusters in the cluster labeling process reduces FPR significantly without sacrificing TPR much. At the same FPR, OCCAM achieves among the highest TPR compared to existing methods. It also has the highest accuracy for UMN and PETS2009 videos compared to existing methods that reported only accuracy. In applications where high FPR is tolerable, OCCAM can run as OCCAM− without ambiguous clusters. Then, OCCAM− achieves TPR of close to 1.0 for all test cases. With ambiguous clusters, OCCAM's spatial precision of detecting abnormal pixels is also very high. In general, OCCAM and existing unsupervised methods can perform as well as or better than supervised methods. Therefore, our research results can serve as a useful benchmark for testing new algorithms and for developing more advanced algorithms that require features other than motion speed and direction.

References

1. Xu, D., Ricci, E., Yan, Y., Song, J., Sebe, N.: Learning deep representations of appearance and motion for anomalous event detection. In: Proceedings of the BMVC, pp. 1–12 (2015)
2. Wu, S., Wong, H.S., Yu, Z.: A Bayesian model for crowd escape behavior detection. IEEE Trans. Circ. Syst. Video Technol. **24**(1), 85–98 (2014)
3. Wu, S., Moore, B.E., Shah, M.: Chaotic invariants of lagrangian particle trajectories for anomaly detection in crowded scenes. In: Proceedings of the CVPR, pp. 2054–2060 (2010)
4. Cheng, K., Chen, Y., Fang, W.: Video anomaly detection and localization using hierarchical feature representation and Gaussian process regression. In: Proceedings of the CVPR, pp. 2909–2917 (2015)
5. Li, W., Mahadevan, V., Vasconcelos, N.: Anomaly detection and localization in crowded scenes. IEEE Trans. PAMI **36**(1), 18–32 (2014)
6. Boiman, O., Irani, M.: Detecting irregularities in images and in video. IJCV **74**(1), 17–31 (2007)
7. Cui, X., Liu, Q., Gao, M., Metaxas, D.N.: Abnormal detection using interaction energy potentials. In: Proceedings of the CVPR, pp. 3161–3167 (2011)
8. Adam, A., Rivlin, E., Shimshoni, I., Reinitz, D.: Robust real-time unusual event detection using multiple fixed-location monitors. IEEE Trans. PAMI **30**(3), 555–560 (2008)
9. Saligrama, V., Chen, Z.: Video anomaly detection based on local statistical aggregates. In: Proceedings of the CVPR, pp. 2112–2119 (2012)
10. Feng, Y., Yuan, Y., Lu, X.: Deep representation for abnormal event detection in crowded scenes. In: Proceedings of the ACM MM, pp. 591–595 (2016)
11. Mahadevan, V., Li, W., Bhalodia, V., Vasconcelos, N.: Anomaly detection in crowded scenes. In: Proceedings of the CVPR, vol. 249, p. 250 (2010)
12. Kim, J., Grauman, K.: Observe locally, infer globally: a space-time MRF for detecting abnormal activities with incremental updates. In: Proceedings of the CVPR, pp. 2921–2928 (2009)
13. Mehran, R., Oyama, A., Shah, M.: Abnormal crowd behavior detection using social force model. In: Proceedings of the CVPR, pp. 935–942 (2009)
14. Sabokrou, M., Fathy, M., Hoseini, M., Klette, R.: Real-time anomaly detection and localization in crowded scenes. In: Proceedings of the CVPR Workshops, pp. 56–62 (2015)
15. Cong, Y., Yuan, J., Liu, J.: Sparse reconstruction cost for abnormal event detection. In: Proceedings of the CVPR, pp. 3449–3456 (2011)

16. Cong, Y., Yuan, J., Tang, Y.: Video anomaly search in crowded scenes via spatio-temporal motion context. IEEE Trans. Inform. Forensics Secur. 8(10), 1590–1599 (2013)
17. Chen, C.Y., Shao, Y.: Crowd escape behavior detection and localization based on divergent centers. IEEE Sens. J. 15(4), 2431–2439 (2015)
18. Chen, D.Y., Huang, P.C.: Motion-based unusual event detection in human crowds. J. Vis. Commun. Image Representation 22(2), 178–186 (2011)
19. Yuan, Y., Fang, J., Wang, Q.: Online anomaly detection in crowd scenes via structure analysis. IEEE Trans. Cybern. 45(3), 548–561 (2015)
20. Lin, H., Deng, J.D., Woodford, B.J., Shahi, A.: Online weighted clustering for real-time abnormal event detection in video surveillance. In: Proceedings of the ACM MM, pp. 536–540 (2016)
21. Roshtkhari, M.J., Levine, M.D.: Online dominant and anomalous behavior detection in videos. In: Proceedings of the CVPR, pp. 2611–2618 (2013)
22. UCSD: Anomaly Detection Dataset. www.svcl.ucsd.edu/projects/anomaly/dataset.htm
23. UMN: Unusual Crowd Activity Dataset. www.mha.cs.umn.edu/proj_events.shtml
24. PETS2009: Event Recognition Dataset. www.cvg.reading.ac.uk/PETS2009/a.html#s3
25. Wang, H., Kläser, A., Schmid, C., Liu, C.L.: Action recognition by dense trajectories. In: Proceedings of the CVPR, pp. 3169–3176 (2011)

Deep Boltzmann Machines Using Adaptive Temperatures

Leandro A. Passos Júnior[1], Kelton A.P. Costa[2], and João P. Papa[2(✉)]

[1] Department of Computing, UFSCar - Federal University of São Carlos,
São Carlos 13565-905, Brazil
leandropassosjr@gmail.com
[2] School of Sciences, UNESP - São Paulo State University,
Bauru 17033-360, Brazil
{kelton,papa}@fc.unesp.br

Abstract. Deep learning has been considered a hallmark in a number of applications recently. Among those techniques, the ones based on Restricted Boltzmann Machines have attracted a considerable attention, since they are energy-driven models composed of latent variables that aim at learning the probability distribution of the input data. In a nutshell, the training procedure of such models concerns the minimization of the energy of each training sample in order to increase its probability. Therefore, such optimization process needs to be regularized in order to reach the best trade-off between exploitation and exploration. In this work, we propose an adaptive regularization approach based on temperatures, and we show its advantages considering Deep Belief Networks (DBNs) and Deep Boltzmann Machines (DBMs). The proposed approach is evaluated in the context of binary image reconstruction, thus outperforming temperature-fixed DBNs and DBMs.

1 Introduction

In the last years, deep learning-driven techniques have been the foremost feature learner tools for a number of applications, that range from object detection to speech recognition, just to name a few. Such techniques are based on the hierarchical-oriented mechanism of the human brain, which learns different levels of information at each processing step. Convolutional Neural Networks (CNNs) [1], Deep Belief Networks (DBNs) [2], and Deep Boltzmann Machines (DBMs) [3] appear to be the most used techniques concerning the deep learning paradigm.

Deep Boltzmann Machines and Deep Belief Networks extend the well-known Restricted Boltzmann Machines (RBMs) to deeper representations, since they are composed of RBMs stacked on top of each other. In a nutshell, RBMs are stochastic neural networks composed of an input and a latent (i.e. hidden) layer, being the latter one in charge of learning the probability distribution of the input data. Roughly speaking, DBNs and DBMs differ in the way the upper layers interact, thus leading to slightly different formulations.

M. Felsberg et al. (Eds.): CAIP 2017, Part I, LNCS 10424, pp. 172–183, 2017.
DOI: 10.1007/978-3-319-64689-3_14

The main problem related to deep architectures concerns the large amount of data that is required for learning purposes; otherwise, the technique may over-fit the data. As a consequence, a number of works have focused on mitigating such drawback, such as regularizing techniques [4,5] and parameter fine-tuning [6–10]. An interesting approach related to RBM-based techniques concerns working on the "stability" of the convergence process to prevent overfitting. Recently, Li et al. [11] studied the influence of different temperatures during DBN learning procedure, and later on Passos and Papa [12] conducted a similar work, tough in the context of Deep Boltzmann Machines. Both studies agreed that temperature helps preventing overfitting, where the lower the temperature values, the better the results. The aforementioned works concluded that low temperature values lead to higher sparsity levels, thus contributing to the regularization of the network. As a matter of fact, sparsity is somehow analogous to dropping out neurons, i.e. one can switch neurons "on" or "off", forcing the network to adapt under such circumstances.

Basically, the problem of learning weights in the RBM training procedure aims at minimizing the energy of each training sample, which leads us to increasing its probability. Therefore, the training procedure of RBMs and related approaches is nothing more than an optimization process. In this work, we borrow the idea from meta-heuristic-based optimization processes, which aim at finding the best trade-off between exploitation and exploration. The first term refers to improving the solutions around the neighborhood of a given sample (local search), meanwhile exploration focuses on improving the solution in far away locations (e.g. global search). At the very beginning of the optimization process, meta-heuristic techniques usually converge faster (high exploration), thus decreasing the step-size (high exploitation) along the iterations in order to avoid overshooting the global/near-global optimum.

Therefore, we propose to use an adaptive temperature-based schema, where the temperature (step-size) decreases along the training procedure, thus simulating the behaviour of exploitation and exploration found out in many meta-heuristic techniques. We showed the proposed approach can outperform temperature-fixed DBNs and DBMs in the context of binary image reconstruction for some situations, or it can be at least competitive to them. Additionally, the adaptive-driven approach does not need a fine-tuning step since it requires the minimum and maximum temperature values only, which are considerably less sensitive and easy to set than the temperature value itself.

In this paper, we also considered two different formulations to control the temperature values. The remainder of this paper is organized as follows. Section 2 presents the theoretical background related to RBMs, DBNs and DBMs, while Sect. 3 discusses the temperature-based approaches used in this work. The methodology and experiments are presented in Sects. 4 and 5, respectively, and Sect. 6 states conclusions and future works.

2   Deep Boltzmann Machines

In this section, we briefly explain the theoretical background related to RBMs and DBMs.

2.1   Restricted Boltzmann Machines

Restricted Boltzmann Machines are energy-based stochastic neural networks composed of two layers of neurons (visible and hidden), in which the learning phase is conducted by means of an unsupervised fashion. A naïve architecture of a Restricted Boltzmann Machine comprises a visible layer $\mathbf{v}$ with m units and a hidden layer $\mathbf{h}$ with n units. Additionally, a real-valued matrix $\mathbf{W}_{m \times n}$ models the weights between the visible and hidden neurons, where w_{ij} stands for the weight between the visible unit v_i and the hidden unit h_j.

Let us assume both $\mathbf{v}$ and $\mathbf{h}$ as being binary-valued units. In other words, $\mathbf{v} \in \{0,1\}^m$ e $\mathbf{h} \in \{0,1\}^n$. The energy function of a Restricted Boltzmann Machine is given by:

$$E(\mathbf{v}, \mathbf{h}) = -\sum_{i=1}^{m} a_i v_i - \sum_{j=1}^{n} b_j h_j - \sum_{i=1}^{m} \sum_{j=1}^{n} v_i h_j w_{ij}, \tag{1}$$

where $\mathbf{a}$ e $\mathbf{b}$ stand for the biases of visible and hidden units, respectively.

The probability of a joint configuration $(\mathbf{v}, \mathbf{h})$ is computed as follows:

$$P(\mathbf{v}, \mathbf{h}) = \frac{1}{Z} e^{-E(\mathbf{v}, \mathbf{h})}, \tag{2}$$

where Z stands for the so-called partition function, which is basically a normalization factor computed over all possible configurations involving the visible and hidden units. Similarly, the marginal probability of a visible (input) vector is given by:

$$P(\mathbf{v}) = \frac{1}{Z} \sum_{\mathbf{h}} e^{-E(\mathbf{v}, \mathbf{h})}. \tag{3}$$

Since the RBM is a bipartite graph, the activations of both visible and hidden units are mutually independent, thus leading to the following conditional probabilities:

$$P(\mathbf{v}|\mathbf{h}) = \prod_{i=1}^{m} P(v_i|\mathbf{h}), \tag{4}$$

and

$$P(\mathbf{h}|\mathbf{v}) = \prod_{j=1}^{n} P(h_j|\mathbf{v}), \tag{5}$$

where

$$P(v_i = 1|\mathbf{h}) = \phi \left(\sum_{j=1}^{n} w_{ij}h_j + a_i \right), \tag{6}$$

and

$$P(h_j = 1|\mathbf{v}) = \phi \left(\sum_{i=1}^{m} w_{ij}v_i + b_j \right). \tag{7}$$

Note that $\phi(\cdot)$ stands for the logistic-sigmoid function.

Let $\theta = (W, a, b)$ be the set of parameters of an RBM, which can be learned through a training algorithm that aims at maximizing the product of probabilities given all the available training data $\mathcal{V}$, as follows:

$$\arg \max_{\Theta} \prod_{v \in \mathcal{V}} P(\mathbf{v}). \tag{8}$$

One can solve the aforementioned equation using the following derivatives over the matrix of weights $\mathbf{W}$, and biases $\mathbf{a}$ and $\mathbf{b}$ at iteration t as follows:

$$\mathbf{W}^{t+1} = \mathbf{W}^t + \underbrace{\eta(P(\mathbf{h}|\mathbf{v})\mathbf{v}^T - P(\tilde{\mathbf{h}}|\tilde{\mathbf{v}})\tilde{\mathbf{v}}^T) + \Phi}_{=\Delta \mathbf{W}^t}, \tag{9}$$

$$\mathbf{a}^{t+1} = \mathbf{a}^t + \underbrace{\eta(\mathbf{v} - \tilde{\mathbf{v}}) + \alpha\Delta\mathbf{a}^{t-1}}_{=\Delta\mathbf{a}^t} \tag{10}$$

and

$$\mathbf{b}^{t+1} = \mathbf{b}^t + \underbrace{\eta(P(\mathbf{h}|\mathbf{v}) - P(\tilde{\mathbf{h}}|\tilde{\mathbf{v}})) + \alpha\Delta\mathbf{b}^{t-1}}_{=\Delta\mathbf{b}^t}, \tag{11}$$

where η stands for the learning rate, and α denotes the momentum. Notice the terms $P(\tilde{\mathbf{h}}|\tilde{\mathbf{v}})$ and $\tilde{\mathbf{v}}$ can be obtained by means of the Contrastive Divergence [13] technique, which basically ends up performing Gibbs sampling using the training data as the visible units. Roughly speaking, Eqs. 9, 10 and 11 employ the well-known Gradient Descent as the optimization algorithm. The additional term Φ in Eq. 9 is used to control the values of matrix $\mathbf{W}$ during the convergence process, and it is formulated as follows:

$$\Phi = -\lambda\mathbf{W}^t + \alpha\Delta\mathbf{W}^{t-1}, \tag{12}$$

where λ stands for the weight decay.

Hinton et al. [2] proposed a learning algorithm concerning DBNs, which are essentially a collection of RBMs stacked on top of each other. The algorithm is pretty straightforward, and it consists into performing the tradicional RBM learning procedure for each layer, being the output of the current layer the input to the next. Once you reach the top, you can perform a fine-tuning step by means of the well-known backpropagation algorithm using an extra layer with the labels of the training samples.

2.2 Deep Boltzmann Machines

Learning more complex and internal representations of the data can be accomplished by using stacked RBMs, such as DBNs and DBMs. In this paper, we are interested in the DBM formulation, which is slightly different from DBN one. Suppose we have a DBM with two layers, where $\mathbf{h}^1$ and $\mathbf{h}^2$ stand for the hidden units at the first and second layer, respectively.

The energy of a DBM can be computed as follows:

$$E(\mathbf{v}, \mathbf{h}^1, \mathbf{h}^2) = -\sum_{i=1}^{m^1}\sum_{j=1}^{n^1} v_i h_j^1 w_{ij}^1 - \sum_{i=1}^{m^2}\sum_{j=1}^{n^2} h_i^1 h_j^2 w_{ij}^2, \tag{13}$$

where m^1 and m^2 stand for the number of visible units in the first and second layers, respectively, and n^1 and n^2 stand for the number of hidden units in the first and second layers, respectively. In addition, we have the weight matrices $\mathbf{W}_{m^1 \times n^1}^1$ and $\mathbf{W}_{m^2 \times n^2}^2$, which encode the weights of the connections between vectors $\mathbf{v}$ and $\mathbf{h}^1$, and vectors $\mathbf{h}^1$ and $\mathbf{h}^2$, respectively. For the sake of simplification, we dropped the bias terms out.

The marginal probability the model assigns to a given input vector $\mathbf{v}$ is given by:

$$P(\mathbf{v}) = \frac{1}{Z} \sum_{\mathbf{h}^1, \mathbf{h}^2} e^{-E(\mathbf{v}, \mathbf{h}^1, \mathbf{h}^2)}. \tag{14}$$

Finally, the conditional probabilities over the visible and the two hidden units are given as follows:

$$P(v_i = 1 | \mathbf{h}^1) = \phi\left(\sum_{j=1}^{n^1} w_{ij}^1 h_j^1\right), \tag{15}$$

$$P(h_z^2 = 1 | \mathbf{h}^1) = \phi\left(\sum_{i=1}^{m^2} w_{iz}^2 h_i^1\right), \tag{16}$$

and

$$P(h_j^1 = 1 | \mathbf{v}, \mathbf{h}^2) = \phi\left(\sum_{i=1}^{m^1} w_{ij}^1 v_i + \sum_{z=1}^{n^2} w_{jz}^2 h_z^2\right). \tag{17}$$

After learning the first RBM using Contrastive Divergence, for instance, the generative model can be written as follows:

$$P(\mathbf{v}) = \sum_{\mathbf{h}^1} P(\mathbf{h}^1)P(\mathbf{v}|\mathbf{h}^1), \qquad (18)$$

where $P(\mathbf{h}^1) = \sum_{\mathbf{v}} P(\mathbf{h}^1, \mathbf{v})$. Further, we shall proceed with the learning process of the second RBM, which then replaces $P(\mathbf{h}^1)$ by $P(\mathbf{h}^1) = \sum_{\mathbf{h}^2} P(\mathbf{h}^1, \mathbf{h}^2)$. Roughly speaking, using such procedure, the conditional probabilities given by Eqs. 15–17 and the Contrastive Divergence algorithm, one can learn DBM parameters one layer at a time [3].

3   Temperature-Based Deep Boltzmann Machines

Li et. al. [11] showed that a temperature parameter T controls the sharpness of the logistic-sigmoid function. In order to incorporate the temperature effect into the RBM context, they introduced this parameter to the joint distribution of the vectors $\mathbf{v}$ and $\mathbf{h}$ in Eq. 2, which can be rewritten as follows:

$$P(\mathbf{v}, \mathbf{h}, T) = \frac{1}{Z}e^{\frac{-E(\mathbf{v}, \mathbf{h})}{T}}. \qquad (19)$$

When $T = 1$, the aforementioned equation degenerates to Eq. 2. In addition, Eq. 7 can be rewritten in order to accommodate the temperature parameter as follows:

$$P(h_j = 1|\mathbf{v}) = \phi\left(\frac{\sum_{i=1}^{m} w_{ij}v_i}{T}\right). \qquad (20)$$

Notice the temperature parameter does not affect the conditional probability of the input units (Eq. 6).

In order to apply the very same idea to DBMs, the conditional probabilities over the two hidden layers given by Eqs. 16 and 17 can be derived and expressed using the following formulation, respectively:

$$P(h_z^2 = 1|\mathbf{h}^1) = \phi\left(\frac{\sum_{i=1}^{m^2} w_{iz}^2 h_i^1}{T}\right), \qquad (21)$$

and

$$P(h_j^1 = 1|\mathbf{v}, \mathbf{h}^2) = \phi\left(\frac{\sum_{i=1}^{m^1} w_{ij}^1 v_i}{T} + \sum_{z=1}^{n^2} w_{jz}^2 h_z^2\right). \qquad (22)$$

3.1   Adaptive Temperature-Based Model

In this paper, we study the influence of two different functions during the convergence process:

- $f_1(t) = L - \frac{t}{t_{max}}(L - U)$; and
- $f_2(t) = L \exp((\log\left(\frac{U}{L}\right)/t_{max})t)$.

In the above functions, $L = 0.1$ and $U = 2.0$ stand for the lower and upper temperature boundaries, respectively. Also, $t_{max} = 200$ denotes the maximum number of iterations concerning DBN/DBM learning procedure. Figures 1a and b display the behaviour of functions f_1 and f_2, respectively. In a nutshell, f_1 stands for a bounded linear function, meanwhile f_2 represents a bounded exponential function. The reason for using functions bounded in $[0.1, 2.0]$ concerns the fact that lower temperatures lead to better results [11, 12].

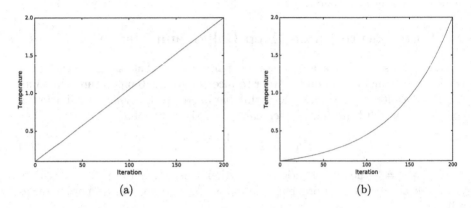

(a)   (b)

Fig. 1. Function F_1 and F_2 for (a) and (b), respectively. Used to update temperature values along the convergence process.

Additionally, we used 100 iterations with step-size of 10 for convergence purposes (i.e. the temperature changes every 10 iterations). Although such number may not be enough to achieve state-of-the-art results, we would like to emphasize we are interested into showing the proposed approach can outperform temperature-fixed ones even using a small number of iterations.

4 Methodology

In this section, we present the methodology employed to evaluate the proposed approach, as well the datasets and the experimental setup. Notice the approach used in this paper is based on the one employed by Passos et al. [12].

4.1 Datasets

We propose to evaluate the behavior of DBNs and DBMs under adaptive temperatures in the context of binary image reconstruction using three public datasets, as described below:

– MNIST dataset[1]: it is composed of images of handwritten digits. The original version contains a training set with 60, 000 images from digits '0'–'9', as well

---

[1] http://yann.lecun.com/exdb/mnist/.

as a test set with 10, 000 images[2]. Due to the high computational burden for DBM model selection, we decided to employ the original test set together with a reduced version of the training set[3].

– CalTech 101 Silhouettes Dataset[4]: it is based on the former Caltech 101 dataset, and it comprises silhouettes of images from 101 classes with resolution of 28 × 28.

Figure 2 displays some training examples from the above datasets.

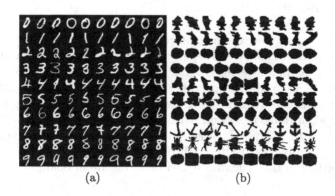

(a)   (b)

Fig. 2. Some training examples from (a) MNIST and (b) Caltech 101 Silhouettes.

4.2   Experimental Setup

We employed a 3-layered architecture for all datasets as follows: I-500-500-2, 000, where I stands for the number of pixels used as input for each dataset, i.e., 196 (14 × 14 images) and 784 (28 × 28 images) considering MNIST and Caltech 101 Silhouettes datasets, respectively. Therefore, we have a first and a second hidden layers with 500 neurons each, followed by a third hidden layer with 2, 000 neurons[5]. The remaining parameters used during the learning steps were chosen empirically and fixed for each layer as follows: $\eta = 0.1$ (learning rate), $\lambda = 0.1$ (weight decay), $\alpha = 0.00001$ (penalty parameter).

In order to provide a statistical analysis by means of the Wilcoxon signed-rank test with significance of 0.05 [14], we conducted a cross-validation procedure with 20 runnings. In regard to the fixed-temperature experiment, we considered a set of values within the range $T \in \{0.1, 0.2, 0.5, 0.8, 1.0, 1.2, 1.5, 2.0\}$ for the sake of comparison purposes.

---

[2] The images are originally available in grayscale with resolution of 28 × 28, but they were reduced to 14 × 14 images.

[3] The original training set was reduced to 2% of its former size, which corresponds to 1, 200 images.

[4] https://people.cs.umass.edu/~marlin/data.shtml.

[5] Since this architecture has been commonly employed in several works in the literature, we opted to employ it in our work either.

Finally, we employed 100 epochs for DBM and DBN learning weights procedure with mini-batches of size 20. In order to provide a more precise experimental validation, we trained both DBMs and DBNs with two different algorithms[6]: Contrastive Divergence (CD) [2] and Persistent Contrastive Divergence (PCD) [15]. Also, in order to evaluate the techniques considered in this work, we computed the mean square error (MSE) error over the training set. Therefore, the smaller the MSE, the better the technique is.

5   Experiments

This section presents the experimental results concerning DBN and DBM optimization by means of adaptive temperatures. Two different adaptive functions, i.e., f_1 and f_2, as well as eight constant temperatures were used for the baseline approach (i.e., fixed-size temperature): 0.1, 0.2, 0.5, 0.8, 1.0, 1.2, 1.5 and 2.0. Furthermore, DBM results were compared against DBN using two different learning algorithms, i.e., Contrastive Divergence and Persistent Contrastive Divergence in a three-layered model. Table 1 presents the average MSE results for DBMs and DBNs over Caltech 101 Silhouettes datasets. The most accurate results according to the Wilcoxon signed rank test are in bold.

Table 1. Average DBM/DBN MSE over the test set considering Caltech 101 Silhouettes dataset with 200 iterations.

	0.1	0.2	0.5	0.8	1.0	1.2	1.5	2.0	Linear	Curve
DBM-CD	0.16048	0.16048	0.16048	0.16049	0.16048	0.16049	0.16049	0.15983	**0.15822**	0.16053
DBM-PCD	0.16049	0.16049	0.16050	0.16048	0.16049	0.16048	0.16049	0.15983	**0.15929**	0.16039
DBN-CD	0.16049	0.16050	0.16049	0.16050	0.16049	0.16058	0.16249	0.17040	**0.15822**	0.16523
DBN-PCD	0.16048	0.16049	0.16049	0.16049	0.16048	0.16049	0.16081	0.16120	**0.15929**	0.16321

Clearly, the best results were obtained using the linear adaptive function for both DBMs and DBNs. A closer look may suggest that adaptive temperature optimization works well for challenging datasets, such as Caltech 101 Silhouettes. Also, one can observe that both DBMs and DBNs obtained pretty much similar results, which can be explained by the fact we are not fine-tuning DBMs with the mean-field learning process. However, it is beyond the scope of this work to show that DBMs may be more accurate than DBNs, since we are interested to show the robustness in using adaptive temperatures for both models.

Table 2 presents the behavior of adaptive temperatures concerning DBMs and DBNs considering the MNIST dataset. Despite the adaptive linear function did not achieve the best results according to Wilcoxon signed-rank test, the difference between fixed- and adaptive-temperature is pretty much irrelevant. With respect to DBNs, both models evaluated in this work, i.e., fixed and adaptive temperatures, obtained quite close results. Additionally, in regard to DBMs, one

---

[6] One sampling iteration was used for all learning algorithms.

can observe the best results were obtained with smaller temperatures, as discussed by Passos et al. [12]. In this case, it is expected that adaptive models will not outperform fixed ones, since the temperature values in these dynamic approaches increase along the iterations.

Table 2. Average DBM/DBN MSE over the test set considering MNIST dataset with 200 iterations.

	0.1	0.2	0.5	0.8	1.0	1.2	1.5	2.0	Linear	Curve
DBM-CD	**0.08642**	**0.08642**	**0.08674**	0.08745	0.08753	0.08750	0.08751	0.08752	0.08747	0.08751
DBM-PCD	**0.08674**	**0.08659**	0.08681	0.08744	0.08752	0.08751	0.08750	0.08752	0.08747	0.08751
DBN-CD	0.08760	0.08771	0.08763	0.08752	**0.08751**	**0.08751**	**0.08751**	0.08749	0.08752	0.08775
DBN-PCD	0.08760	0.08769	0.08762	0.08751	**0.08751**	0.08751	**0.08750**	**0.08751**	0.08752	0.08775

We performed an extra round of experiments to analyze the impact of adaptive temperatures during the convergence process. For comparison purposes, we considered both the temperature value and learning algorithm that achieved the best results concerning the fixed-temperature approach. Figure 3 depicts the MSE of the first layer during the learning process across the iterations for both DBMs and DBNs considering the Caltech 101 Silhouettes dataset. Therefore, we compared DBM-CD with $T = 2.0$ against the proposed approach in Fig. 3a, as well as we compared DBN-PCD with $T = 1.0$ against the proposed approach in Fig. 3b.

Clearly, one can observe the adaptive temperatures converged faster during the first 50 iterations, and for DBN (Fig. 3b) they did not get stuck in local optima, as one can observe in the experiment with the fixed temperature, which stabilized after 75 iterations. Also, it seems there is no difference in using the linear or exponential function to update the temperature values considering DBMs, while the exponential model seemed to fit better for DBNs, but for a very small difference.

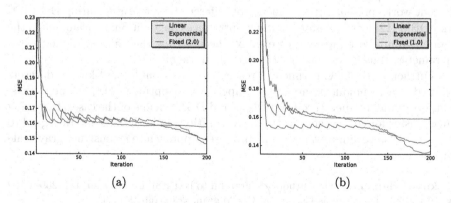

(a)   (b)

Fig. 3. MSE during the learning step of the first layer considering Caltech 101 Silhouettes dataset for (a) DBM and (b) DBN.

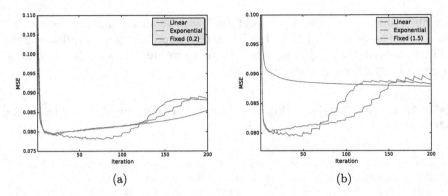

Fig. 4. MSE during the learning step of the first layer considering MNIST dataset for (a) DBM and (b) DBN.

Figure 4 shows the very same procedure for MNIST dataset. Once again, the fast convergence of the proposed approaches can be evidenced. Notice that adaptive-temperatures achieve by far the lower MSE since the beginning, but the model starts to "unlearn" and moves back to a point where the MSE is higher than the one achieved by the fixed-temperature after a long period of training.

Roughly speaking, the proposed approaches can benefit in situations where higher temperatures lead to the better results. However, since the adaptive model always increases the temperature, one may not get suitable results at the very end of the convergence process, which means one can halt the process much earlier.

6   Conclusions

In this work, we dealt with the problem of hastening the DBM learning step using adaptive temperatures, as well as we also evaluated them in the context of DBNs. Recent works presented the influence of different temperatures during DBN [11] and DBM [12] learning process, which introduces an additional parameter to the model. Adaptive temperatures exempt the need for the aforementioned extra parameter, thus becoming easier to handle those models.

Furthermore, the experimental results over two public well-known datasets showed the technique is at least competitive to optimize DBMs and DBNs, outperforming temperature-fixed DBNs and DBMs in one of the cases, but being much faster for convergence at the early iterations in both datasets. In regard to future works, we aim to validate the proposed approach to reconstruct gray-scale images either.

Acknowledgments. The authors are grateful to FAPESP grants #2014/16250-9 and #2014/12236-1, as well as Capes and CNPq grant #306166/2014-3.

References

1. Lecun, Y., Bottou, L., Bengio, Y., Haffner, P.: Gradient-based learning applied to document recognition. Proc. IEEE **86**(11), 2278–2324 (1998)
2. Hinton, G.E., Osindero, S., Teh, Y.-W.: A fast learning algorithm for deep belief nets. Neural Comput. **18**(7), 1527–1554 (2006)
3. Salakhutdinov, R., Hinton, G.E.: An efficient learning procedure for deep Boltzmann machines. Neural Comput. **24**(8), 1967–2006 (2012)
4. Wan, L., Zeiler, M., Zhang, S., LeCun, Y., Fergus, R.: Regularization of neural networks using dropconnect. In: Dasgupta, S., Mcallester, D. (eds.) Proceedings of the 30th International Conference on Machine Learning. JMLR Workshop and Conference Proceedings, ICML 2013, vol. 28, no. 3, pp. 1058–1066 (2013)
5. Srivastava, N., Hinton, G.E., Krizhevsky, A., Sutskever, I., Salakhutdinov, R.: Dropout: a simple way to prevent neural networks from overfitting. J. Mach. Learn. Res. **15**(1), 1929–1958 (2014)
6. Papa, J.P., Rosa, G.H., Costa, K.A.P., Marana, A.N., Scheirer, W., Cox, D.D.: On the model selection of Bernoulli restricted Boltzmann machines through harmony search. In: Proceedings of the Genetic and Evolutionary Computation Conference, GECCO 2015, pp. 1449–1450. ACM, New York (2015)
7. Papa, J.P., Rosa, G.H., Yang, X.-S.: Quaternion-driven deep belief networks fine-tuning. Applied Soft Computing (2016) (submitted)
8. Papa, J.P., Rosa, G.H., Marana, A.N., Scheirer, W., Cox, D.D.: Model selection for discriminative restricted Boltzmann machines through meta-heuristic techniques. J. Comput. Sci. **9**, 14–18 (2015)
9. Rosa, G.H., Papa, J.P., Marana, A.N., Scheirer, W., Cox, D.D.: Fine-tuning convolutional neural networks using harmony search. In: Pardo, A., Kittler, J. (eds.) IARP 2015. LNCS, vol. 9423, pp. 683–690. Springer, Cham (2015)
10. Papa, J.P., Scheirer, W., Cox, D.D.: Fine-tuning deep belief networks using harmony search. Appl. Soft Comput. **46**, 875–885 (2016)
11. Li, G., Deng, L., Xu, Y., Wen, C., Wang, W., Pei, J., Shi, L.: Temperature based restricted Boltzmann machines. Sci. Rep. **6**, 1–12 (2016)
12. Passos, L.A., Papa, J.P.: Temperature-based deep Boltzmann machines, arXiv (2016). http://arxiv.org/abs/1608.07719
13. Hinton, G.E.: Training products of experts by minimizing contrastive divergence. Neural Comput. **14**(8), 1771–1800 (2002)
14. Wilcoxon, F.: Individual comparisons by ranking methods. Biometrics Bull. **1**(6), 80–83 (1945)
15. Tieleman, T.: Training restricted Boltzmann machines using approximations to the likelihood gradient. In: Proceedings of the 25th International Conference on Machine Learning, pp. 1064–1071. ACM, New York (2008)

Twin Deep Convolutional Neural Network for Example-Based Image Colorization

Domonkos Varga[1,2(✉)] and Tamás Szirányi[1,3]

[1] MTA SZTAKI, Institute for Computer Science and Control, Budapest, Hungary
{varga.domonkos,sziranyi.tamas}@sztaki.mta.hu
[2] Department of Networked Systems and Services,
Budapest University of Technology and Economics, Budapest, Hungary
[3] Department of Material Handling and Logistics Systems,
Budapest University of Technology and Economics, Budapest, Hungary

Abstract. This paper deals with the colorization of grayscale images. Recent papers have shown remarkable results on image colorization utilizing various deep architectures. Unlike previous methods, we perform colorization using a deep architecture and a reference image. Our architecture utilizes two parallel Convolutional Neural Networks which have the same structure. One CNN, which uses the reference image, helps the other CNN in color prediction for the input image. On the other hand, the second CNN, which uses the input image, helps to identify the areas which holds essential information about the color scheme of the scene. Comprehensive experiments and qualitative and quantitative evaluations were conducted on the images of SUN database and on other images. Quantitative evaluations are based on Peak Signal-to-Noise Ratio (PSNR) and on Quaternion Structural Similarity (QSSIM).

Keywords: Image colorization · Deep learning · Convolutional Neural Network

1 Introduction

Automatic image colorization examines the problem how to add realistic colors to grayscale images without any user intervention. It has some useful applications such as colorizing old photographs or movies, artist assistance, visual effects and color recovering. On the other hand, colorization is a heavily ill-posed problem. In order to effectively colorize any images, the algorithm or the user should have enough information about the scene's semantic composition.

As pointed out in [16], image colorization is also a good model for a huge number of applications where we want to take an arbitrary image and predict values or different distributions at each pixel of the input image, exploiting information only from this input image. This is a very common task in the image processing and pattern recognition community.

M. Felsberg et al. (Eds.): CAIP 2017, Part I, LNCS 10424, pp. 184–195, 2017.
DOI: 10.1007/978-3-319-64689-3_15

To date, deep learning techniques have shown impressive results on both high-level and low-level vision problems including image classification [1], removing phantom objects from point clouds [2], pedestrian detection [3], face detection [4], handwritten character classification [5], photo adjustment [6], etc. In recent years, deep learning based approaches appeared to address the colorization problem.

Main contributions. Image colorization algorithms can be divided into three classes: scribble-based, example-based, and learning-based. In this paper, we show a possible solution that utilizes the advantages of example-based and learning-based approaches. Unlike previous methods, we perform colorization using a deep architecture and a reference image.

Paper organization. This paper is organized as follows. In Sect. 2, the related and previous works are reviewed primarily focused on learning-based approaches. We describe our algorithm in Sect. 3. Section 4 shows experimental results and analysis. The conclusions are drawn in Sect. 5.

2   Related Works

Image colorization has been intensively studied since 1970's. Broadly speaking, the existing algorithms can be divided into three groups: scribble-based, example-based, and learning-based approaches. In this section, we mainly concentrate on reviewing learning-based approaches.

Scribble-based approaches interpolate colors in the grayscale image based on color scribbles produced by a user or an artist. Levin et al. [7] presented an interactive colorization method which can be applied to still images and video sequences as well. The user places color scribbles on the image and these scribbles are propagated through the remaining pixels of the image. Huang et al. [8] improved further this algorithm in order to reduce color blending at image edges. Yatziv et al. [9] developed the algorithm of Levin et al. [7] in another direction. The user can provide overlapping color scribbles. Furthermore, a distance metric was proposed to measure the distance between a pixel and the color scribbles. Combinational weights belong to each scribbles which were determined based on the measured distance.

Example-based approaches require two images. These algorithms transfer color information from a colorful reference image to a grayscale target image. Reinhard et al. [10] applied simple statistical analysis to impose one image's color characteristics on another. Welsh et al. [11] utilized on pixel intensity values and different neighborhood statistics to match the pixels of the reference image with the pixels of grayscale target image. On the other hand, Irony et al. [12] determine first for each pixel which example segment it should learn its color from. This carried out by applying a supervised classification algorithm that considers the low-level feature space of each pixel neighborhood. Then each color assignment is treated as color micro-scribbles which were the inputs to Levin et al.'s [7] algorithm. Charpiat et al. [13] predicted the expected variation of color at each pixel, thus defining a non-uniform spatial coherency criterion. Then graph cuts were applied to maximize the probability of the whole colored image at the global level. Gupta et al. [14] extracted features from the target and reference images at the resolution of superpixels. Based on different kind of features, the superpixels of the reference image were matched with the superpixels of the target image and the color information was transfered to the center of the superpixels of the target image with the help of micro color-scribbles. Then these micro-scribbles were propagated through the target image.

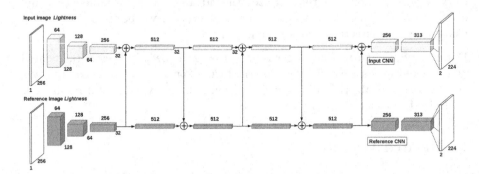

Fig. 1. The architecture of the proposed method. The input and the reference CNN have the same structure. First, only the reference CNN is trained then the input CNN and the reference CNN are trained simultaneously. Information is transmitted from input CNN to reference CNN and vica versa using element-wise addition operator to certain convolutional blocks.

Learning-based approaches model the variables of the image colorization process by applying different machine learning techniques and algorithms. Bugeau and Ta [15] introduced a patch-based image colorization algorithm that takes square patches around each pixel. Patch descriptors of luminance features were extracted in order to train a model and a color prediction model with a general distance selection strategy was proposed. Deshpande et al. [16] colorize an image by optimizing a linear system that considers local predictions of color, spatial consistency, and consistency with an overall histogram. Cheng et al. [17] introduced a fully-automatic method based on a deep neural network which was trained by hand-crafted features. Three levels of features were extracted from each pixel of the training images: raw grayscale values, DAISY features [18], and high-level semantic features.

In recent years, Convolutional Neural Network based approaches appeared to tackle the colorization problem. Iizuka et al. [19] elaborated a colorization method that jointly extracts global and local features from an image and then merge them together. In [20], the authors proposed a fully automatic algorithm based on VGG-16 [21] and a two-stage Convolutional Neural Network to provide richer representation by adding semantic information from a preceding layer. Furthermore, the authors proposed Quaternion Structural Similarity [22] for quality evaluation. Zhang et al. [23] trained a Convolutional Neural Network to map from a grayscale input to a distribution of quantized color values. This algorithm was evaluated with the help of human participants asking them to distinguish between colorized and ground-truth images. In [24], the authors introduced a patch-based colorization model using two different loss functions in a vectorized Convolutional Neural Network framework. During colorization patches are extracted from the image and are colorized independently. Guided image filtering [25] is applied as postprocessing. Larsson et al. [26] processed a grayscale image through VGG-16 [21] architecture and obtained hypercolumns [27] as feature vectors. The system learns to predict *hue* and *chroma* distributions for each pixel from its hypercolumn. Deshpande et al. [28] proposed a conditional model for predicting multiple colorizations. The low dimensional embedding of color fields was learned by a Variational Autoencoder. Similarly, Cao et al. [29] worked with a conditional model but a Conditional Generative Adversarial Network was utilized to model the distribution of real-world colors. Limmer and Lensch [30] proposed a method for transferring the RGB color spectrum to near-infrared images using deep multi-scale convolutional neural networks. The transfer between RGB and near-infrared images is trained.

3   Our Approach

The objectiveness of our framework is to combine example-based and learning-based approaches in order to produce more realistic and plausible colors. To capitalize on the advantages of example-based and learning-based methods as well, we propose a novel architecture which is shown in Fig. 1. Our architecture consists of two parallel CNNs which are called Input CNN and Reference CNN. These have the same structure. In the following, this structure is firstly described and then the co-operation of the two networks is discussed.

We reimplemented the algorithm of [23] using Keras [31] deep learning library. This algorithm has some appealing properties. First of all, the authors elaborated a *class rebalancing* method because the distribution of ab values in natural images is biased towards low ab values. Second, colorization is treated as multinomial classification instead of regression. This means that the ab output space is quantized into bins with grid size 10 and keep the $Q = 313$ values which are in gamut. For all details, we refer to [23].

We used SUN database [32] to compile our training database. We denote a reference image by R and an input image by I. Formally, our database can be defined as $\mathcal{L}_i = \{(I_i, R_i)|i = 1, ..., N\}$ where N is the number of image pairs and reference image R_i is semantically similar to input image I_i. That is why we opted to utilize SUN database [32] since this dataset contains images grouped by their semantic information. Figure 2 shows the empirical distribution of pixels in ab space gathered from our database. Figure 3 illustrates the empirical and smoothed empirical distribution of ab pairs in the quantized space. These curves were determined and were applied in the training process based on the algorithm of [23].

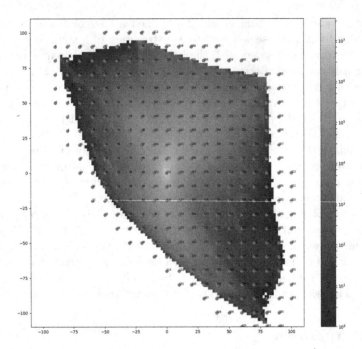

Fig. 2. Empirical probability distribution of ab values in our database, shown in *log* scale. The horizontal axis represents the b values and the vertical axis represents the a values. The green dots denote the quantized ab value pairs. (Color figure online)

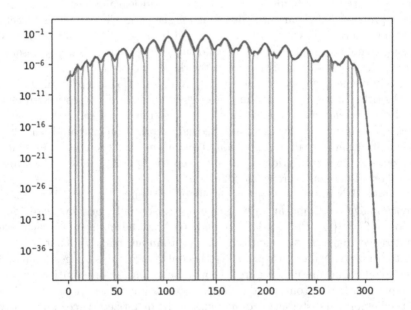

Fig. 3. Empirical (blue curve) and smoothed empirical distribution (red curve) of ab pairs in the quantized space of our cartoon database. (Color figure online)

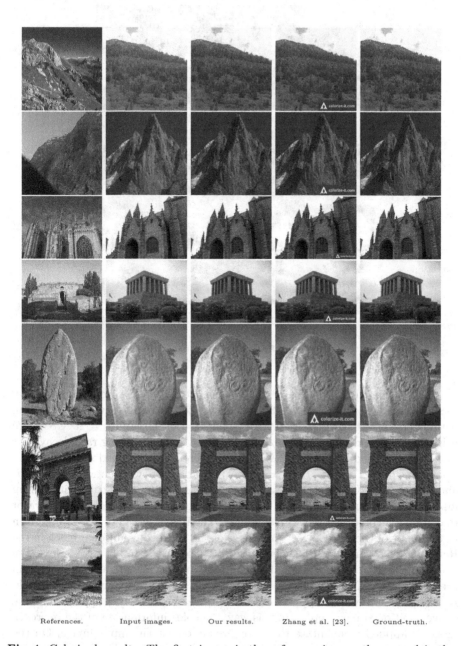

References.   Input images.   Our results.   Zhang et al. [23].   Ground-truth.

Fig. 4. Colorized results. The first image is the reference image, the second is the grayscale input, the third is our colorized result, and fourth is the result of [23], and the fifth is the ground-truth image. Digital watermarks in the lower right corners were embedded by the application of [23] (available: http://demos.algorithmia.com/colorize-photos). (Color figure online)

References.   Input images.   Our results.   Ground-truth.

Fig. 5. Colorized results.

First, we train only the Reference CNN using only the R_i's from our database. We utilize *ADAM* optimizer [33] and early stopping [34] with the following parameters: $\alpha = 0.0001$, $\beta_1 = 0.9$, $\beta_2 = 0.999$, $d = 0.0$, and $\varepsilon = 1e - 8$ where α is the learning rate, ε is the fuzz factor, and d is the learning rate decay over each update. Then the input CNN and the reference are trained simultaneously using the whole $\mathcal{L}_i = \{(I_i, R_i)|i = 1, ..., N\}$ database. As we mentioned the input and the reference CNN have the same structure. Information is transmitted from input CNN to reference CNN and vica versa using element-wise addition operator to certain convolutional blocks (see Fig. 1). The image pairs $(I_i, R_i)_{i=1}^N$ are given to the input of the two CNNs. The values of the third convolutional block in the Reference CNN are added element-wise to those in the Input CNN. Next, the values of the fourth convolutional block in the input CNN are added to those in the Reference CNN. This process repeats to the second last convolutional block. In this process, we also applied ADAM optimizer and early stopping with the above mentioned parameters. In this way, the color information of the reference image is applied to facilitate the color prediction for the input image. On the other hand, information from the input image helps to identify the areas which holds essential information about the color scheme of the scene. The proposed framework was trained on 60.000 image pairs of the SUN database.

As pointed out in many papers [20, 23, 24, 26], Euclidean loss function is not an optimal solution because it will result in the so-called averaging problem. Namely, the system will produce grayish sepia tone effects. That is why we use a cross-entropy like loss function to compare predicted $\hat{\mathbf{Z}} \in [0,1]^{H \times W \times Q}$ against the ground truth $\mathbf{Z} \in [0,1]^{H \times W \times Q}$:

$$L(\hat{\mathbf{Z}}, \mathbf{Z}) = -\sum_{h=1,w=1}^{H,W} v(\mathbf{Z}_{h,w}) \sum_{q=1}^{Q=313} \mathbf{Z}_{h,w,q} \cdot log(\hat{\mathbf{Z}}_{h,w,q}), \tag{1}$$

where $Q = 313$ is the number of quantized ab values (see Fig. 2), $v(\cdot)$ is a weighting term used to rebalance the loss based on color-class rarity, and H and W denote the height and the width of the training images. The weighting term $v(\cdot)$ is obtained using the smoothed empirical distribution of ab pairs in the quantized space (see Fig. 3). For all details of the weighting term, we refer to [23].

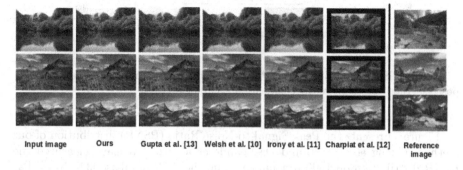

Input image   Ours   Gupta et al. [13]   Welsh et al. [10]   Irony et al. [11]   Charpiat et al. [12]   Reference image

Fig. 6. Comparison with state-of-the-art example-based colorization algorithms.

4   Experimental Results

Figure 4 presents several colorization results obtained by our proposed method with respect to the inputs, the ground-truth colorful images, and the reference images. Figure 4 also illustrates the results of [23] which were obtained using their web application (available: http://demos.algorithmia.com/colorize-photos). Note that the digital watermarks in the lower right corners were embedded by this application. From this qualitative comparison, we can see that our method is able to reduce visible artifacts, especially for detailed scenes, objects with large color variances (*e.g.* building). The color filling is nearly flawless. We could reduce the amount of false edges near the object boundaries. Figure 5 shows further results of our method.

Figure 6 shows a comparison with the major state-of-the-art example-based colorization algorithms such as [11–14]. It can be seen that we could produce more realistic and plausible colors than most state-of-the-art example-based colorization algorithms.

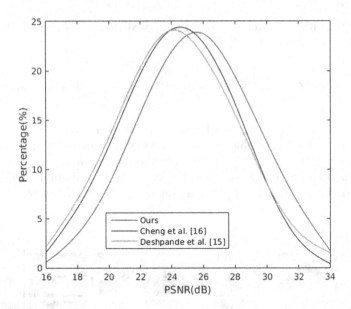

Fig. 7. Peak Signal-to-Noise Ratio (PSNR) distribution. It can be seen that the proposed method can improve the colorization accuracy.

Figure 7 presents the Peak Signal-to-Noise Ratio (PSNR) distribution of our method, Cheng et al. [17], and Deshpande et al. [16]. We have measured the PSNR distribution on 1500 test images from the SUN database [32]. Note that we reimplemented for this experiment the method of [17] using Keras deep learning library [31]. In our experiment, we have applied a 33-dimensional semantic feature vector for [17] and have trained the proposed deep neural network architecture using ADAM optimizer [33] and the images of SUN database. Besides, we have used the source code (available: http://vision.cs.illinois.edu/projects/lscolor) provided by Deshpande et al. [16]. Figure 7 illustrates that the proposed method is able to improve colorization accuracy since it outperforms these two state-of-the-art algorithms.

Unfortunately, there is no widely used quality metrics which clearly indicates the quality of a colorization. Methodical quality evaluation by showing colorized images to human observers is slow, expensive, and subjective. Empirically, we have found that Quaternion Structural Similarity (QSSIM) [22] gives a good base for quantitative evaluation. It is a theoretically well based measure which has been accepted by the colorimetry research community as a potential qualification value. We have measured the QSSIM distribution on 1500 test images from the SUN database. Figure 8 presents the QSSIM distribution of our method, Cheng et al. [17], and Deshpande et al. [16]. It can be seen that the proposed method outperforms the two other state-of-the-art algorithms. A higher QSSIM values indicates better image quality. This experiment was based on the

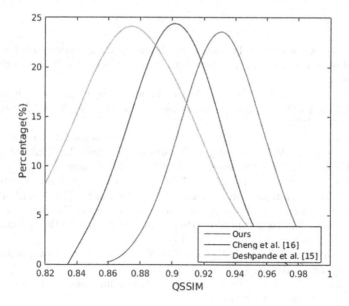

Fig. 8. Quaternion Structural Similarity (QSSIM) distribution. It can be seen that the proposed method can improve the colorization quality. A higher QSSIM value indicates better image quality.

source code (available: http://www.ee.bgu.ac.il/~kolaman/QSSIM) provided by Kolaman et al. [22].

5   Conclusion

In this paper, we have introduced a novel framework which capitalizes on the advantages of example-based and learning-based colorization approaches. Specifically, we have shown a possible solution that combines the information between two CNNs in order to help the input CNN in color prediction for the input image. To this end, we have trained first a reference CNN which facilitates the identification of the specific color scheme of the input scene. We have shown that the semantic enhancement capability of a deep CNN can be switched into a colorization scheme to result in an effective image analysis and interpretation framework. The QSSIM method has been proved a superior measuring method for color modeling. There are many directions for further research. First, it is worth to generalize the proposed method for arbitrary size input images. Another direction of research would be automatizing the search for a suitable reference image to an input image.

Acknowledgment. The research was supported by the Hungarian Scientific Research Fund (No. OTKA 120499). We are very thankful to Levente Kovács for helping us with professional advices in high-performance computing.

References

1. Krizhevsky, A., Sutskever, I., Hinton, G.: Imagenet classification with deep convolutional neural networks. In: Advances in Neural Information Processing Systems, pp. 1097–1105 (2012)
2. Nagy, B., Benedek, C.: 3D CNN based phantom object removing from mobile laser scanning data. In: International Joint Conference on Neural Networks, pp. 4429–4435 (2017)
3. Bochinski, E., Eiselein, V., Sikora, T.: Training a convolutional neural network for multi-class object detection using solely virtual world data. In: IEEE International Conference on Advanced Video and Signal Based Surveillance, pp. 278–285 (2016)
4. Lawrence, S., Giles, C.L., Tsoi, A.C., Back, A.D.: Face recognition: a Convolutional Neural Network approach. IEEE Trans. Neural Networks 8(1), 98–113 (1997)
5. Ciresan, D., Meier, U.: Multi-column deep neural networks for offline handwritten Chinese character classification. In: Proceedings of the International Joint Conference on Neural Networks, pp. 1–6 (2015)
6. Yan, Z., Zhang, H., Wang, B., Paris, S., Yu, Y.: Automatic photo adjustment using deep learning. CoRR, abs/1412.7725 (2014)
7. Levin, A., Lischinski, D., Weiss, Y.: Colorization using optimization. ACM Trans. Graph. 23(3), 689–694 (2004)
8. Huang, Y.C., Tung, Y.S., Chen, J.C., Wang, S.W., Wu, J.L.: An adaptive edge detection based colorization algorithm and its applications. In: Proceedings of the 13th Annual ACM International Conference on Multimedia, pp. 351–354 (2005)
9. Yatziv, L., Sapiro, G.: Fast image and video colorization using chrominance blending. IEEE Trans. Image Process. 15(5), 1120–1129 (2006)
10. Reinhard, E., Ashikhmin, M., Gooch, B., Shirley, P.: Color transfer between images. IEEE Comput. Graph. Appl. 21(5), 34–41 (2001)
11. Welsh, T., Ashikhmin, M., Mueller, K.: Transfering color to greyscale images. ACM Trans. Graph. 21(3), 277–280 (2002)
12. Irony, R., Cohen-Or, D., Lischinski, D.: Colorization by example. In: Eurographics Symposium on Rendering (2005)
13. Charpiat, G., Hofmann, M., Schölkopf, B.: Automatic image colorization via multimodal predictions. In: Forsyth, D., Torr, P., Zisserman, A. (eds.) ECCV 2008. LNCS, vol. 5304, pp. 126–139. Springer, Heidelberg (2008). doi:10.1007/978-3-540-88690-7_10
14. Gupta, R.K., Chia, A.Y.S., Rajan, D., Ng, E.S., Zhiyong, H.: Image colorization using similar images. In: Proceedings of the 20th ACM International Conference on Multimedia, pp. 369–378 (2012)
15. Bugeau, A., Ta, V.T.: Patch-based image colorization. In: Proceedings of the IEEE International Conference on Pattern Recognition, pp. 3058–3061 (2012)
16. Deshpande, A., Rock, J., Forsyth, D.: Learning large-scale automatic image colorization. In: Proceedings of the IEEE International Conference on Computer Vision, pp. 567–575 (2015)
17. Cheng, Z., Yang, Q., Sheng, B.: Deep colorization. In: Proceedings of the IEEE International Conference on Computer Vision, pp. 415–423 (2015)
18. Tola, E., Lepetit, V., Fua, P.: DAISY: an efficient dense descriptor applied to wide-baseline stereo. IEEE Trans. Pattern Anal. Mach. Intell. 32(5), 815–830 (2010)
19. Iizuka, S., Simo-Serra, E., Ishikawa, H.: Let there be color!: joint end-to-end learning of global and local image priors for automatic image colorization with simultaneous classification. ACM Trans. Graph. (TOG) 35(4), 110 (2016)

20. Varga, D., Szirányi, T.: Fully automatic image colorization based on Convolutional Neural Network. In: International Conference on Pattern Recognition (2016)
21. Simonyan, K., Zisserman, A.: Very Deep Convolutional Networks for Large-Scale Image Recognition. CoRR, abs/1409.1556 (2014)
22. Kolaman, A., Yadid-Pecht, O.: Quaternion structural similarity: a new quality index for color images. IEEE Trans. Image Process. **21**(4), 1526–1536 (2012)
23. Zhang, R., Isola, P., Efros, A.A.: Colorful image colorization. In: Leibe, B., Matas, J., Sebe, N., Welling, M. (eds.) ECCV 2016. LNCS, vol. 9907, pp. 649–666. Springer, Cham (2016). doi:10.1007/978-3-319-46487-9_40
24. Liang, X., Su, Z., Xiao, Y., Guo, J., Luo, X., Deep patch-wise colorization model for grayscale images. SIGGRAPH ASIA 2016 Technical Briefs 13 (2016)
25. He, K., Sun, J., Tang, X.: Guided image filtering. In: Daniilidis, K., Maragos, P., Paragios, N. (eds.) ECCV 2010. LNCS, vol. 6311, pp. 1–14. Springer, Heidelberg (2010). doi:10.1007/978-3-642-15549-9_1
26. Larsson, G., Maire, M., Shakhnarovich, G.: Learning representations for automatic colorization. In: Leibe, B., Matas, J., Sebe, N., Welling, M. (eds.) ECCV 2016. LNCS, vol. 9908, pp. 577–593. Springer, Cham (2016). doi:10.1007/978-3-319-46493-0_35
27. Hariharan, B., Arbeláez, P., Girshick, R., Malik, J.: Hypercolumns for object segmentation and fine-grained localization. In: Proceedings of the IEEE Conference on Computer Vision and Pattern Recognition, pp. 447–456 (2015)
28. Deshpande, A., Lu, J., Yeh, M.C., Forsyth, D.: Learning Diverse Image Colorization. arXiv preprint arXiv:1612.01958 (2016)
29. Cao, Y., Zhou, Z., Zhang, W., Yu, Y.: Unsupervised Diverse Colorization via Generative Adversarial Networks. arXiv preprint arXiv:1702.06674 (2017)
30. Limmer, M., Lensch, H.: Infrared Colorization Using Deep Convolutional Neural Networks. arXiv preprint arXiv:1604.02245 (2016)
31. Chollet, F.: Keras (2015). https://github.com/fchollet/keras
32. Xiao, J., Hays, J., Ehinger, K., Oliva, A., Torralba, A.: SUN database: large-scale scene recognition from abbey to zoo. In: Proceedings of IEEE Conference on Computer Vision and Pattern Recognition, pp. 3485–3492 (2010)
33. Kingma, D., Adam, J.B.: A method for stochastic optimization. arXiv preprint arXiv:1412.6980 (2014)
34. Girosi, F., Jones, M., Poggio, T.: Regularization theory and neural networks architectures. Neural Comput. **7**(2), 219–269 (1995)

Optimization of Facade Segmentation Based on Layout Priors

Radwa Fathalla[1]([⊠]) and George Vogiatzis[2]

[1] College of Computing and Information Technology,
Arab Academy for Science and Technology, Alexandria, Egypt
`fathallr@aston.ac.uk`
[2] School of Engineering and Applied Science, Aston University, Birmingham, UK

Abstract. We propose an algorithm that provides a pixel-wise classification of building facades. Building facades provide a rich environment for testing semantic segmentation techniques. They come in a variety of styles affecting appearance and layout. On the other hand, they exhibit a degree of stability in the arrangement of structures across different instances. Furthermore, a single image is often composed of a repetitive architectural pattern. We integrate appearance, layout and repetition cues in a single energy function, that is optimized through the TRW-S algorithm to provide a classification of superpixels. The appearance energy is based on scores of a Random Forrest classifier. The feature space is composed of higher-level vectors encoding distance to structure clusters. Layout priors are obtained from locations and structural adjacencies in training data. In addition, priors result from translational symmetry cues acquired from the scene itself through clustering via the α-expansion graphcut algorithm. We are on par with state-of-the-art. We are able to fine tune classifications at the superpixel level, while most methods model all architectural features with bounding rectangles.

1 Introduction

Generating models of buildings has innumerable applications, such as heritage conservation, disaster management and urban planning. One particular field of interest has been analysis of building facades. Facades capture the architectural essence of the buildings. They are a dense representation of their characteristics in terms of layout and materials used, which translate into surface properties.

Facade parsing is often regarded as a classical case of semantic segmentation. As most scene interpretation approaches, the problem was originally tackled with appearance-based segmentation algorithms, in which weak priors of smoothness assumption are applied. Research was then directed to the incorporation of mid-level and high level cues of translational symmetry and sub-part classifications, based on training data. The challenges for achieving high accuracies rise from imaging artifacts: blur and noise, non-uniform lighting conditions, reflections and, shadows. Also, they include, the existence of irregular lattices of structures, occlusions, intra- and inter- geometric style variations. This lead to investigating

© Springer International Publishing AG 2017
M. Felsberg et al. (Eds.): CAIP 2017, Part I, LNCS 10424, pp. 196–207, 2017.
DOI: 10.1007/978-3-319-64689-3_16

the pattern of arrangement of facade elements rather than their individual visual attributes. We present an algorithm that exploits higher level reasoning about scene entities, suggested by the appearance characteristics. We combine both aspects in a single energy function, to provide optimized solution at the lowest level of image primitives. In contrast, state-of-the-art methods [10,14] apply their optimization steps on formed Bounding Boxes (BB), whose assignments are either rejected or accepted as a whole. As such, their algorithms incorporates layout principles only in the recognition step of pre-segmented regions, resulting from appearance cues phase. Whereas, we carry out segmentation and recognition simultaneously, while exploiting the layout priors to correct preliminary segmentations. We provide an algorithm that minimizes the use of thresholds, prior assumptions except for fronto-parallelism and works in an approximate inference framework. More importantly, it does not require manual specification of architectural rules as in the 3-layered approach [10].

1.1 Related Work

Research is directed towards implementing architectural guidelines in automated flexible form. These guidelines are concerned with alignment, symmetry, similarity, co-occurrence and components layout. In [10], Martinović et al. make use of these architectural principles in their final classification decision. They refine the output of a preceding segmentation step by applying this set of restricting principles in an ad-hoc procedure. Each principle is applied in isolation and in most part, as a matter of fulfilling a certain criterion is exceeding a manually specified threshold. The classification into structures is achieved by an Recurrent Neural Networks RNN [12] fed with an oversegmentation of the image and a Dollar's Integral Channel [5] specialized window and door detector.

[6] is the only reported work that allows a per-pixel final classification. Every pixel is represented by a vector of image features (such as: location, RGB values, and HOG features), in addition to contextual ones (such as: neighbourhood statistics, and bounding box features) obtained from the preliminary predictions based on image features. The drawback is, each feature vector is supplied independently to an ensemble of classifiers. It lacks the concurrency in classification of pixels of the arrangement and hence, it lacks the global optimality in the proper sense. Perhaps the most related work to ours is [14]. In [14], they build a factor graph of higher order cliques on the images, based on structural aspects more sophisticated than spatial proximity. However, their nodes are Bounding Boxes (BBs) of preliminary segmented regions with the pixel assignment done as a region-to-pixel mapping of the chosen label without the capability of fine tuning the results. Also, based on their reported inadequacy in localizing segment borders, the hardwired specification of thresholds on aspects like alignment, size similarity and regular spacing, will fail with inaccuracies in the segments and subsequent BBs formation. The way they handle size variations and the subsequent reliability of relative location priors is unclear, given that they use vertical and horizontal distances in their absolute form. In addition, their algorithm does

not incorporate appearance in determining edges between the BBs, as they rely on purely geometrical properties.

2 Facade Segmentation Optimization

Our proposed algorithm (Fig. 1) receives as input a set of image pixels in the $2d$ space. It is required to provide an interpretation of these data points by assigning them to a predefined set of labels $\mathcal{L} = \{L_i\}_{m=1}^{M}$, such that $\mathcal{L}$ holds indices to M architectural structures. To keep the problem tractable and enhance computational efficiency, we work with superpixels. Thus, the data points for our algorithm is the set $\mathbf{X} = \{\mathbf{x}_i\}_{i=1}^{n}$ of n superpixels. The image is subjected to a watershed transform [15]. The transform aggregates pixels to a region until reaching a peak in the $2d$ space of the gradient image. The result is a severe oversegmentation of the images with color coherent regions, called *basins*. The superpixels are the minima pixels corresponding to the lowest gradient value in each region.

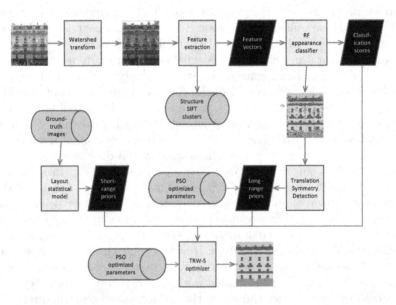

Fig. 1. Diagram showing proposed system modules and their interactions.

We pose our problem as an optimization problem under both appearance and layout constraints, emerging from architecture characteristic patterns. To this end, we define an energy function and minimize it using the sequential tree-reweighted message passing (TRW-S) [7]. We chose this minimization technique due to its ability to handle arbitrary forms of cost function and scalability, while providing state-of-the-art results in some applications. We aim to ensure that the labeling of a pixel is influenced not only by the labeling of its neighbours, but also by that of pixels in other possibly distant regions based on extracted architectural patterns.

A distinctive aspect of our algorithm is imparting structural knowledge on image primitives. The TRW-S operates on the original set of superpixels. The total energy function Ξ of the TRW-S is as follows:

$$\Xi = \Xi_1(L) + \Xi_2(L) \ , \tag{1}$$

where

$$\Xi_1(L) = \sum_{x_i} D(L_i|x_i) \ , \tag{2}$$

is the datacost received from the appearance module. $D(L_i|x_i) = -\log(P(L_i|x_i))$. $P(L_i|x_i)$, are the classification posteriors resulting from a Random Forest (RF) classifier. And, the layout prior

$$\Xi_2(L) = \beta_1 \sum_{x_i} \sum_{x_j \in \Psi_1} Q_1(L_i, L_j|x_i, x_j) + \beta_2 \sum_{x_i} \sum_{x_j \in \Psi_2} Q_2(L_i, L_j|x_i, x_j) \tag{3}$$

is the total energy relayed from the layout statistical model and the translational symmetry modules (Fig. 1). Ψ_1 and Ψ_2 are the neighbourhoods defined based on the short- and long- range edges (Sect. 2.2). $Q_1(\cdot)$ is the prior for the plausible structural adjacencies, while $Q_2(\cdot)$ is the regularizer for the translational symmetry of structures in the architectural scene at hand. The assigned label of a superpixel is mapped to all pixels sharing its basin.

In the following sections, we explain how the appearance and layout priors are established to be incorporated in our energy function for the TRW-S algorithm.

2.1   Appearance Cues

A well-known fact about visual perception is, it is evoked by appearance. Thus, our algorithm is launched by obtaining preliminary classification of the image superpixels that utilizes textural characteristics of the regions. We choose Random Forest (RF) as our classifier [2], which performs a recursive partitioning of the data based on an ensemble of decision trees. But, other efficient classifiers can be used instead.

Another critical choice is the space in which the feature vectors are embedded. We examine 2 spaces. Firstly, the vector $\mathbf{s}_i$ is comprised of the 128 SIFT descriptors [9], calculated densely over the image with a bin size of 8. Secondly, the vector $\mathbf{r}_i$ (4) and (5) is the distances to M predefined clusters, corresponding to M architectural structures. Each cluster consists of the SIFT feature vectors of the superpixels, belonging to a certain structure and acquired from the groundtruth data. The distance is calculated as the mean Euclidean norm between the SIFT vector of the superpixel and the k-nearest neighbours vectors in the cluster after removing the exact match. We preferred this distance over a centroidal one, because the clusters exhibit a high degree of scattering, due to the high degree of appearance variation among instances of the same structure. Hence, the centroid would not be a proper representative of a cluster. We downsampled over-sized clusters to ensure a uniform prior for the RF.

$$\mathbf{r}_i = \begin{bmatrix} r_i^1 r_i^2 \ldots \ldots r_i^M \end{bmatrix} . \tag{4}$$

$$r_i^j = \frac{\sum_{o=1}^k |\mathbf{s}_i - \mathbf{NN}_{ij}^o|}{k} . \tag{5}$$

$\mathbf{NN}_{ij}^o$ is the SIFT vector of the o-th nearest neighbour in cluster j with respect to data point i. And k is the count of neighbours.

In practice the later space was found to outperform the former. In our opinion, it introduced a higher level of semantics over the raw SIFT features, that achieved a substantial dimensionality reduction (from 128 to M features). The challenge for any dimensionality reduction algorithm is, not disturbing the position of a feature vector in its space, relative to label clusters. In the described space, we retain this relative position of the vector, by storing its distances to the clusters in the space, without the overhead of low-level SIFT details. In addition, this space transformation provided better characteristics for the training vectors, namely inter-separability and intra-compactness of the clusters. These characteristics are expected to boost, not only k-nn equivalents but also margin-maximizing hyperplane classifiers. However, further investigation is required to evaluate the proposed idea with other classifiers and clusters of various topologies. Similar approach of using a meta-feature vector can be found in [3]. The resulting segmentations are provided as input to the next phase. We also retain the classification probabilities $P(L_i|x_i)$ computed by the RF for each super-pixel to be used as datacosts in the TRW-S framework.

2.2   Layout Cues

In this module, we make use of 5 architectural principles, namely, spatial coherence, approximate structural location, structure ordering, recurring structural adjacencies, and translational symmetry. In our framework, these principles are expressed in the edge costs of the TRW-S graph. The edge costs are look-up tables giving the penalties for various combinations of labelings for the edge vertices. There are 2 types of edges: short-range and long-range.

Short-Range Edges. They specify neighbours based on spatial proximity, and their edge costs used to establish $Q_1(\cdot)$ for the TRW-S function (3). Super-pixels are connected by an edge if there is a common boundary between their encompassing basins. Hence, each superpixel is allowed a different number of neighbours. During the learning phase, we build a statistical model of the found adjacencies among structures. We argue that the familiar adjacencies is the most stable feature across different architectural scenes. For instance, a door structure can be seen adjacent to a wall, but never next to a sky structure.

The edge costs are $M \times M$ matrices, where M is the number of architectural features encoding the costs for different combinations of labels for adjacent superpixels. We introduce the concept of *location-aware* edges, which entails different costs for edges in different zones of the facade. In POTTs model [1], the diagonal values are set to zero encourage neighbouring nodes get the same label. However, we utilize a non-POTTs model, in which the values on the diagonals of the cost matrices are non-zeros. Therefore, there is a penalty incurred even if nodes are given the same labeling. This penalty is dependent on the frequency by which the label has been seen in this zone of the image in the training samples. The frequencies of the labels with respect to locations are obtained through the following procedure. To account for image size variability, the groundtruth images are transformed to an approximate scale invariant space. This is done by subdividing each image into k horizontal and k vertical stripes of equal width, such that k^2 rectangular patches are formed. The corresponding patch is determined for each labeled pixel and the information is used to update the frequency of the label in the patch. The values are then normalized by dividing by the total pixel density within the patch to get probability $\mathbf{P}_{rc}^m$, such that $r, c \in \{1, 2, \ldots k\}$.

To fill the upper and lower triangles of the cost matrices, we build a $2d$ histogram for the structural tangencies based on the same image subdivision, but this time for a pair of labels (instead of a single label) to encode a transition. The recorded frequencies in each patch, are normalized per structure to reflect the probability $\mathbf{P}_{rc}^{ab}$ that a pair of labels (a and b) exist in adjacency at this location, when a testing sample is introduced. $a, b \in \{M \times M\}$, such that $a \neq b$. Edges and their cost matrices are established in 2 directions corresponding to the directions for tangency: horizontal and vertical. For each structure instance in the ground truth, we record the structures to its east and south. We bypass the west and north directions because they are inverses of the included directions and would only require a transpose of the cost matrix. So, including them will redundantly duplicate the cost. The matrices are non-symmetrical. For instance, a roof structure is more frequently seen to the south of sky than to its north.

In this way, the edge cost matrices (Fig. 2) encode the architectural principles of, vertical and horizontal arrangement ordering of structures, in addition to locations and structural direct adjacencies. At inference time, if basins are tangent in both directions, we choose the direction of the common boundary with the longest length. We convert the probabilities to costs to build labeling penalty matrices, according to the Boltzmann distribution, $\mathbf{E}_{rc}^m = -\log(\mathbf{P}_{rc}^m)$ and $\mathbf{E}_{rc}^{ab} = -\log(\mathbf{P}_{rc}^{ab}) + \xi$. We add ξ, a constant to raise the range of values in the upper and lower triangles of the cost matrices over the diagonal values, to bias the optmization algorithm towards same labeling for the vertices of the edge. As such, spatial coherence is achieved while promoting the frequently encountered label in the training set, at this location. If the algorithm chooses to label the vertices differently, the most frequent adjacencies at this location are preferred.

Some practical adjustments were carried out, because the subdivision of the image is arbitrary and to prevent over-fitting to training data. We apply a Gaussian smoothing filter on the frequency histograms of location and

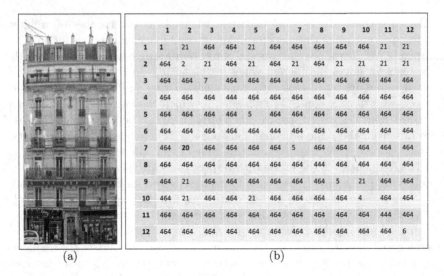

	1	2	3	4	5	6	7	8	9	10	11	12
1	1	21	464	464	21	464	464	464	464	464	21	21
2	464	2	21	464	21	464	21	464	21	21	21	21
3	464	464	7	464	464	464	464	464	464	464	464	464
4	464	464	464	444	464	464	464	464	464	464	464	464
5	464	464	464	464	5	464	464	464	464	464	464	464
6	464	464	464	464	464	444	464	464	464	464	464	464
7	464	20	464	464	464	464	5	464	464	464	464	464
8	464	464	464	464	464	464	464	444	464	464	464	464
9	464	21	464	464	464	464	464	464	5	21	464	464
10	464	21	464	464	21	464	464	464	464	4	464	464
11	464	464	464	464	464	464	464	464	464	464	444	464
12	464	464	464	464	464	464	464	464	464	464	464	6

(a)   (b)

Fig. 2. (a) A sample of long-range edges shown in red. (b) A sample of short-range edge approximated cost matrix for the CMP dataset [14]. Structure 1 incurs the least cost, which signals that it is the most frequently encountered structure in this image patch. The most abundant transition is between structures 7 and 2. Structures 4, 6, 8, and 11 are never seen in this image patch during training. Values on the diagonal are in a lower range than the ones on the lower and upper triangles to promote same labeling. (Color figure online)

structural adjacency. In addition, Inf costs, resulting from zero frequency, are replaced by a relatively high value π, to discourage rather than eliminate the possibility of an assignment. Same goes for Inf values in the appearance datacost, as they are replaced by ρ.

Long-Range Edges. These encode the translational symmetries found in the scene, used for building the $Q_2(\cdot)$ (3). To establish these symmetries we use the α-expansion graphcut algorithm [8], to assign a translation vector to each superpixel in the image. The ultimate goal is to establish a smoothness prior over distant instances of the same structure, in the TRW-S labeling optimization step. It is run separately for each type of putative structure resulting from the appearance classification phase. A Markov Random Field (MRF) is defined over all superpixels belonging to the structure and forming the nodes of the graph. The smoothness prior is based on neighbourhood Ω, detected between superpixels when their basins share a common boundary and belong to the same putative structure. Neighbourhoods are assigned a constant weight. The terminal nodes of the graph of the α-expansion algorithm constitute the labels and they are a set of translational vectors. This set is constructed from the SIFT feature points of the image and their best matches. The matching score is calculated based on Euclidean norm in the SIFT space. The set of translational vectors is refined by preserving only the ones that exhibit a translation in either the x and y directions but not both.. As such the long-range cliques promote the vertical and horizontal alignment of facade structures. The energy function E, to be minimized by the graphcut, is as follows:

$$E\left(Y\right) = \sum_{x_i} D_Y\left(y_i|x_i\right) + \mu \sum_{x_i} \sum_{x_j \in \Omega} F_Y\left(y_i, y_j|x_i, x_j\right) + \theta \cdot |Y_T| \ . \tag{6}$$

The unary term $D\left(\cdot\right)$ is the dissimilarity score between an examined superpixel x_i and the superpixel of the watershed basin, to which the destination belongs. The destination is obtained when applying translation y_i ($\in T$) on the examined superpixel. We constraint the translations to result in destinations being within image boundaries, but not necessarily belonging to the same structure as the source superpixel, to minimize the propagation of errors from the previous appearance-based stage. The pairwise term $F\left(\cdot\right)$ follows a POTTS model, in which a pair of neighbouring superpixels labeled differently, is penalized with a constant value. θ is a constant label cost that penalizes the assignment of x_i to new redundant labels. Redundancy in the sense that they can be replaced by one of the already utilized labels without drastically increasing the datacost. Afterwards, the edges that will be relayed to the TRW-S algorithm are found by applying on each superpixel its preferred translation vector in the specified, in addition to the reverse direction (a 180° rotated variant). In effect, this extends the putative structures into a loci of points that complete their contained grids. An outcome of this phase is shown in Fig. 2.

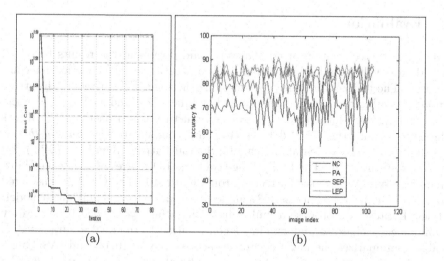

(a)   (b)

Fig. 3. (a) Semi-log scale plot of the cost against PSO iterations. (b) Accuracy plots for the images in ECP dataset when different options for IASC are activated.

2.3   Learning the Weight Parameters

For learning the parameters of the energy functions, we use the Particle Swarm Optimization [11] (PSO). A meta-heuristic technique, that relies on a user-specified range of values for the parameters. The algorithm initializes a swarm of vectors randomly. Each vector U_i holds values for the parameters and is named a particle. Iteratively, it updates the vectors based on their best previous position U_{i_pbest} and the best position in the swarm U_{global_best}. The quality of the particle is evaluated based on a cost function. In all our experiments, the cost function is single objective. The position update rule for the ith particle is

$$U_i = U_i + V_i \ . \tag{7}$$

The velocity V_i of the particle is given by,

$$V_i = \omega \times V_i + c_1 \times \text{rand}\,() \times (U_{i\_pbest} - U_i) + c_2 \times \text{rand}\,() \times (U_{global\_best} - U_i) \ . \tag{8}$$

The rule guarantees that the procedure yields non-increasing cost values in each iteration Fig. 3, thus leading to convergence. First, we use the PSO in learning the α-expansion parameters (θ and μ). In this case, the objective is minimizing the number of erroneous edges that link superpixels belonging to different genre of structures. In the second setting, it is used for optimizing β_1, β_2, ξ, π, and ρ in the TRW-S framework. The objective is minimizing the errors in the final labeling of the superpixels, when compared to ground truth data.

3   Evaluation

We follow the convention of related work, and document the results based on 5-fold cross validation and using pixel-based accuracy as the criterion for comparison. The training folds are used for constructing SIFT clusters of the structures, collecting the layout statistics and training the Random Forest. We test our model IASC (Integrated Appearance Structure Cues) on the *ECP-Monge* dataset [13] and the *CMP* dataset [14], and compare to the *state-of-the-art* results from the 3-layered approach [10], Spatial Pattern Templates (SPT) [14] and Auto-Context [6]. The ECP-Monge contains 104 images of facades in Haus-mannian style. We use the corrected groundtruth [10]. The CMP is considered more challenging as it contains 378 samples from various (often difficult to model) styles. Because, we propose a multi-phase algorithm, we needed to separately examine each phase to understand its contribution to the final accuracy value. Table 1 summarizes the mean accuracies achieved by [6,10,14] and IASC algorithm in various stages. We include the results of the commonly used POTTS model for spatial smoothness (PA), as a variant of our algorithm, and use the same datacosts of the IASC. We follow the naming conventions of the original papers [6,10,14] in reporting results. Per-image accuracies are shown in Fig. 3, for the different factors affecting the performance of our model. In Fig. 4, we display results of a selection of samples. We can conclude from experiments, for IASC,

Table 1. Average accuracies on datasets. NC: No context (appearance only), AP: Aligned Pairs, APRT: Aligned Pairs Regular Triplets, SH: Structural Heuristics, PA: POTTS Adjacency, ST3: Auto-Context classified, PW3: POTTS Smoothed Auto-Context, SEP: Short-range Edges Prior, and LEP: Layout Edges Prior (short- and long- range).

	SPT [14]			3-layers [10]			Auto-context [6]		IASC (our method)			
	(NC)	(AP)	(APRT)	(NC)	(PA)	(SH)	(ST3)	(PW3)	(NC)	(PA)	(SEP)	(LEP)
ECP	59.6	79.0	84.2	82.6	85.1	84.2	90.8	91.4	68.9	79.9	86.3	87.8
CMP	33.2	54.3	60.3	-	-	-	66.2	68.1	41.4	55.5	60.3	64.4

each phase consistently improved accuracy over the preceding one. Despite our efforts to minimize the propagation of errors, across the system modules, it is evident that appearance classification failures remain a limiting factor for subsequent improvements. It is evident for [10], the incorporation of the structural heuristics (such as: the existence of a running balcony on the second and fifth floor) degraded the accuracy of their smoothed appearance classifications. As for [14], the fact that their neighbourhoods of pairs and triplets were based on a manually assigned threshold was a severe limitation. The reported result for ECP-Monge in [6] is based on 7 classes of structures, whereas we include the result using the updated groundtruth which added the chimney structure. In IASC, we record one of the highest accuracy net gains when incorporating layout cues in the problem of facade parsing, even when starting with severely damaged results based on appearance. This is attributed to the generalization ability of our optimization function that relies only on persistent architectural guidelines without being style specific.

We use the Davies Bouldin (DB) index [4] to shed light on the characteristics of the proposed feature space of distance-to-cluster, against the raw SIFT feature space. The clustering is predefined from the groundtruth and we normalized the 2 spaces. It was found that the proposed space transformation increased both separability and compactness of the clusters, thus, favorably lowering the average DB on the training folds from 8.4616 to 1.4497. As for classification accuracy, raw SIFT vectors achieved 63.3% on ECP-Monge in the No Context setting. For the distance vectors, the figure was 68.9%. In both settings we use the PSO to learn the parameters, the number of iterations was set to 75. The swarm size was 10 when optimizing the parameters for finding long-range edges and 40 for the TRW-S function. The parameters ranges were based upon our observations during experiments, but we provided a much wider range to lower the risk of a local minimum. In evaluating the objective functions, 10 samples were selected randomly for each dataset. The objective function yields the highest calculated cost based on the 10 samples.

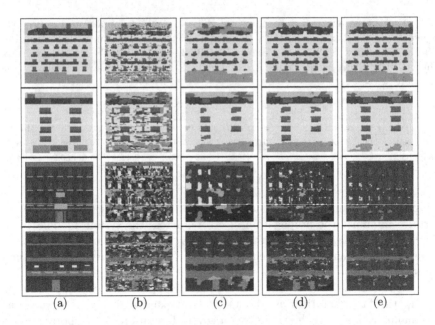

(a)   (b)   (c)   (d)   (e)

Fig. 4. Sample outcomes in tabular format. Row (1) ECP-Monge sample with accuracy 91.72%; (2) ECP-Monge sample with accuracy 83.18%; (3) CMP sample with accuracy 74.21%; (4) CMP sample with accuracy 72.00%. Column (a) Ground truth; results of (b)NC; (c) PA; (d) SEP; (e) LEP.

4   Conclusion

We present an algorithm for handling semantic segmentation of architectural scenes. The algorithm relies on the output of a Random Forest classifier on SIFT-based meta-feature vectors. We carry out a feature space transformation from raw SIFT to distance-to-cluster vectors. Also, we incorporate layout principles in the form of labeling costs for superpixel long-range cliques resulting from translation vectors, detected by α-expansion. Other labeling costs are based on location and structural adjacencies, defined on short range neighbourhoods. We report competitive results. We believe our method offers significant advantages over competitors in terms of algorithm elegance. The priors are automatically learned from training samples and its weight parameters are deduced via the single objective PSO algorithm. At inference time, the labeling is efficiently optimized using the TRW-S algorithm, while including no heuristics or manually determined thresholds. Our future work is intended towards boosting the accuracy figures, by plugging in the state-of-the-art Convolutional Neural Networks in the appearance module and relaying its resulting posteriors to our layout optimization function.

References

1. Boykov, Y., Veksler, O., Zabih, R.: Fast approximate energy minimization via graph cuts. IEEE Trans. Pattern Anal. Mach. Intell. **23**(11), 1222–1239 (2001)
2. Breiman, L.: Random forests. Mach. Learn. **45**(1), 5–32 (2001)
3. Dang, K., Yuan, J.: Location constrained pixel classifiers for image parsing with regular spatial layout. In: Proceedings of the British Machine Vision Conference. BMVA Press (2014)
4. Davies, D.L., Bouldin, D.W.: A cluster separation measure. IEEE Trans. Pattern Anal. Mach. Intell. **1**(2), 224–227 (1979)
5. Dollar, P., Tu, Z., Perona, P., Belongie, S.: Integral channel features. In: Proceedings of the BMVC, pp. 91.1–91.11 (2009)
6. Jampani, V., Gadde, R., Gehler, P.V.: Efficient 2D and 3D facade segmentation using auto-context. In: Proceedings of the 2015 IEEE Winter Conference on Applications of Computer Vision, WACV 2015, Washington, DC, USA, pp. 1038–1045. IEEE Computer Society (2015)
7. Kolmogorov, V.: Convergent tree-reweighted message passing for energy minimization. IEEE Trans. Pattern Anal. Mach. Intell. **28**(10), 1568–1583 (2006)
8. Kolmogorov, V., Zabih, R.: What energy functions can be minimized via graph cuts? In: Heyden, A., Sparr, G., Nielsen, M., Johansen, P. (eds.) ECCV 2002. LNCS, vol. 2352, pp. 65–81. Springer, Heidelberg (2002). doi:10.1007/3-540-47977-5_5
9. Lowe, D.G.: Distinctive image features from scale-invariant keypoints. Int. J. Comput. Vision **60**(2), 91–110 (2004)
10. Martinović, A., Mathias, M., Weissenberg, J., Gool, L.: A three-layered approach to facade parsing. In: Fitzgibbon, A., Lazebnik, S., Perona, P., Sato, Y., Schmid, C. (eds.) ECCV 2012. LNCS, vol. 7578, pp. 416–429. Springer, Heidelberg (2012). doi:10.1007/978-3-642-33786-4_31
11. Poli, R., Kennedy, J., Blackwell, T.: Particle swarm optimization. Swarm Intell. **1**(1), 33–57 (2007)
12. Socher, R., Lin, C.C., Ng, A.Y., Manning, C.D.: Parsing natural scenes and natural language with recursive neural networks. In: Proceedings of the 26th International Conference on Machine Learning (ICML) (2011)
13. Teboul, O., Kokkinos, I., Simon, L., Koutsourakis, P., Paragios, N.: Shape grammar parsing via reinforcement learning. In: CVPR, pp. 2273–2280. IEEE Computer Society (2011)
14. Tyleček, R., Šára, R.: Spatial pattern templates for recognition of objects with regular structure. In: Weickert, J., Hein, M., Schiele, B. (eds.) GCPR 2013. LNCS, vol. 8142, pp. 364–374. Springer, Heidelberg (2013). doi:10.1007/978-3-642-40602-7_39
15. Vincent, L., Soille, P.: Watersheds in digital spaces: an efficient algorithm based on immersion simulations. IEEE Trans. Pattern Anal. Mach. Intell. **13**(6), 583–598 (1991)

Development of a New Fractal Algorithm to Predict Quality Traits of MRI Loins

Daniel Caballero[1](✉), Andrés Caro[1], José Manuel Amigo[2], Anders B. Dahl[3], Bjarne K. Ersbøll[4], and Trinidad Pérez-Palacios[5]

[1] Computer Science Department, Research Institute of Meat and Meat Product, University of Extremadura, Av/ Universidad S/N, 10003 Cáceres, Spain
dcaballero@unex.es

[2] Department of Food Science, Quality and Technology, Faculty of Life Science, University of Copenhagen, Rolighedsvej 30, 1958 Fredriksberg C, Denmark

[3] Department of Applied Mathematics and Computer Science, Technical University of Denmark, Richard Petersen Plads, Building 324, 2800 Kongens Lyngby, Denmark

[4] Department of Informatics and Mathematical Modelling, Technical University of Denmark, Richard Petersen Plads, Building 324, 2800 Kongens Lyngby, Denmark

[5] Food Technology Department, Research Institute of Meat and Meat Product, University of Extremadura, Av/ Universidad S/N, 10003 Cáceres, Spain

Abstract. Traditionally, the quality traits of meat products have been estimated by means of physico-chemical methods. Computer vision algorithms on MRI have also been presented as an alternative to these destructive methods since MRI is non-destructive, non-ionizing and innocuous. The use of fractals to analyze MRI could be another possibility for this purpose. In this paper, a new fractal algorithm is developed, to obtain features from MRI based on fractal characteristics. This algorithm is called OPFTA (One Point Fractal Texture Algorithm). Three fractal algorithms were tested in this study: CFA (Classical fractal algorithm), FTA (Fractal texture algorithm) and OPFTA. The results obtained by means of these three fractal algorithms were correlated to the results obtained by means of physico-chemical methods. OPFTA and FTA achieved correlation coefficients higher than 0.75 and CFA reached low relationship for the quality parameters of loins. The best results were achieved for OPFTA as fractal algorithm (0.837 for lipid content, 0.909 for salt content and 0.911 for moisture). These high correlation coefficients confirm the new algorithm as an alternative to the classical computational approaches (texture algorithms) in order to compute the quality parameters of meat products in a non-destructive and efficient way.

Keywords: MRI · Fractal · Algorithms · Quality traits · Iberian loin

1 Introduction

The traditional methods established for determining physico-chemical parameters related to loin quality are laborious, time and solvent consuming and require

© Springer International Publishing AG 2017
M. Felsberg et al. (Eds.): CAIP 2017, Part I, LNCS 10424, pp. 208–218, 2017.
DOI: 10.1007/978-3-319-64689-3_17

the destruction of the meat piece. Magnetic Resonance Imaging (MRI) and computer vision techniques have been proposed as an alternative, since they are non-destructive, non-invasive, non-intrusive, non-ionizing and innocuous. Several works have been carried out to determine quality characteristics of dry-cured products by MRI, most of them focused on hams, allowing to monitor the ripening process of Iberian [1], Parma [2] and San Daniele [3] hams.

Indeed, the extraction of textural information from images is very common to explore parameters related to food quality. Cernadas [4,5] allowed the prediction of some sensory and physico-chemical traits. Ávila [6] analyzed marbling and fat level in Iberian loin based on texture features of MRI. Recently, Pérez-Palacios applied texture analysis to predict moisture and lipid of hams [7] and loins [8]. Jackman et al. [9,10] have proved the efficiency of texture feature methods to solve problems related to food technology.

In recent years, there is a growing interest in the use of fractal analysis techniques instead of classical texture analysis methods. Mainly because the image texture seeks to compress image information while the use of fractals allows the identification of recurring patterns, removing the possibility of image compression. The fractal concept studies the degree of symmetry or self-similarity found in a structure at all scales [11,12]. In relation to the use of fractals in food technology, mainly, fractal techniques have been applied to characterize the microstructure of different fruit and vegetables [13], fish [14] and meat [15]. For prediction issues, only two studies have applied fractal techniques. Tsuta [16] used them to predict the sugar content of melons and Polder [17] measured the chlorophyll of tomato by applying fractals. However, as our knowledge, the use of fractal analysis has not been carried out to predict quality parameters on meat products.

This paper aims (1) to develop an algorithm for studying texture features based on fractal and second order statistics and (2) its application on MRI images of Iberian loins in order to predict some physico-chemical parameters.

This paper is organized as follows: Sect. 2 presents the Materials and Methods used in this work. Section 3 describe the obtained results and their discussion. Section 4 draws the main conclusions and their implications.

2 Materials and Methods

The prediction of quality attributes was carried out with 5220 MRI images from fresh and dry-cured loins. In addition, the quality attributes of fresh and dry-cured loins were determined by means of traditional physico-chemical methods in order to obtain values for moisture [18], salt [18] and lipid content [19].

2.1 Image Acquisition

MRI from the loins were generated at the Animal Source Foodstuffs Innovation Services (SiPA) at Faculty of Veterinary Science of University of Extremadura

(Cáceres, Spain). A low-field MRI scanner (ESAOTE VET-MR E-SCAN XQ 0.18 T) was used, with a hand/wrist coil, with nine different configurations on echo time (TE) and repetition time (TR). Sequences of Spin Echo (SE) weighted on T1 were applied with a field of view (FOV) of $150 \times 150\,\text{mm}^2$, slice thickness 4 mm, a matrix size of 256×204 and 29 slices per loin were obtained.

All images were acquired in DICOM format, with a 512×512 resolution and 256 grey levels. The MRI acquisition was performed at $23\,^{\circ}\text{C}$.

2.2   Computer Vision Algorithms

Once the MRI of loins were obtained, three Computer Vision algorithms based on fractal were applied to extract numerical data from the images.

Table 1. Texture features equations of FTA algorithm

	Equation	
Uniformity (UNI)	$\sum_i F_i^2$	(1)
Entropy (ENT)	$\sum_i F_i * log_{10}(F_i)$	(2)
Correlation (COR)	$\sum_i (i - \mu) * F_i$	(3)
Inverse Difference Moment (IDM)	$\sum_i \frac{F_i}{1+i^2}$	(4)
Inertia (INE)	$\sum_i F_i * i^2$	(5)
Contrast (CON)	$\sum_i F_i^2 * i^2$	(6)
Emphasis (EMP)	$\sum_i \frac{F_i}{i^2}$	(7)
Jorna Correlation (JC)	$\sum_i (i - \mu)^2 * F_i$	(8)
Cluster Shade (CS)	$\sum_i (i - \mu)^3 * F_i$	(9)
Cluster Prominence (CP)	$\sum_i (i - \mu)^4 * F_i$	(10)

The first one, classical fractal algorithm (CFA) [20] studies the repetition of patterns in the MRI. The method measures the number of boxes (small fractions of the image depending on the size of the original image) needed to cover an area occupied by the object as a function of the size of boxes. This is calculated by computing the so-called local exponent with different box sizes. The local exponent, D, is the variation of the number of objects (N) depending on the box size (R).

$$D = -\frac{\triangle \ln N}{\triangle \ln R} \qquad (11)$$

The fractal dimension is the value of the number of D when it remains constant respect to the box size. Quantification of the fractal dimension of each simple image was calculated using the compression box counting package (tool-box downloaded from http://www.mathworks.com/matlabcentral/fileexchange/ 13063-boxcount-last accessed November 2016, for MATLAB (The Mathworks Inc., Natick, Massachusetts, U.S.A.)). Eight features were obtained from CFA algorithm.

The second one, the fractal texture algorithm (FTA) [21] is a novelty texture algorithm which is not based on texture features of images. In fact, it is based on fractals characteristics, obtained from a two dimensional variation of Minkowski-Bouligand algorithm [20]. Algorithm 1 shows the new version of the Minkowski-Bouligand algorithm, proposed to compute the local exponents for FTA. These fractal characteristics reflect the number of times that a pattern is repeated for each image depending of the box size calculated in each case. These fractal characteristics were gathered in a vector. The features of this algorithm were computed applying second order statistics [22] on these vectors, a total of ten features were calculated and they were the following: Uniformity (UNI), Entropy (ENT), Correlation (COR), Inverse Difference Moment (IDM), Inertia (INE), Contrast (CON), Emphasis (EMP), Jorna's Correlation (JC), Cluster shade (CS) and Cluster Prominence (CP). Table 1 shows the equation to compute each feature from the values of the previously calculated vector, where i is the index of the vector, F_i is the value of the cell with the position i of the vector, and μ is the average value of the vector.

2.3 One Point of Fractal Texture Algorithm (OPFTA)

The third algorithm studied was our proposal One Point of Fractal Texture Algorithm (OPFTA). Figure 1 summarizes the flow chart of this algorithm.

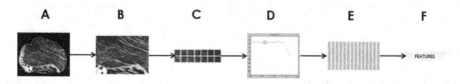

Fig. 1. The proposed computational texture algorithm (OPFTA algorithm). (A) Input image. (B) Largest area rectangle inside of loin contour (C) Calculating ROIs (D) Selecting fractal value (E) Input fractal value in the matrix (F) Calculating features

First, the image acquisition process obtained sets of MRI, in a high resolution (pixel resolution 0.23×0.23) (Fig. 1A). When the images were acquired, the largest rectangle inscribed in the contour of the loin was selected (Fig. 1B) [23]. Then, each rectangle was divided into smaller rectangles of 32×32 pixels, so called region of interest (ROI) (Fig. 1C). At this point, a two dimensional variation of the Minkowski-Bouligand algorithm [20] was applied on each one of the ROI in order to obtain local exponents with the different box sizes (powers of 2). Again, algorithm 1 was used to compute the local exponents for OPFTA.

These local exponents reflect the number of times that a pattern is repeated for each ROI depending of the size of the boxes calculated in each case. From all local exponents, we select the local exponent with the box size equal to eight (Fig. 1D), since this local exponent is very representative, because reflect the patterns inside of the ROI of medium size. After that, we gathered one value for each ROI in order to create a matrix with the fractal values. Each cell of the matrix represents one ROI from the image (Fig. 1E).

Finally, seven texture features were computed on each matrix (Fig. 1F). These features were calculated based on second order statistics [22], and were the following: Uniformity (UNI), Entropy (ENT), Correlation (COR), Homogeneity (HOM), Inertia (INE), Contrast (CON) and Efficency (EFI). Table 2 shows the equation to calculate each feature from the values of the previously computed matrix. Algorithm 2 shows the pseudocode of the OPFTA algorithm.

Table 2. Texture features equations of OPFTA algorithm

	Equation	
Uniformity (UNI)	$\sum_i \sum_j P(i,j)^2$	(12)
Entropy (ENT)	$\sum_i \sum_j P(i,j) * log_{10}(P(i,j))$	(13)
Correlation (COR)	$\dfrac{\sum_i \sum_j \mu_x * \mu_y * P(i,j)}{\sigma_x/\sigma_y}$	(14)
Homogeneity (HOM)	$\dfrac{\sum_i \sum_j P(i,j)}{1+(i-j)^2}$	(15)
Inertia (INE)	$\sum_i \sum_j P(i,j) * (i-j)^2$	(16)
Contrast (CON)	$\sum_i \sum_j P(i,j)^2 * (i-j)^2$	(17)
Efficency (EFI)	$\dfrac{\sigma_x}{\mu_x} + \dfrac{\sigma_y}{\mu_y}$	(18)

2.4 Prediction Analysis

The prediction of physico-chemical parameters is made as a function of computer vision features from CFA, FTA and OPFTA algorithms. To achieve the prediction, data mining techniques were carried out, specifically multiple linear regression (MLR) [24]. MLR models the linear relationship between a target variable and more independent prediction variables, to produce a linear regression equation that can be used to predict future values. For this purposal the free software WEKA was used (http://www.cs.waikato.ac.nz/ml/weka - last accessed November 2016). The M5 method was applied to select attributes, this method is based on stepping though the attributes, being the one with the smallest standardized coefficient removed until no improvement is observed in the estimation error [25] and a ridge value of 1.0×10^{-4} was applied in the linear regression in this study.

Algorithm 1. Method for obtaining fractals dimensions: Obtain fractals()

Input: img: image
Output: datavector: vector
 1: Begin
 2: $(height, width)$ ←obtain dimensions (img)
 3: **if** height > width **then**
 4:     max ←height
 5: **else**
 6:     max ←width
 7: **end if**
 8: p ←log(max) / log(2)
 9: $size$ ←2^p
10: $ampliedimg$ ←create image (size,size)
11: **for** $i = 0$ to $height$ **do**
12:     **for** $j = 0$ to $width$ **do**
13:         $ampliedimg$ ←img
14:     **end for**
15: **end for**
16: **for** $i = 0$ to $height$ **do**
17:     **for** $j = width$ to $size$ **do**
18:         $ampliedimg$ ←0
19:     **end for**
20: **end for**
21: **for** $i = height$ to $size$ **do**
22:     **for** $j = 0$ to $size$ **do**
23:         $ampliedimg$ ←0
24:     **end for**
25: **end for**
26: **for** $g = p - 1$ to 0 decrease: -1 **do**
27:     aux ←2^{p-g}
28:     $auxil$ ←aux/2
29:     **for** $i = 1$ to aux increase: $i + max - aux$ **do**
30:         **for** $j = 1$ to aux increase: $j + max - aux$ **do**
31:             **if** ampliedimg(i,j) = ampliedimg(i+auxil,j) **then**
32:                 cont++
33:             **else if** ampliedimg(i,j) = ampliedimg(i,j+auxil) **then**
34:                 cont++
35:             **else if** ampliedimg(i,j) = ampliedimg(i+auxil,j+auxil) **then**
36:                 cont++
37:             **end if**
38:         **end for**
39:     **end for**
40:     Update datavector
41: **end for**
42: End

Algorithm 2. Main method of OPFTA

Input: img: image
Output: featurevector: vector
1: Begin
2: $(height, width)$ ←obtain dimensions (img)
3: $columns$ ←width/32
4: $rows$ ←height/32
5: mat ←create matrix (rows,columns)
6: **for** $i = 0$ to $i <= rows * 32$ increment: $i + 32$ **do**
7:   **for** $j = 0$ to $j <= columns * 32$ increment: $j + 32$ **do**
8:     $cell$ ←cut image (img,i*32,j*32,(i+1)*32,(j+1)*32)
9:     /* applying algorithm 1 */
10:    obtain fractals (cell,datavector)
11:    /* Select the position of datavector with the box size equal to 8 */
12:    $fractalvalue$ ←datavector (3)
13:    $mat(i, j)$ ←fractal value
14:   **end for**
15: **end for**
16: /* Computing equations from Table 2 */
17: compute features(mat,featurevector)
18: End

3 Results and Discussion

Table 3 shows the computational complexity of some of most used computer vision algorithms. In this table, the fractal-based algorithms tested in this study are shown first and then, other classical computer vision algorithms traditionally applied to analyze images. OPFTA performs a computational complexity $O(n^2)$ lower than other fractal algorithms ($O(n^2*\log(n))$ and $O(n^3)$) and similar to the classical texture algorithms. GLCM and GLRLM obtained a computational complexity of $O(n^2)$. Low computational complexities are mandatory from an industrial point of view.

Table 3. Computational cost of the most used computer vision algorithms

Algorithm	Authors	Reference	Computational cost
CFA	Mandelbrot	[20]	$O(n^3)$
FTA	Caballero et al.	[21]	$O(n^2 * \log(n))$
OPFTA	Caballero et al.	This study	$O(n^2)$
GLCM	Haralick et al.	[26]	$O(n^2)$
NGLDM	Sun and Wee	[27]	$O(n^3)$
GLRLM	Siew et al.	[28]	$O(n^2)$
GLCM + NGLDM + GLRLM	Durán et al.	[29]	$O(n^3)$

The predicted values based on the three fractals algorithms were correlated to the real values obtained by physico-chemical analysis. Thus, the correlation coefficient (R) of equations were calculated (Table 4), and was used to evaluate the accuracy in the predictions. These results were analyzed taking into account the rules given by Colton [30], who considered correlation values between 0 and 0.25 as little degree of relationship, from 0.25 to 0.50 as a fair degree of relationship, from 0.50 to 0.75 as moderate to good relationship and between 0.75 and 1 as very good to excellent relationship.

Table 4. Correlation coefficients between some physico-chemical traits and computer vision algorithms

	CFA	FTA	OPFTA
Moisture (%)	0.289	0.832	0.911
Lipid (%)	0.201	0.835	0.837
Salt (%)	0.507	0.795	0.909

As can be seen in Table 4, according to Colton [30], for physico-chemical parameter, CFA reached little relationship, whereas FTA and OPFTA achieved very good to excellent correlations. OPFTA obtained slightly higher R values for all physico-chemical attributes than FTA, 0.911 versus 0.832 for moisture, 0.837 versus 0.835 for lipid content and 0.909 versus 0.795 for salt content. The fact of the OPFTA obtained higher correlations than FTA could be related to the better perform in terms to simulate textures from the images. In addition, FTA was previously validated in order to predict some physico-chemical parameters of loin [21]. Therefore, these facts could validate the use of OPFTA to predict quality traits from the loins.

Table 5. Prediction equations obtained applying OPFTA algorithm

Equation

Moisture(%) = $136.970 * ENT - 33.214 * COR + 24.320 * HOM + 82.487 * INE + 33.350 * CON + 10.610 * EFI - 83.966$

Lipid(%) = $-52.016 * UNI - 52.642 * ENT + 8.502 * COR - 13.734 * INE + 11.566 * CON - 10.309 * EFI + 67.576$

Salt(%) = $-11.033 * ENT + 2.333 * COR - 1.812 * HOM - 6.675 * INE - 2.615 * CON - 0.628 * EFI + 11.986$

Table 5 shows the prediction equations of quality parameters of loin as a function of features obtained from OPFTA. As can be seen, there are six independent variables of the prediction equations for the OPFTA. In addition, only seven features of OPFTA need to be computed, while FTA require the computation of ten features. Besides, the computational cost for OPFTA is lower than FTA, as Table 3 shown. All these facts point out the suitability of OPFTA for MRI analysis in order to predict some physico-chemical characteristics of loin.

4   Conclusion

In this study, a new texture algorithm based on fractals and second order statistics has been proposed, developed and validated. The prediction of moisture, lipid and salt content of loins by applying the proposed algorithm on MRI have also been tested. Therefore, the use of this approach could be suitable for the meat industries in order to characterize meat products in a non-destructive, effective, efficient and accurate way.

Acknowledgments. The authors wish to acknowledge the funding received from the FEDER-MICCIN Infrastructure Research Project (UNEX-10-1E-402), Junta de Extremadura economic support for research group (GRU15173 and GRU15113) and the COST association, Farm Animal Imaging action (FAIM) (COST-FA1102) (COST-STSM-FA1102-26642). We also wish to thank the Animal Source Foodstuffs Innovation Service (SiPA, Cáceres, Spain) from the University of Extremadura.

References

1. Antequera, T., Caro, A., Rodríguez, P.G., Pérez-Palacios, T.: Monitoring the ripening process of Iberian ham by computer vision on magnetic resonance imaging. Meat Sci. **76**, 561–567 (2007)
2. Fantazzini, P., Gombia, M., Schembri, P., Simoncini, N., Virgili, R.: Use of magnetic Resonance Imaging for monitoring Parma dry-cured ham processing. Meat Sci. **82**, 219–227 (2009)
3. Manzoco, L., Anese, M., Marzona, S., Innocente, N., Lazaglio, C., Nicoli, M.C.: Monitoring dry-curing of San Daniele ham by magnetic resonance imaging. Food Chem. **141**, 2246–2252 (2013)
4. Cernadas, E., Antequera, T., Rodríguez, P.G., Durán, M.L., Gallardo, R., Villa, D.: Magnetic resonance imaging to classify loin from Iberian pig. In: Webb, G.A., Belton, P.S., Gil, A.M., Delgadillo, I. (Eds.) Magnetic Resonance Imaging in Food Science: A View to the Future. The Royal Society of Chemistry. Cambridge (2001)
5. Cernadas, E., Carrión, P., Rodríguez, P.G., Muriel, E., Antequera, T.: Analyzing magnetic resonance images of Iberian pork loin to predict its sensorial characteristics. Comput. Vis. Image Underst. **98**, 345–361 (2005)
6. Ávila, M.M., Durán, M.L., Antequera, T., Palacios, R., Luquero, M.: 3D reconstruction on mri to analyse marbling and fat level in iberian loin. In: Martí, J., Benedí, J.M., Mendonça, A.M., Serrat, J. (eds.) IbPRIA 2007. LNCS, vol. 4477, pp. 145–152. Springer, Heidelberg (2007). doi:10.1007/978-3-540-72847-4_20
7. Pérez-Palacios, T., Caballero, D., Caro, A., Rodríguez, P.G., Antequera, T.: Applying data mining and computer vision techniques to MRI to estimate quality traits in Iberian hams. J. Food Eng. **131**, 82–88 (2014)

8. Pérez-Palacios, T., Caballero, D., Caro, A., Antequera, T.: Magnetic resonance imaging and computational texture features to predict moisture and lipid content of loins. In: IV Farm Animal Imaging Conference, Edinburgh, UK (2015)
9. Jackman, P., Sun, D.W., Allen, P.: Recent advances in the use of computer vision technology in the quality assessment of fresh meat. Trends Food Sci. Technol. **22**(4), 185–197 (2011)
10. Jackman, P., Sun, D.W.: Recent advances in image processing using image texture features for food quality assessment. Trends Food Sci. Technol. **29**(1), 35–43 (2013)
11. Celigueta-Torres, I., Amigo-Rubio, J.M., Ipsen, R.: Using fractal image analysis to characterize microstructure of low-fat stirred yogurt manufactured with microparticulated whey protein. J. Food Eng. **109**, 721–729 (2012)
12. Sun, J., Zhang, Y.B., Dahl, A.B., Conradsen, K., Juul Jensen, D.: Boundary fractal analysis of two cube-oriented grains in partly recrystallized copper. In: XVII International Conference on Texture of Materials, ICOTOM 2017, Dresden, Germany (2014)
13. Quevedo, R., Pedreschi, F., Bastías, J.M., Díaz, O.: Correlation of the fractal enzymatic browning rate with the temperature in mushroom, pear and apple slices. LWT-Food Sci. Technol. **65**, 406–413 (2016)
14. Manera, M., Giari, L., De Pasquale, J.A., Dezfuli, B.S.: Local connected fractal dimmension analysis in gill of fish experimentally exposed to toxicants. Aquat. Toxicol. **175**, 12–19 (2016)
15. Zapotoczny, P., Szczypinski, P.M., Daszkiewicz, T.: Evaluation of the quality of cold meats by computer-assisted image analysis. LWT-Food Sci. Technol. **67**, 37–49 (2016)
16. Tsuta, M., Sugiyama, J., Sagara, Y.: Near-infrared imaging spectroscopy based on sugar absorption band for melons. J. Agric. Food Chem. **50**(1), 48–52 (2002)
17. Polder, G., Van Der Heijden, G.W.A.M., Van Der Hoet, H., Young, I.T.: Measuring surface distribution of caretones and chlorophyll in ripening tomatoes using imaging spectrometry. Postharvest Biol. Technol. **34**, 117–129 (2004)
18. Association of Official Analytical Chemist (AOAC): Official Methods of Analysis of AOAC International, 17th edn. AOAC International. Gaithersburg, Maryland, U.S.A
19. Pérez-Palacios, T., Ruiz, J., Martín, D., Muriel, E., Antequera, T.: Comparison of different methods for total lipid quantification. Food Chem. **110**, 1025–1029 (2008)
20. Mandelbrot, B.B.: The Fractal Geometry of Nature. W.H. Freeman and Co., New York (1982)
21. Caballero, D., Caro, A., Antequera, T., Pérez-Palacios, T.: Non destructive analysis of loin by magnetic resonance imaging and fractal. In: IX Symposium of Mediterranean Pig, Portalegre, Portugal (2016)
22. Peckinpaugh, S.: An improved method for computing gray-level coocurrence matrix based texture measured. Comput. Vis. Graph. Image Process. **53**, 574–580 (1991)
23. Molano, R., Rodríguez, P.G., Caro, A., Durán, M.L.: Finding the largest area rectangle of arbitrary orientation in a closed contour. Appl. Math. Comput. **218**(19), 9866–9874 (2012)
24. Witten, I.H., Frank, E.: Data Mining: Practical Machine Learning Tools and Techniques with Java Implementations. Morgan-Kauffmann, San Francisco (2005)
25. Kira, K., Rendell, L.A.: A practical approach to feature selection. In: IX International Conference on Machine Learning, Aberdeen, UK (1992)
26. Haralick, R.M., Shanmugam, K., Dinstein, I.: Texture features for image classification. IEEE Trans. Man Cybern. **3**(6), 610–621 (1973)

27. Sun, C., Wee, G.: Neighboring gray level dependence matrix. Comput. Vis. graph. Image Proc. **23**, 341–352 (1982)
28. Siew, L.H., Hodgson, R.M., Wood, E.J.: Texture measures for carpet wear assessment. IEEE Trans. Pattern Anal. Mach. Intell. **10**(1), 92–104 (1988)
29. Durán, M.L., Rodríguez, P.G., Arias-Nicolas, J.P., Martín, J., Disdier, C.: A perceptual similarity method by pairwise comparison in a medical image case. Mach. Vis. Appl. **21**(6), 865–877 (2010)
30. Colton, T.: Statistics in Medicine. Little Brown and Co., New York (1974)

A Visual Inspection Method
Based on Periodic Feature for Wheel Mark
Defect on Wafer Backside

YangSub Park$^{(\boxtimes)}$, KilBum Kang, and SeongSoo Kim

Advanced Technology Inc, 112 Gaetbeol-Ro, Yeonsu-Gu, Incheon,
Republic of Korea
{yspark,kbkang,sskim}@ati2000.co.kr

Abstract. In this paper, we propose a method to inspect wheel mark defect on the wafer backside. The aim of this method is to detect wheel mark defects through back side film and back grinding tape on back side of the wafer. To reduce noise from both films, we used a vignetting correction to eliminate vignetting effects from line scan and Gaussian smoothing filter to reduce noises from back side films. Then, we used a Circle Curve Fitting to find the center point of the wafer and extracted the periodic feature in polar coordinates using the Fourier transform. And we also measured the noise signal on the background to calculate SNR and parameterize it. A sample test result shows that the proposed method is effective to control the quality of products in the factory.

Keywords: Wheel mark · Wafer backside · TSV

1 Introduction

Wafer backgrinding(Backlap) is an essential process in semiconductor device fabrication to reduce wafer thickness to allow for stacking and high density packaging of integrated circuits(IC) [1]. Usually, the active part of the die is just a few microns thick and the other part is silicon, which can be removed via the wafer backgrinding procedure [2]. However, this procedure has been faced with increasing challenges due to a reduction in die thickness in the process of incorporating multiple dies to make high-integration packages. Today, a die can be reduced to a 10-micron thickness with a cutting-edge wafer backgrinding system [3] while wafers thinned down to 30 to 50 μms are common now.

It is necessary to control the quality of surface backside; particularly, prevention of cracks is important to create through silicon via (TSV) 3D packages and TSV 3D integrated circuits [4, 5]. A broken die would spoil a package containing whole dies [6]. For the same reason, detecting a wheel mark signal is very important to monitor the grinding process and prevent cracks. A die with wheel marks has the potential to develop into the crack. But it is hard to detect them because the thinned wafer is attached to some film such as die attach film (DAF) with wafer ring frame. Unfortunately, the light transmission rate of these films is too low. For that reason, many manufacturers rely on sampling inspection by human eyes.

© Springer International Publishing AG 2017
M. Felsberg et al. (Eds.): CAIP 2017, Part I, LNCS 10424, pp. 219–227, 2017.
DOI: 10.1007/978-3-319-64689-3_18

Therefore, it is necessary to find a method to inspect wheel mark defect through backside film to ensure the quality of the backgrinding process.

2 Method

Although the ultra-thin wafer handling is very difficult, it is a widely acknowledged technology. Generally, a thin wafer having an edge support ring is called as metal film frame or wafer ring frame. A wafer ring frame is used to hold the tape and carry a wafer as non-contact handling. This kind of step is wafer mounting process. This step is necessary for not only wafer handling but also wafer dicing. After wafer backgrinding, die is separated from a wafer. There are many methods for wafer dicing such as mechanical sawing (dicing saw) and laser cutting.

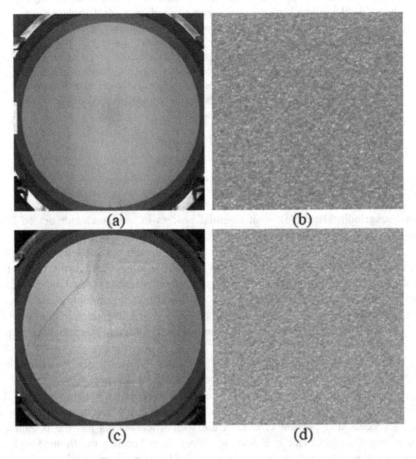

Fig. 1. (a): a normal wafer, (b): a close-up of surface of (a), (c): a wafer with wheel mark defect, (d): a close-up of wheel mark defect.

During dicing, a wafer is mounted on a dicing tape. The dicing tape has different properties depending on the dicing application. The dicing tape can be made of PVC, polyolefin, or polyethylene backing material with an adhesive [7]. And some tapes have UV curable properties. UV tapes are made to reduce adhesive strength by exposure to ultraviolet light after dicing. There is a wide variety of tapes and films to choose from. Unfortunately, most of them are not made for visible properties. The backside films have rough particles and opaque properties, which makes it hard to detect wheel mark defects by a visual system.

Figure 1(a) is an image of a good wafer, which was captured by the proposed system. The pixel resolution was 20 um. Figure 1(c) is an image with wheel mark defects. Figure 1(d) is an original scale image of the position of wheel mark defect. It is very hard to find defect signals because the contrast of defect is too low while the noise level is quite high. Figure 1(b) is an original scale image of normal position. There is no big difference between two images. This is the reason why detecting wheel mark defects through backside film is a challenge. Wheel mark defects appear like a wave from the center of the wafer. The interval in concentric circles between each wheel mark is almost the same. It is a key characteristic. We focused on this signature and here, we propose a method to find wheel mark signals in polar coordinates.

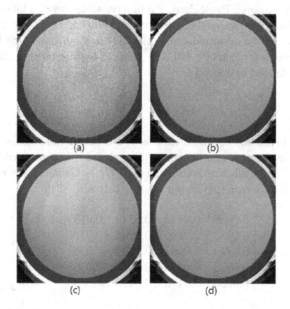

Fig. 2. (a): an original image, (b): a vignetting corrected image, (c): a image with Gaussian smoothing filter ($\sigma = 1$, kernel size = 5 × 5), (d): a vignetting corrected image with Gaussian smoothing filter

2.1   Pre-processing

We used a wide-line scanning camera to grab the image of the wafer at one time-point. The number of pixels is 16,000 and field of view(FOV) is 320 mm. The uniformity of illumination is important in this system because the wheel mark signal is too low while the noise level is too high. Unfortunately, it was impossible to get an image with good uniformity for full FOV. Therefore, vignetting correction is an essential part of pre-processing to improve wheel mark signals. To correct for illumination vignetting effects in an image, often the easiest way is to get an empty field in advance [8]. A key characteristic of illumination vignetting effects in our system is the same effects for each vertical line. To characterize a vertical vignetting, we used the average intensity of vertical line in the center to make an empty field.

Pixel greyscale values are defined as $I(x, y)$. $I_{mean}(x)$ is the mean intensity profile and I_{median} is the median value across the profile. The correction function is determined as follows

$$I_{correction}(x) = I_{median} - I_{mean}(x) \tag{1}$$

The vignetting-free image is $I_{correction(x,y)}$ as

$$I_{correction}(x, y) = I(x, y) + I_{correction}(x) \tag{2}$$

Figure 2(b) is vignetting-free image given by (2). Figure 2(b) has noise signals such as an electronic noise. Particularly, the film on wafer makes a lot of noise signals. Normally, a thin wafer is mounted on the back side film and DAF. The textures of films are rough. So, we need a noise reduction. The 2D Gaussian smoothing filter is widely used to reduce noises for years [9, 10]. The value of each element in the 2D Gaussian function can be defined by (3) [9]

$$G(x, y) = e^{\frac{-(x^2 + y^2)}{2\sigma^2}} \tag{3}$$

It is possible to use this 2D distribution as a point spread function for image and it is replaced by convolution. A noisy image is $T(x, y)$ and it is convolved with Gaussian function. $G(x, y)$ is impulse response function. G_1 is Gaussian smoothing filter of 5×5 kernel size with $\sigma = 1$. (varying x and y between -2 to 2)

$$G_1 = \begin{bmatrix} 0.0030 & 0.0133 & 0.0219 & 0.0133 & 0.0030 \\ 0.0133 & 0.0596 & 0.0983 & 0.0596 & 0.0133 \\ 0.0219 & 0.0983 & 0.1621 & 0.0983 & 0.0219 \\ 0.0133 & 0.0596 & 0.0983 & 0.0596 & 0.0133 \\ 0.0030 & 0.0133 & 0.0219 & 0.0133 & 0.0030 \end{bmatrix} \tag{4}$$

T_1 is an image matrix made by $I_{corrected}(x, y)$ then the convolution is given by:

$$Z = T_1 \otimes G_1 \tag{5}$$

Z is an image matrix of a corrected image from noise and vignetting. Figure 2(d) is a result image of it.

2.2   Finding Center of Wafer

It is necessary to find the exact center point of the wafer because defects of wheel mark generally occur from the center of the wafer. But wafers under 200 mm diameter have flat cuts and wafers of 200 mm diameter and above have a single small notch. Both are made to convey wafer orientation. And a thin wafer has an edge that is prone to cracks. These factors make it difficult to calculate the center point of the wafer. Therefore, a method of Circle Curve Fitting [11] is needed to find the exact center point of the wafer without error points.

To find the fine edge of the wafer by Circle Curve Fitting, at first, we need to detect a candidate circle. We searched for the edges from the center of the image to the edge of the image with the edge gradient direction. Finding the edge with every direction is the best way but it is good enough to find 0 and 90° only with changing position to reduce the complexity of calculating direction. W is a data set to represent wafer edge. The least square method is a widely used method to get the approximate solution of overdetermined systems [11]. The best fitting ellipse is defined as a collection of points (x, y) of satisfying the following implicit Eq. (7).

$$x^2 + ay^2 + bxy + cx + dy + e = 0 \tag{6}$$

To simplify the following analysis, we subtract x^2 from both sides of the equality sign, which can be reduced to

$$ay^2 + bxy + cx + dy + e = -x^2 \tag{7}$$

The analysis method of least square is a standard approach to the approximate solution of overdetermined systems. But it is complicated for some equations which have many parameters. So, we use an algebraic method using pseudo inverse method [12]. The matrix equation for (8) is (9)

$$\begin{pmatrix} y_1^2 & x_1y_1 & y_1 & x_1 & 1 \\ \vdots & \vdots & \vdots & \vdots & \vdots \\ y_n^2 & x_ny_n & y_n & x_n & 1 \end{pmatrix} \begin{pmatrix} a \\ b \\ c \\ d \\ e \end{pmatrix} = \begin{pmatrix} -x_1^2 \\ \vdots \\ -x_n^2 \end{pmatrix} \tag{8}$$

The pseudo inverse has the following properties.

$$X = pinv(A)B$$
$$= (A^T A)^{-1} A^T B \tag{9}$$

(8) can be solved by using pseudo inverse (9) [12]. Following this process, we were able to find the best fitting ellipse of the wafer and C is the center point of the wafer from that.

2.3   Extracting Periodic Feature in Polar Coordinates

Wheel mark defects have an arc shape from the center of the wafer to outside. And this defect is repeated on the wafer with similar intervals in a circle. A signal of individual defects is very weak, so it is difficult to detect with a typical surface inspection method. Therefore, we used the Fourier transform in polar coordinates to observe the periodic features and check that there are signals repeated at a specific frequency.

r is a radius of a wafer which was calculated in the previous step. K_r is a circle which is centered at C and has a radius of r. K_n is a concentric circle with K_r and has a radius of n. $\mathbb{R}_n$ is a data set of points on the circle K_n. $R_n(t)$ is an intensity of the image at the position which is given by (10)

$$R_n(t) = p(r \cdot \cos(t), r \cdot \sin(t)) + c \tag{10}$$

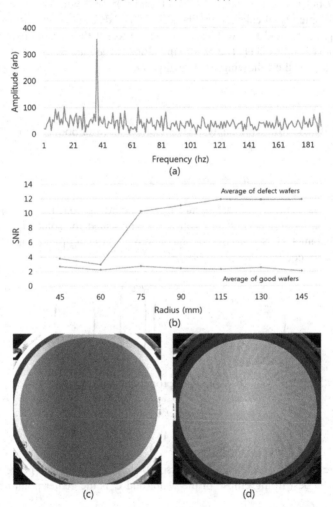

Fig. 3. (a): R_{75} of (d), (b): blue SNR of defect wafer and orange is SNR of good wafer, (c): good wafer for test, (d): defect wafer for test (color figure online)

We can get the amplitude in the frequency domain using Fourier transform. T_n is a result of Fourier transform of R_n. R_n is a data set of points on concentric circles with the wafer with different radius. So, T_n means a result of Fourier transform of points on concentric circles with a radius of n.

Figure 3(a) is a result of Fourier transform of R_{75} (the radius is 75 mm). S_n is a maximum value of T_n. S_n means a signal of wheel mark. N_n is a noise signal of T_n. The signal of wheel mark only exists in low frequency. T_{nh} is a high frequency part of T_n. (over 150 Hz) N_n is a mean value of T_{nh}. N_n means a noise signal of background. We can analyze wheel mark defects by calculating Signal to Noise Ratio (SNR)

$$S_n = MAX(T_n(x))$$
$$N_n = MEAN(T_n(x)) \ (if \ x > 150hz) \tag{11}$$

$$SNR_n = \frac{S_n}{N_n} \tag{12}$$

Figure 3(b) shows that the wheel mark signal at 75 mm position in defect wafer is 10.24. To calculate SNR for the whole wafer, we measure SNR from 45 mm to 145 mm at intervals of 15 mm. A SNR of a defect sample wafer is 9.08 and a SNR of the normal wafer is 2.41. The result shows that the proposed method amplifies the wheel mark signals to distinguish between the good and defects.

3 Experiment and Result Analysis

Today, most production lines for frontend processes are fully automated. In contrast to frontend process, the backend process is not yet automated enough. Products are still carried by people and examined by human eyes in some backend lines. Because there is no proper inspection equipment to inspect many kinds of defects, sometimes human inspectors make different judgments about the same defects. As the standards for judgment are always changing according to person and time, critical and huge defects are often overlooked. In addition, new defects may occur during human handling of the products. In order to solve these problems, we made the first inspection application to the production line to inspect backside of the wafer with a consistent condition.

In order to test its reliability for cases of mass production, we made an inspection zone in Equipment Front End Module (EFEM). EFEM is used to transfer wafers in a clean environment. It is located in front of the process modules and passes wafers between the carriers such as FOUP and FOSB. Figure 4 is a blueprint of our system. The purpose of EFEM is to transfer wafers from FOUP to process module using Wafer Transfer Robot (WTR). We integrated an inspection module into EFEM to eliminate the need for additional steps and space. This EFEM, which gets the back side inspection zone has the same throughput as the original version. This is an advantage in terms of efficiency.

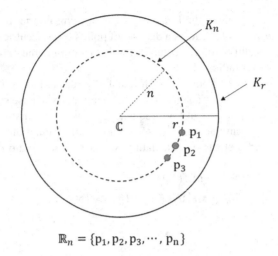

$$\mathbb{R}_n = \{p_1, p_2, p_3, \cdots, p_n\}$$

Fig. 4. A diagram about the data set of concentric circles in a wafer

We tested a variety of defect sample wafers for a while. Actually, the inspection module was made to inspect many kinds of defects. The number of wheel mark defects found during the evaluation period was only 13. The lack of samples was a big factor that made this problem difficult. The number of other defects found the same period was 103. We compared these two groups. Figure 5 is the result of the evaluation. The average SNR of defects is 7.51 and the average SNR of the normal wafer is 2.80. The maximum value of SNR of the normal wafer is 3.4. And standard deviation of it is 0.34. This means that only the wheel mark signal could be amplified by the proposed method.

The results from this application to a mass-production line resulted in the satisfactory management of backgrinding wheel defects. After some refinements, this method can contribute for the engineering of backend packaging process.

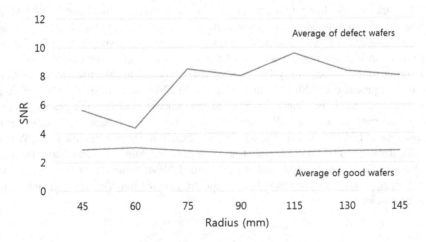

Fig. 5. A result of evaluation

4   Conclusions

To detect wheel mark defects on the back side of the wafer through films, we reduced the noise from backside films using vignetting correction and Gaussian smoothing. And we extracted the edge of the wafer and found the center point of the wafer by Circle Curve Fitting using a pseudo inverse method. And we got an intensity of points, which are on the concentric circle with the wafer. Then we measured a signal of wheel mark using Fourier transform in polar coordinates. And we also measured a noise signal, which represents the background. We used both signals to define SNR of wheel mark. The average SNR of defect wafers is 7.51 and the average SNR of good wafers is 2.80. The ratio of defects to normal is 2.68. The test result showed that the proposed method to find wheel mark signal in polar coordinates seems to have enough potential for applications to the actual production of products in factories.

References

1. Patti, R.S.: Three-dimensional integrated circuits and the future of system-on-chip designs. Proc. IEEE **94**(6), 1214–1224. doi: 10.1109/JPROC.2006
2. Wikipedia: Wafer backgrinding, the free encyclopedia 2016. Accessed 19 Jan 2017. https://en.wikipedia.org/wiki/Wafer_backgrinding
3. Micross components: What is the difference between wafer thinning, wafer backgrinding and wafer backlapping?. Accessed 12 Jan 2017. http://www.micross.com/articles-what-is-wafer-backgrinding.aspx
4. Topol, A.W., Jr. La Tulipe, D.C., Shi, L., Frank, D.J., Bernstein, K., Steen, S.E., Kumar, A., Singco, G.U., Young, A.M., Guarini, K.W., Ieong, M.: Three-dimensional integrated circuits. IBM J. Res. Develop. **50**(4/5), 491–506 (2006). doi:10.1147/rd.504.0491
5. Chong, D.Y.R., Lee, W.E., Pang, J.H.L., Low, T.H., Lim, B.K.: Mechanical characterization in failure strength of silicon dice. In: Proceedings of the 5th EPTC 2003. doi:10.1109/EPTC.2003.1271591
6. Wikipedia: Dicing tape, the free encyclopedia (2016). Accessed 24 Jan 2017. https://en.wikipedia.org/wiki/Dicing_tape
7. Gareau, D., Patel, Y., Li, Y., Aranda, I., Halpern, A., Nehal, K., Rajadhyaksha, M.: Confocal mosaicing microscopy in skin excisions: a demonstration of rapid surgical pathology. J. Microscopy **233**(1), 149–159 (2009). doi:10.1111/j.1365-2818.2008.03105.x
8. Hsiao, P.-Y., Chen, C.-H., Chou, S.-S., Li, L.-T., Chen, S.-J.: A parameterizable digital-approximated 2D gaussian smoothing filter for edge detection in noisy image. In: Proceedings of the 2006 IEEE International Symposium on Circuits and Systems. ISCAS 2006. p. 4 May 2006. doi:10.1109/ISCAS.2006.1693303
9. Basu, M.: Gaussian-based edge-detection methods-a survey. IEEE Trans. Syst. Man Cybern. c **32**, 252–260 (2002). doi:10.1109/TSMCC.2002.804448
10. Gander, W., Golub, G.H., Strebel, R.: Least-squares fitting of circles and ellipse, 1994 unpublished
11. Moore, E.H.: On the reciprocal of the general algebraic matrix. Bull. Am. Math. Soc. **26**(9), 394–395 (1920)

Retinal Vessel Segmentation Using Matched Filter with Joint Relative Entropy

Mohsin Challoob$^{(\boxtimes)}$ and Yongsheng Gao

School of Engineering, Griffith University, Nathan, Australia
mohsin.challoob@griffithuni.edu.au,
yongsheng.gao@griffith.edu.au

Abstract. The matched filter is an effective method for the detection of retinal vessels when combined with other processing techniques. This paper presents a segmentation method to improve the extraction of retinal vessels based on the matched filter. The method combines a morphological approach to enhance retinal vessels before applying the matched filter and a modified joint relative entropy (MJRE) thresholding method to segment the matched filter response. The morphological approach is designed to suppress irregular bright regions and noise while preserving the information of vessel edges, and to improve the contrast of vessels, especially thin ones. The joint relative entropy thresholding is modified to provide an optimal threshold value for segmenting the retinal vessel tree properly. The proposed method is tested on the DRIVE dataset, yielding an average accuracy, specificity and sensitivity of 0.9546, 0.9742 and 0.7527 respectively. Experimental results demonstrate that the proposed method achieved better performance than the state-of-the-art methods.

Keywords: Retinal vessel · Mathematical morphology · Matched filter · Joint relative entropy

1 Introduction

Retinal blood vessels have been widely used in the medical society for pathological diagnoses. The physical appearance of blood vessels plays an important role in the diagnosis and treatment of many diseases such as hypertension, glaucoma, arteriosclerosis and diabetic retinopathy. Detecting blood vessels in retinal image is performed manually in some cases. The manual segmentation of retinal blood vessels is a time-consuming process and requires remarkable skills [1, 2]. Hence, an automated segmentation of retinal blood vessels using computer algorithms would be highly desirable for medical diagnoses. However, the development of an automated method for retinal vessel segmentation faces several challenges such as low intensity contrast between thin vessels and the background, and the presence of noise and other retinal structures [1]. Many methods have been presented in the literature [1, 3–14] for the segmentation of blood vessels from retinal image. These methods have achieved great progress in the segmentation of retinal vessels. However, the following aspects are required to be improved for better segmentation results.

© Springer International Publishing AG 2017
M. Felsberg et al. (Eds.): CAIP 2017, Part I, LNCS 10424, pp. 228–239, 2017.
DOI: 10.1007/978-3-319-64689-3_19

- Some of the enhanced methods are associated with one or both of the following two problems. The first problem is that noise is exaggerated when the contrast of vessels is improved. The second is losing small and weak vessels, or some part of a vessel, when noise is removed. These problems are because the intensity contrast between vessels, especially thin ones, and background is relatively low. It is important to use an enhanced method that can reduce the effects of these problems on the vessel detection.
- False detection might result from presenting irregular bright regions such as exudates and central light reflexes. Exudates can have higher contrast and be misclassified as vessel pixels. Central light reflex may present as a bright strip along the center of a wider vessel, which leads to detecting the vessel as two small ones. Also, when these bright regions are removed, some vessel pixels are misclassified as background. The segmentation method should be concerned with the influence of these regions.
- A thresholding method should provide the optimal value that distinguishes vessel pixels from background pixels properly. At the same time, the resultant retinal vessels should be linked in order to provide a good visual appearance. This is crucial to enable the underlying shape of vessels to be identified.

The contribution of this paper is to improve the mentioned aspects for obtaining better segmentation results. This is achieved by combining a morphologically enhanced approach and a modified joint relative entropy (MJRE) thresholding method with the matched filter. The morphological approach is applied before using the matched filter to remove noise and irregular bright regions, preserve the information of vessel edges, and improve the contrast of vessels. The joint relative entropy thresholding is modified in a way to provide an optimal threshold value for extracting the retinal vessel tree from the matched filter response properly. Also, the MJRE thresholding method can provide a better threshold value than other thresholding methods used to segment the matched filter response.

The rest of the paper is organized as follows. Section 2 presents the proposed vessel segmentation method. Results are discussed in Sect. 3 and conclusion is given in Sect. 4.

2 The Proposed Vessel Segmentation Method

The proposed vessel segmentation method includes several operations. The green channel of the retinal image is firstly extracted for its higher contrast than the other channels (red and blue). Then, a morphological approach is applied to the green channel for the removal of noise and irregular bright regions, and vessel contrast enhancement while preserving the information of vessel edges. The matched filter is the next operation to detect retinal vessels in the morphologically enhanced image. After that, the matched filter response is segmented by a modified joint relative entropy thresholding method. The final step is area filtering operation to suppress small unwanted and isolated regions that do not belong to the vessel tree.

2.1 Morphological Approach

Dilation and erosion are the basic morphological operations that are used for detecting, modifying and manipulating features presented in an image based on their shapes. The dilation and erosion of an image I by a structuring element SE are defined as $I \oplus SE$ and $I \ominus SE$ respectively. Other morphological operations such as opening and closing are built by performing a series of dilation and erosion operations. The opening of an image I by a structuring element SE (denoted by $I \circ SE$) is an erosion operation followed by a dilation operation. The closing of an image I by a structuring element SE (denoted by $I \bullet SE$) is a dilation followed by an erosion [15, 16].

The green channel of retinal image includes the vessel structure, optic disk, heavy background noise and occasionally irregular bright regions. As the intensity contrast between vessels, especially narrow and small ones, and background is relatively lower, the enhancement of vessel structure is associated with problems of exaggerating noise when improving contrast, and losing some vessel pixels when removing noise. Irregular bright regions might also appear in the retinal image because of the presence of exudates or the reflection of artefacts, which can result in false detection. Exudate lesions are higher intensity pixels that may complicate the detection process of vessels and present in the final segmented vessel tree. The central light reflex may present as a bright strip along the center of a wider vessel. This strip can result in detecting a vessel as two small ones when it is strong. Moreover, if two vessels lie two close to each other, they might be merged in the detection process as one wider vessel. Therefore, there is a need to design an enhanced method that can reduce the effects of these problems on the vessel detection. This can be achieved by an opening-by-reconstruction operation and a bottom hat transformation.

The opening-by-reconstruction operation is applied to the green channel to remove irregular bright regions, and reduce the risk of merging close vessels together and missing small and narrow vessels. Applying the opening-by-reconstruction involves two steps. The first one, the green channel is eroded to suppress bright regions that have a smaller size than the structuring element. The next step is reconstructing image by dilating eroded image iteratively to restore the contours of components that have not been completely removed. The eroded image is used as the marker image and the green channel is the mask image for reconstruction. Assume Gr be the green channel of a retinal image and SE be a structuring element, the opening-by-reconstruction operation is defined as:

$$Fr = R_G(Gr \ominus SE) \qquad (1)$$

where Fr is the resultant image from opening-by-reconstruction operation (reconstructed image). The notation $\ominus$ denotes the opening-by-reconstruction operation.

Bottom hat transformation can improve the contrast of dark objects on a light background. It extracts dim regions that are smaller than the structuring element [16]. Therefore, since retinal blood vessels appear darker compared to the background, this transformation can enhance the contrast of retinal vessels. Bottom hat transformation is built by subtracting input image from its closing, where the input image is the image Fr. The bottom hat transformation can be expressed as

$$Bd = ((Fr \bullet SE) - Fr) \tag{2}$$

where Bd is the resultant image having dim regions. Then, these dim regions (Bd) are subtracted from the reconstructed image (Fr) to improve the contrast of vessels and remove small components of noise that are associated with vessels. The enhanced image is given as:

$$I_{enhanced} = Fr - Bd \tag{3}$$

where $I_{enhanced}$ is the enhanced image. By using Eqs. (1) and (2) in Eq. (3), the morphologically enhanced image can be rewritten as:

$$I_{enhanced} = (R_G(Gr \ominus SE)) - \{((R_G(G_r \ominus SE)) \bullet SE) - (R_G(Gr \ominus SE))\} \tag{4}$$

Moreover, selecting the appropriate structuring element (SE) is important into the performance of the used morphological approach. Since the vessels have uncertain direction and are symmetric, the disk structuring element is the optimal type to be used. Its radius is determined experimentally and set to 4 to preserve the information of major vessels when noise and bright regions are removed.

2.2 Matched Filter

The matched filter (MF) is a template matching method which is used in the detection of blood vessels in retinal images. The MF was firstly designed by Chaudhuri et al. [9] based on the assumption that the cross-sectional intensity profile of a retinal vessel might be approximated by the Gaussian shaped curve. The two-dimensional matched filter kernel is built based on the Gaussian function to detect blood vessels in the enhanced image ($I_{enhanced}$). According to [9], such a kernel can be expressed as:

$$K(x, y) = -e^{\left(\frac{-x^2}{2\sigma^2}\right)} \ \forall |y| \leq \frac{Ls}{2} \tag{5}$$

where σ is the spread of the intensity profile and Ls is the length of the vessel segment assumed to have a fixed orientation. The values of Ls and σ are set to 9 and 1.2 respectively. To detect retinal vessel in all possible orientations, the kernel is rotated from (0° to 180°) by angular resolution of 15° using the rotation matrix, which is shown in Eq. (6). Then, a set of twelve kernels with a size of 16 × 15 is convolved with the enhanced image and only the maximum response is taken for each pixel.

$$r_i = \begin{bmatrix} \cos \theta_i & -\sin \theta_i \\ \sin \theta_i & \cos \theta_i \end{bmatrix} \tag{6}$$

Once the MF response is obtained, it is multiplied by a mask. The mask is used to label pixels that belong to the region of interest (ROI) in the retinal image, removing pixels outside the ROI. The mask is generated by two steps. The first step is thresholding the green channel; in the next, erosion operation with a disk structuring element of radius 3 is subsequently applied to the thresholded image. Let Ms be a mask, the matched filter response (MF) may be defined as:

$$MF = CONV(I_{enhanced}, K) \otimes Ms \tag{7}$$

where $\otimes$ is the notation for element-wise multiplication.

2.3   Modified Joint Relative Entropy Thresholding

Grey level co-occurrence matrix (GLCM) is one of a widely used feature extraction techniques in image processing. It was introduced by Haralick et al. [17] to acquire the spatial dependence of grey level values. The values of GLCM demonstrate related frequencies of pij in which two neighbouring pixels with constant distance of d, one of them with grey level of i and the other with grey level of j occur on the image [18]. Assume an image of size $M \times N$ with L grey levels denoted by $G = \{0,1,2,...., L-1\}$ and consider $f(x, y)$ be the grey level of the pixel at the spatial location (x, y). Then, the image can be defined as $F = [f(x, y)]_{MXN}$, where $f(x, y) \in G$. A co-occurrence matrix of an image is an $L \times L$ square matrix, which is denoted by $W = [t_{ij}]_{LXL}$, where t_{ij} represents the number of transitions from grey level i to grey level j. In other words, each entry in the matrix t_{ij} indicates the number of times that the pixel grey level j follows the grey level i in some pattern [11, 19, 20]. The co-occurrence matrix of the MF response is computed to acquire the special distribution of the grey levels of retinal vessels. In this paper, the value of t_{ij} is calculated as follows, where the angle between two pixels is considered as $0°$ and the distance is set to 1.

$$t_{ij} = \sum_{m=1}^{M} \sum_{n=1}^{N} \begin{cases} 1 & if\, f(m,n) = i\, and\, f(m,n+1) = j \\ 0, & otherwise \end{cases} \tag{8}$$

The desired transition probability P_{ij} from grey level i to j can be obtained by normalizing the total number of transitions in the co-occurrence matrix as:

$$P_{ij} = t_{ij}/(\sum_{k=0}^{L-1} \sum_{l=0}^{L-1} t_{kl}) \tag{9}$$

Assume t be the value used to threshold an image. Then, the co-occurrence matrix is divided by t into four quadrants: A, B, C and D. The quadrants A and C represent grey level transitions within background and foreground respectively. The quadrants B and D represent the grey level transitions across the boundaries of background and foreground. In other words, these quadrants (B and D) include edge information on transitions from background to foreground and foreground to background. The four quadrants are grouped into two classes. The first class is local quadrants that are A and C. The second class is B and D, which are known as joint quadrants [19–21]. In this work, the joint quadrants (B and D) are only considered. Figure 1 shows an example to calculate the co-occurrence matrix and its quadrants of an image, where t is set to 2.

a

1	1	5	6	8
2	3	5	7	1
4	5	7	1	2
8	5	1	2	5

b

		1	2	3	4	5	6	7	8	
1		1	2	0	0	1	0	0	0	
2	A	0	0	1	0	1	0	0	0	B
3		0	0	0	0	1	0	0	0	
4		0	0	0	0	1	0	0	0	
5		1	0	0	0	0	1	2	0	
6	D	0	0	0	0	0	0	0	1	C
7		2	0	0	0	0	0	0	0	
8		0	0	0	0	1	0	0	0	

Fig. 1. (a) An original image, (b) the co-occurrence matrix.

The probability of each quadrant can be defined as follows [21]:

$$P_A^t = \sum_{i=0}^{t} \sum_{j=0}^{t} P_{ij} \qquad P_B^t = \sum_{i=0}^{t} \sum_{j=t+1}^{L} P_{ij}$$
$$P_C^t = \sum_{i=t+1}^{L-1} \sum_{j=t+1}^{L-1} P_{ij} \qquad P_D^t = \sum_{i=t+1}^{L-1} \sum_{j=0}^{t} P_{ij} \tag{10}$$

The relative entropy has been used to measure the information distance between two probability distributions. The relative entropy is smaller when the two probability distributions are closer to each other. Assume two sources with L grey levels be described by probability distributions (p and h). The relative entropy between p and h is defined by

$$J(p; h) = \sum_{j=0}^{L-1} p_j \log \frac{p_j}{h_j} \tag{11}$$

The entropy $J(p; h)$ is computed as h relative to p, where p is considered as the original image, while h is the processed image which tries to match p [19, 21]. A second order joint relative entropy is defined as:

$$J\left(\{p_{ij}\}; \left\{h_{ij}^t\right\}\right) = \sum_{i=0}^{L-1} \sum_{j=0}^{L-1} p_{ij} \log \frac{p_{ij}}{h_{ij}^t} \tag{12}$$

Then, the conditional probabilities of the quadrants B and D are expressed as:

$$h_{ij|B}^t = q_B^t = \frac{P_B^t}{(t+1)(L-t-1)} \qquad h_{ij|D}^t = q_D^t = \frac{P_D^t}{(L-t-1)(t+1)} \tag{13}$$

This paper utilizes the studies in [19, 21, 22] to propose a modified thresholding method. The modified joint relative entropy (MJRE) thresholding method for segmenting the MF response is designed as follows:

(1) The probabilities of the quadrants B and D (P_B^t and P_D^t) are calculated as shown in Eq. (10):
(2) The probabilities of these quadrants (B and D) are multiplied by two constant values ($f1$ and $f2$ respectively) as follows.

$$P_B^{\prime t} = P_B^t * f1 \qquad P_D^{\prime t} = P_D^t * f2 \tag{14}$$

where $f1$ and $f2$ are experimentally examined and set to 0.168 and 0.01 respectively.

(3) The conditional probabilities of the joint quadrants are obtained as:

$$\hat{h}^t_{ij|B} = q'^t_B = \frac{P'^t_B}{(t+1)(L-t-1)} \quad \hat{h}^t_{ij|D} = q'^t_D = \frac{P'^t_D}{(L-t-1)(L+1)} \tag{15}$$

Note that the term $(L+1)$ is used in Eq. (15) rather than the term $(t+1)$, which is used in Eq. (13) when calculated the q'_D.

(4) The maximum entropy is taken instead of the minimum one as the optimal threshold value.

$$t'_{jre} = arg\left[\max_{t\in L}\hat{H}_{jre}(t)\right] \tag{16}$$

where $\hat{H}_{jre}(t) = -(P'^t_B \log q'^t_B + P'^t_D \log q'^t_D)$. The value t'_{jre} is the threshold value used to segment retinal vessels from the MF response. The segmented image contains small unwanted and isolated regions caused by noise and pathological changes. These regions are wrongly classified as vessels. They can be removed based on the connectivity of retinal vessels. Each region connected to an area less than 40 pixels is removed and reclassified as background.

3 Results and Discussion

The proposed method is tested on a well-known database, DRIVE. This database includes 40 retinal images, which were taken from a diabetic retinopathy screening program in the Netherlands. The images were captured by a Canon CR5 3CCD camera with a FOV (field of view) of 45°, with a resolution of 565*584 pixels and 8 bits per color channel. The database is divided into two sets: a training set and test set, each consisting of 20 images. They were manually segmented by three observers trained by an experienced ophthalmologist. The images in the test dataset were segmented twice by two different observers, resulting into two sets: A and B [1, 2]. The performance of the proposed method is only evaluated on the test dataset using the segmented images in the set A (1st_manual) as a ground truth. To evaluate the performance of the proposed method, sensitivity, specificity and accuracy measures are computed as follows.

$$Sensitivity = TP/(TP+FN) \tag{17}$$

$$Specificity = TN/(TN+FP) \tag{18}$$

$$Accuracy = (TP+TN)/(TP+TN+FP+FN) \tag{19}$$

where TP (True Positive) represents the number of correctly classified vessel pixels. TN (True Negative) is the number of correctly classified non-vessel pixels. FP (False Positive) represents the number of non-vessel pixels incorrectly classified as vessel. FN (False Negative) is the number of vessel pixels incorrectly classified as non-vessel. Sensitivity shows the ability of the proposed method to detect the vessel pixels. Specificity demonstrates the ability to detect the non-vessel pixels. Accuracy represents the ratio of the correctly detected vessel and non-vessel pixels to the number of pixels in the image FOV [2]. Figure 2 shows the segmentation results of the proposed method. Sensitivity, specificity and accuracy measures are calculated for each retinal image in the test dataset as shown in Table 1.

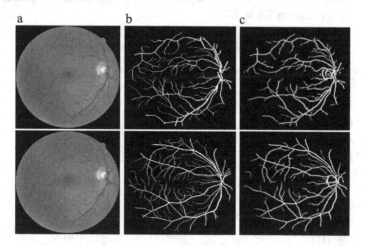

Fig. 2. (a) The green channel, (b) the ground truth, (c) segmentation results by the proposed method.

Table 1. Segmentation results on the DRIVE test dataset.

Image	Sensitivity	Specificity	Accuracy
1	0.8222	0.9688	0.9558
2	0.8270	0.9659	0.9517
3	0.6710	0.9839	0.9527
4	0.7782	0.9723	0.9545
5	0.6976	0.9866	0.9595
6	0.6497	0.9870	0.9542
7	0.7577	0.9655	0.9465
8	0.6923	0.9737	0.9495
9	0.6925	0.9837	0.9601
10	0.7125	0.9826	0.9604
11	0.7595	0.9578	0.9400
12	0.7403	0.9774	0.9570
13	0.7212	0.9774	0.9524
14	0.8003	0.9644	0.9511
15	0.8423	0.9519	0.9441
16	0.7668	0.9765	0.9576
17	0.7382	0.9772	0.9570
18	0.7889	0.9746	0.9599
19	0.8257	0.9791	0.9664
20	0.7700	0.9768	0.9616
Avg.	**0.7527**	**0.9742**	**0.9546**

As shown in Table 1, the maximum accuracy, specificity and sensitivity are 0.9664, 0.9870 and 0.8423 respectively. In contrast, the minimum results are 0.9400, 0.9519 and 0.6497 respectively. The average accuracy, specificity and sensitivity of the proposed method are 0.9546, 0.9742 and 0.7527 respectively when applied on the test dataset (20 images). For comparative purposes, the average accuracy, specificity and sensitivity are reported in Table 2.

Table 2. Comparison of different vessel segmentation methods on the DRIVE test dataset.

Method	Average accuracy	Average specificity	Average sensitivity
Zhao et al. [1]	0.9477	0.9789	0.7354
Marín et al. [3]	0.9452	0.9801	0.7067
You et al. [4]	0.9434	0.9751	0.7410
Soares et al. [5]	0.9446	0.9762	0.7230
Lam et al. [6]	0.9472	–	–
Nguyen et al. [7]	0.9407	–	–
Fraz et al. [8]	0.9430	0.9768	0.7152
Classical MF [9]	0.8773	–	–
Kande et al. [10]	0.9437	–	–
Singh et al. [11]	0.9459	0.9721	0.6735
Kumar et al. [12]	0.9626	–	0.7006
Zhang et al. [13]	0.9382	–	0.7120
Odstrcilik et al. [14]	0.9340	0.9693	0.7060
The proposed method	**0.9546**	**0.9742**	**0.7527**

The performance of the proposed method is compared with 13 different segmentation methods in Table 2 using the same dataset in terms of average accuracy, specificity and sensitivity. These methods are based on region growing [1], supervised learning [3–5], modelling [6], multi-scale line detection [7], morphological bit plane slicing [8], and matched filter [9–14] methods. The matched filter studies [9–14] used different thresholding methods to segment the MF response. The vessel tree of the classical MF [9] and Kumar et al. [12] methods were obtained by filtering. Kande et al. [10] and Singh et al. [11] used the local entropy thresholding. Zhang et al. [13] and Odstrcilik et al. [14] applied a global thresholding and the Kittler minimum error thresholding methods respectively. The segmentation results of Soares et al. [5] and classical MF [9] methods are obtained from the method [1, 3] respectively. The results of the other methods are taken from their original studies.

As shown in Table 2, the proposed method obtains a better performance than other vessel segmentation methods. It achieves a higher average sensitivity than the reported methods, which means that the proposed method detects more vessel pixels than them. In terms of average accuracy, the proposed method is higher than the other methods except the method in [12]. The method [12] produced an average accuracy of 0.9626, while the average accuracy of the proposed method is 0.9546. However, the method [12] misclassified some vessel pixels as non-vessel and increased the FN. This resulted in an average sensitivity of 0.7006, which is lower than the sensitivity of the proposed method (0.7527). Regarding specificity, the methods [1, 3–5, 8] produced higher specificity than the proposed method. Nevertheless, the proposed method obtains higher accuracy and sensitivity than these methods, and the specificity of the proposed method is also higher (0.9742).

According to the experimental results, the proposed segmentation method provides encouraging results and a better performance compared to the other vessel segmentation methods. The proposed method is able to segment both thin and wide vessels. Suppressing noise and irregular bright regions while preserving the information of vessel edges results in making the contrast between vessels and the background relatively higher, helping to extract more vessel pixels, and decreasing false detection. Designing the matched filter kernel with Ls of 9 and σ of 1.2 contributes to detecting different vessel sizes and producing better detection results. Also, using the MJRE thresholding method for segmenting the MF response provides a better threshold value than the local entropy thresholding, the global thresholding and the Kittler minimum error methods, which were also used to segment the MF response in the literature. This leads to increasing the TP and TN results, decreasing the FN and FP results, and producing a well linked structure with more vessels. The parameter $f1$ in the used thresholding method can be set between the range of (0.167-0.173). It is set to 0.168 to produce better thresholding results on the DRIVE database. It can be adjusted to provide an optimal threshold value when applied to other databases. The proposed method does not require any training data for the extraction of vessels. It is an unsupervised method. Regarding the execution time, the proposed method runs on a PC Intel ® Xeon® 3.10 GHz CPU and 4 GB RAM using MATLAB software. It takes 1.14 s to segment a DRIVE retinal image. The main drawback of the proposed method is that false detection can occur around the optic disk and in some pathological regions.

4 Conclusion

This paper proposes a segmentation method by combining a morphological approach and a modified joint relative entropy thresholding method with the matched filter. The morphological approach is applied to the green channel to suppress noise and irregular bright regions, to preserve the information of vessel edges, and to improve the contrast between the vessels and background. It is built using an opening-by-reconstruction operation and a bottom hat transformation. The opening-by-reconstruction operation is used to remove irregular bright areas, and reduce the risk of merging close vessels together and missing of small and fine vessels. The bottom hat transformation removes noise and improves the contrast of vessels, especially thin ones. This is achieved by subtracting small components which are smaller than the structuring element from the reconstructed image. The matched filter is then applied to the enhanced image. The MF response is segmented by the proposed MJRE thresholding method. The MJRE thresholding method can provide a better threshold value for extracting retinal vessels from the MF response than the other thresholding methods used to segment the MF. The proposed segmentation method is tested on the DRIVE database, producing an average accuracy, specificity and sensitivity of 0.9546, 0.9742 and 0.7527 respectively. The method obtains encouraging results and a better performance compared to the existing segmentation methods. Combining the proposed morphological approach and the MJRE thresholding method with the matched filter leads to detecting both thin and wide vessels with a well-connected vessel structure. The main disadvantage of the proposed method is its false detection around the optic disk and in some pathological regions. The focus for future work is implementing a segmentation method that can avoid the false detection produced from these regions.

Acknowledgment. The first author would like to thank the Higher Committee for Education Development in Iraq (HCED) for providing scholarship.

References

1. Zhao, Y., Wang, X., Wang, X., Shih, F.: Retinal vessels segmentation based on level set and region growing. Pattern Recogn. **47**, 2437–2446 (2014). doi:10.1016/j.patcog.2014.01.006
2. Fraz, M., Remagnino, P., Hoppe, A., Uyyanonvara, B., Rudnicka, A., Owen, C., Barman, S.: Blood vessel segmentation methodologies in retinal images – a survey. Comput. Meth. Programs Biomed. **108**, 407–433 (2012). doi:10.1016/j.cmpb.2012.03.009
3. Marín, D., Aquino, A., Gegundez-Arias, M., Bravo, J.: A new supervised method for blood vessel segmentation in retinal images by using gray-level and moment invariants-based features. IEEE Trans. Med. Imaging **30**, 146–158 (2011). doi:10.1109/tmi.2010.2064333
4. You, X., Peng, Q., Yuan, Y., Cheung, Y., Lei, J.: Segmentation of retinal blood vessels using the radial projection and semi-supervised approach. Pattern Recogn. **44**, 2314–2324 (2011). doi:10.1016/j.patcog.2011.01.007
5. Soares, J., Leandro, J., Cesar, R., Jelinek, H., Cree, M.: Retinal vessel segmentation using the 2-D Gabor wavelet and supervised classification. IEEE Trans. Med. Imaging **25**, 1214–1222 (2006). doi:10.1109/tmi.2006.879967
6. Lam, B., Gao, Y., Liew, A.: General retinal vessel segmentation using regularization-based multiconcavity modeling. IEEE Trans. Med. Imaging **29**, 1369–1381 (2010). doi:10.1109/tmi.2010.2043259
7. Nguyen, U., Bhuiyan, A., Park, L., Ramamohanarao, K.: An effective retinal blood vessel segmentation method using multi-scale line detection. Pattern Recogn. **46**, 703–715 (2013). doi:10.1016/j.patcog.2012.08.009
8. Fraz, M., Barman, S., Remagnino, P., Hoppe, A., Basit, A., Uyyanonvara, B., Rudnicka, A., Owen, C.: An approach to localize the retinal blood vessels using bit planes and centerline detection. Comput. Meth. Programs Biomed. **108**, 600–616 (2012). doi:10.1016/j.cmpb.2011.08.009
9. Chaudhuri, S., Chatterjee, S., Katz, N., Nelson, M., Goldbaum, M.: Detection of blood vessels in retinal images using two-dimensional matched filters. IEEE Trans. Med. Imaging **8**, 263–269 (1989). doi:10.1109/42.34715
10. Kande, G., Savithri, T., Subbaiah, P.: Retinal vessel segmentation using local relative entropy thresholding. In: 2008 IEEE International Conference on Systems, Man and Cybernetics. pp. 3448–3453. IEEE (2008). doi:10.1109/ICSMC.2008.4811831
11. Singh, N., Kumar, R., Srivastava, R.: Local entropy thresholding based fast retinal vessels segmentation by modifying matched filter. In: International Conference on Computing, Communication & Automation. pp. 1166–1170. IEEE (2015). doi:10.1109/CCAA.2015.7148552

12. Kumar, D., Pramanik, A., Kar, S., Maity, S.: Retinal blood vessel segmentation using matched filter and laplacian of gaussian. In: 2016 International Conference on Signal Processing and Communications (SPCOM). pp. 1–5. IEEE (2016). doi:10.1109/SPCOM. 2016.7746666

13. Zhang, B., Zhang, L., Zhang, L., Karray, F.: Retinal vessel extraction by matched filter with first-order derivative of Gaussian. Comput. Biol. Med. **40**, 438–445 (2010). doi:10.1016/j. compbiomed.2010.02.008

14. Odstrcilik, J., Kolar, R., Kubena, T., Cernosek, P., Budai, A., Hornegger, J., Gazarek, J., Svoboda, O., Jan, J., Angelopoulou, E.: Retinal vessel segmentation by improved matched filtering: evaluation on a new high-resolution fundus image database. IET Image Process. **7**, 373–383 (2013). doi:10.1049/iet-ipr.2012.0455

15. Hassan, G., El-Bendary, N., Hassanien, A., Fahmy, A., Shoeb, A., Snasel, V.: Retinal blood vessel segmentation approach based on mathematical morphology. Procedia Comput. Sci. **65**, 612–622 (2015). doi:10.1016/j.procs.2015.09.005

16. Liao, M., Zhao, Y., Wang, X., Dai, P.: Retinal vessel enhancement based on multi-scale top-hat transformation and histogram fitting stretching. Opt. Laser Technol. **58**, 56–62 (2014). doi:10.1016/j.optlastec.2013.10.018

17. Haralick, R., Shanmugam, K., Dinstein, I.: Textural features for image classification. IEEE Trans. Syst. Man Cybern. **SMC-3**, 610–621 (1973). doi:10.1109/tsmc.1973.4309314

18. Rahebi, J., Hardalaç, F.: Retinal blood vessel segmentation with neural network by using gray-level co-occurrence matrix-based features. J. Med. Syst. **38**, (2014). doi:10.1007/s10916-014-0085-2

19. Yang, C., Ma, D., Chao, S., Wang, C., Wen, C., Lo, C., Chung, P., Chang, C.: Computer-aided diagnostic detection system of venous beading in retinal images. Opt. Eng. **39**, 1293–1303 (2000). doi:10.1117/1.602487

20. Villalobos-Castaldi, F., Felipe-Riverón, E., Sánchez-Fernández, L.: A fast, efficient and automated method to extract vessels from fundus images. J. Vis. **13**, 263–270 (2010). doi:10. 1007/s12650-010-0037-y

21. Chang, C., Chen, K., Wang, J., Althouse, M.: A relative entropy-based approach to image thresholding. Pattern Recogn. **27**, 1275–1289 (1994). doi:10.1016/0031-3203(94)90011-6

22. Zhang, Y., Zhang, Y.: Another method of building 2D entropy to realize automatic segmentation. J. Phys: Conf. Ser. **48**, 303–307 (2006). doi:10.1088/1742-6596/48/1/056

Sparse-MVRVMs Tree for Fast and Accurate Head Pose Estimation in the Wild

Mohamed Selim$^{(\boxtimes)}$, Alain Pagani, and Didier Stricker

Augmented Vision Research Group, German Research Center for Artificial Intelligence (DFKI), Technical University of Kaiserslautern, Tripstaddterstr. 122, 67663 Kaiserslautern, Germany
{mohamed.selim,alain.pagani,didier.stricker}@dfki.uni-kl.de
http://www.av.dfki.de

Abstract. Head pose estimation is an important problem in the field of computer vision and facial analysis. We model the problem of head pose estimation as a regression problem, where the three rotation angles (yaw, pitch, roll) are functions of the face appearance. We make use of that fact and learn the appearance of the face using a tree cascade of sparse Multi-Variate Relevance Vector Machines (MVRVM). Our method is fast and suitable for real-time applications as it is not computationally expensive. Our method learns the face appearance to estimate the head rotation angles. We evaluated our approach on two challenging datasets, the YouTube Faces and the Point and Shoot Challenging (PaSC) dataset. We achieved results of head pose estimation (yaw, pitch, roll) with mean error less than 5o and with error tolerance less than ±4 on the PaSC dataset. In terms of speed, one prediction takes around 6 milliseconds, which is suitable for real-time applications and also with high frame rate.

Keywords: Head pose estimation · MVRVM · Cascade · YouTube Faces · PaSC

1 Introduction

Due to many potential applications, head pose estimation has become one of the most active and important topics in computer vision [12]. The problem can be considered as a sole problem to be solved and tackled, or as an important part of a bigger system. For example, it can help in gaze estimation problems. Valenti *et al.* [17] combined head pose with eye location to solve gaze estimation problem.

As outlined in [12], the problem of head pose estimation has been framed as a crucial factor in the field of facial analysis, in case robustness to pose is required in an application. For example, in implementing a gender classifier, a pose estimator can be an important pre-processing step in the system.

The problem can be addressed as a classification problem, where the system can try to classify the face in one of the main rotations, like left profile, right profile, semi profile on both sides and frontal face. An SVM could be sufficient

© Springer International Publishing AG 2017
M. Felsberg et al. (Eds.): CAIP 2017, Part I, LNCS 10424, pp. 240–250, 2017.
DOI: 10.1007/978-3-319-64689-3_20

in that application. However, if we add more possibilities, the number of classes will be very big in a way a classifier can fail at. Thus, to predict a wide range of angles, we model the problem as a regression problem, where we provide data in the training phase, and our approach can learn the data, and use it to estimate the three rotation angles at a time.

Our approach builds a tree cascade of regressors, where each node in the tree is trained in subset of the training dataset. We estimate the three rotation angles with the cascade tree of Multi-Variate Relevance Vector Machines [16].

Fig. 1. Sample frames from the PaSC dataset [3]. The top images are from the control subset videos (steady camera). The bottom frames are from the hand-held videos. Hand-held have lower quality and resolution. The dataset have videos captured indoor and outdoor. The persons walk during the video, thus we have different, continuous head poses

Although we build over previous work where MVRVM was used for head pose estimation [15], we significantly improve over this work by building a more complex structure of MVRVMs that yields less error. The work in [15] was limited to single subject only. However, with our new tree structure, we generalized the method for faces of unseen subjects. Moreover, We trained MVRVM models with better input angles generated by state of the art head pose estimation algorithm by [2]. Moreover, we validated our new approach for generalization and worked with more challenging datasets, the PaSC [3].

2   Related Work

In recent time, head pose estimation attracted more interest in the computer vision community. Different approaches have been investigated in solving this problem. Some researchers work on 2D facial images [5,10,13], and others work on 3D data [4,8]. For the methods that use AAMs [5] or any specific facial feature, their estimation is dependent on specific features detection, like facial landmarks, thus, making that error prune. In case an error exists in the features detection, it propagates to the head pose estimation.

In the approach proposed by [7], 3D data is used. The 3D data requires special hardware for capturing. In fact, that makes their approach limited to this type of data and cannot be applied on 2D video sequences. Besides, the work done by [11] uses both color data and depth data. They base their estimation on the point cloud data, they achieve very good results. However, comparing to these approaches is not possible as we work with 2D images from video sequences.

Our work deals with 2D facial images. This problem was addressed before in the work done by [19], however, they have a high error tolerance of ±15°. The problem of head pose was addressed in the work by [2], and they depends on landmark detection. Thus, making head pose estimation depending on the landmark detection. Having this constraint in their approach, they are limited to angles of about −60 to +60°, where enough landmarks are still visible. Our proposed approach doesn't rely on landmarks.

Previously, using MVRVMs in solving the problem in head pose estimation was introduced in [15]. The idea was tested on videos of single subjects from the YouTube faces dataset [18]. It was limited to one subject in the training, in other words, it wasn't generalized to work with any unseen subjects. In this work, we go deeper into the MVRVMs by testing different kernel types. We also work with larger dataset, the Point and Shoot Challenging dataset [3].

3   Methodology

As introduced before in the introduction and related work sections, we build our approach on previous work by [15]. We used MVRVMs for head pose estimation, where it was trained on a single subject. In this work, we want to reduce the error in the estimated head rotation angles and validate the generalization of our approach for unseen faces. For that we introduce the idea of building a more complex structure, that doesn't only have one single MVRVM to make the head pose estimation, but the structure has a tree of MVRVMs. As we build upon previous work, we use the same feature extraction method, which is a vector of normalized pixel intensities extracted from the facial image.

Figure 2 shows an abstract overview of the proposed method. The detected faces are fed into the feature extraction step, then the features and corresponding angles are fed into the Root node of the cascade tree. The cascade is discussed in details in the next subsection.

3.1   Cascade of Sparse Regressors

The cascade of the regressors is built in a tree structure. The yaw angle is used in branching the tree, as it is the rotation angle of the head that has the widest

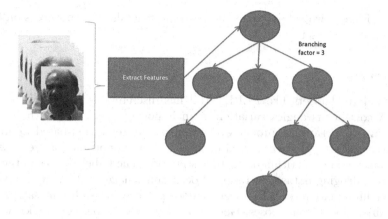

Fig. 2. Overview of the input faces and the cascade tree of sparse MVRVMs

range. The yaw angle can start from $-90°$ (left profile face) to $+90°$ (right profile face). At the root node in the tree, the MVRVM is trained on the input samples, which consists of the features of each face and its corresponding three rotation angles. Going to the next level of the tree, the number of children of the node is determined by the branching factor b. If $b = 3$, the yaw angle range is split into three ranges, and the data is filtered such that each node has the samples that lie in the corresponding yaw angle range. The branching goes on untill we reach the maximum depth of the cascade, or a node does not have sufficient data to be learnt.

The resulting tree of MVRVMs, is used in the prediction process. The prediction process starts from the root node. The root node is designed to give a rough estimation of the head pose. Based on the predicted yaw angle, the child node is chosen to be the next node used in the path while traversing the tree of the cascade. The longest prediction path is predicting and improving the estimation by d predictions, where d is the maximum depth of the cascade.

Although, free variables available in using MVRVM for solving the problem of head pose estimation were optimized [15], we now introduce new free variable that needs to be optimized. We carried out experiments to optimize those parameters, the tree branching factor and its depth. We also investigated different MVRVM kernel types.

4   Evaluation and Results

In this section, we present the evaluation of our approach on challenging datasets of persons captured in different conditions. These datasets have continuous head pose variation. They vary in the background, illumination, indoor and outdoor locations, resolution, etc. We show the results of the experiments on the datasets to optimize the free variables in our approach. In general we optimize the kernel width and type of the RVM. Moreover, we optimize the branching factor of the tree. Finally, we validate our approach for generalization purposes on large subsets of the dataset.

4.1 Datasets

The standard datasets like FERET [14] has discrete specific values for head pose. A continuous angles variation is an important feature that the dataset must have to perform a proper evaluation of our regression-based approach. Moreover, using real data captured in the wild is an important feature to assure the validity of our algorithm on real-life scenarios. The Labeled faces in the wild [9] is a challenging dataset in terms of occlusion, image quality, varying poses, different illumination, etc. However, it does not provide sufficient samples for each subject in different poses. Good candidates to the best of our knowledge are the video datasets, YouTube faces [18] and the PaSC [3].

YouTube Faces Dataset. The YouTube faces dataset [18] is a challenging dataset that has 3425 videos of 1595 different people. The authors of the dataset provided the rotation angles of each frame in the dataset. They used face.com API to provide the head rotation angles.

Point and Shoot Face Recognition Challenge (PaSC). In 2013, Beveridge *et al.* produced the PaSC dataset [3]. They used inexpensive "point-and-shoot" cameras. They collected 9376 still images and 2802 videos of 293 people. The videos were recorded in different locations, outdoors and indoors, with varying illumination and backgrounds. The authors provided meta-data with the dataset that contains the face detection in the video frames. The head rotation angle was provided by the PittPatt detector. The scenarios they had in the videos shows the face from the right profile to the left profile in continuous motion, where the yaw angles changes widely along the videos. Two video types were provided in the dataset, hand-held and controlled subsets. In the hand-held videos, the frames are very shaky and challenging. The controlled videos, have a stable background. Both video types are challenging. Figure 1 shows sample images from the dataset.

The rotation angles in the datasets were produced using the face.com API for the YouTube Faces, and PittPatt for the PaSC (yaw angle only) dataset. However, the work done by [2] focuses on facial landmarks detection, and can estimate the head pose. We used their approach to generate estimations of the head rotation angles. However, even this approach was challenged, as it was unable to detect and track the landmarks in some hard frames of the detected faces in the PaSC dataset. It worked with about 72% of the faces provided by the meta-data in the dataset. However, we don't need all the frames, as our approach learns the head pose from appearance and can estimate it for any detected face.

4.2   Parameters Optimization

Choice of the Kernel and Its Size. As mentioned in Sect. 3, the kernel type is chosen in a way that suits the data provided to the RVM. Kernel type affects the accuracy of the training as it is the metric mapping the input to the output of the RVM. We evaluated four kernel types (Gaussian, Linear, Bubble, and Cubic) on the PaSC dataset subject videos and on the YouTube faces dataset. The Gaussian kernel yields the least error in the yaw angle estimation, hence it is the kernel that we used in the next evaluations. The kernel width has a strong effect on the accuracy of the cascade (Fig. 3). In [15], the kernel was optimized for the head pose estimation problem. It was varied starting from value 3 up to 50. The optimal value found was 13.

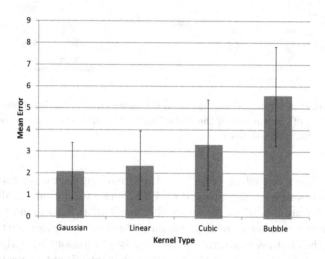

Fig. 3. Comparing different kernel types on the PaSC dataset. Average errors in the yaw angle estimation are shown with the standard deviation among all dataset videos. Gaussian kernel yields the least error.

Grid Size. In the work [15], the result of optimizing the face grid size to was reported to be 15 × 15. Here we build upon these results, and proceed to build the cascade tree for solving the problem of head pose estimation.

4.3   Branching Factor

The tree branching factor affects the accuracy of the estimations. We variate the branching factor of the cascade tree from 2 splits up to 5 splits. Having more than 5 splits makes the range very small in the child nodes. We start the optimization at branching factor of value 2, which is the minimum number of splits possible. When the branching factor was more than 4, the number of input samples decreased quickly in the tree, h ence, resulting in a shallow tree. The branching factor with the least error was 3. We set the root node to have only two children, thus classifying right or left profile faces. Further children in the tree have branching factor of 3. Based on that setup, the leaf nodes of the tree get very small range of angles after depth of 3. Consequently, we set the maximum depth to be 3 levels.

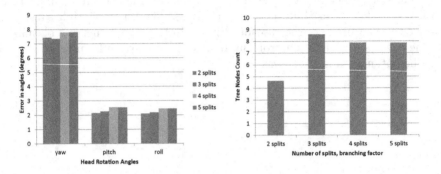

Fig. 4. The effect of the branching factor on the mean error in the head rotation angles (left). The effect of the number of the branching factor on the tree nodes count (right), using the YouTube faces dataset

Figure 4 shows the effect of varying the branching factor on the YouTube dataset. The results are the average of 4-fold cross validation on all the 1595 subjects in the dataset. In Fig. 4, we see that the 3 splits has the most number of nodes, which means a better representation of the data in the cascade tree. It also follows that the least average error on the main head rotation angle, the yaw, is at 3 splits. Considering the presented evaluations, we optimized the free parameters in our approach by 4-fold cross validation experiments which considered all the videos of one of the subjects. Thus, the next step is validating the approach with as many samples from the dataset as possible, which is discussed in Subsect. 4.5.

4.4   Single RVM vs. Cascade Tree

It is important to compare the single MVRVM [15] to the Cascade tree of MVRVMs. Table 1 shows the mean accuracy of 4-fold cross validation test on the PaSC dataset. The cascade tree approach yields smaller errors in all head rotation angles.

4.5   Validation

Based on the findings so far, the kernel width optimal value is 13, and the optimal number of grid divisions is 15. The final step is approach validation. We generated better head pose estimations for the PaSC dataset using a newer

Table 1. Comparing the Single with the Tree Cascade of MVRVMs. The Cascade shows smaller mean error - PaSC. (Mean error ±std). Validation experiment of chunks of 5000 samples.

Method	Yaw	Pitch	Roll
Single MVRVM [15]	5.4 ±4	5.4 ±4	3 ±2.5
Cascade Tree	**4.6 ±3.32**	**5 ±4**	**2.3 ±2.1**

method proposed by [2] (compared to PittPatt used in the dataset metadata). Their method deals with landmark localization and tracking, and it can be used in head pose estimation. To validate our method for generic use, we first shuffle all video frames from all subjects. Then we divide the frames into sets of 5000 frames each. We run 4 fold cross-validation on each set. The number of validation sets in the PaSC is 25, each having 5000 random frames from different subjects. Table 2 shows the average mean error with standard deviation reported on all the sets.

Table 2. Validation results, reported on the PaSC dataset. Average errors in the angles with the standard deviations are reported.

Dataset	Yaw	Pitch	Roll
PaSC [15]	5.4 ±4	5.4 ±4	3 ±2.5
PaSC (Our work)	**4.6 ±3.32**	**5 ±4**	**2.3 ±2.1**
HPEG (Work [6])	4.25 ±3.04	3.83 ±2.72	-
HPEG (75,25)	**2.6 ±1.2**	**1.9 ±2**	-
HPEG (25,75)	**3.45 ±1.9**	**2.15 ±2.25**	-

The MVRVMs can learn the head pose by the appearance of the face with high accuracy. Less the 4.6° error in the Yaw angle in the validation tests are reported on very challenging uncontrolled datasets. Regarding the pitch and roll angles, the MVRVM reported errors less than 5° on PaSC. Figure 5 shows the distribution of the errors in the angles on the dataset frames. We can see that the error is below 5° in the yaw angle for about 66% of the data. and below 10° for about 98% of the data. The errors in the pitch and roll are a bit higher which could be due to the fact that the video frames did not have as big variations as in the yaw angle. Finally, our method is suitable for real-time applications as the time taken by the computation of one single prediction of the three head rotation angles is only 6 milliseconds, with no need of complex landmark detection or model fitting or tracking.

We compared our approach with the work in [6] using the same dataset, the HPEG dataset [1]. The dataset contains 10 video sequences of people rotating their head, the groundtruth angles were provided by the dataset. They acquired it using markers attached to the subject's head (outside the face). The work in [6] requires tracking and doesn't require training, We require training on a subset of the dataset. We used different training and testing percentages of the dataset, either 25% or 75%. Results are shown Table 2.

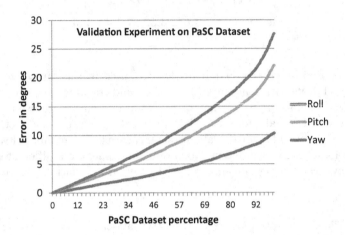

Fig. 5. Results of our new approach in the validation experiment on the PaSC dataset. Yaw error is below 5° in the yaw angle for about 66% of the data. and below 10° for about 98% of the data

The architecture of the machine used in the evaluations is a 6-core Intel Xeon CPU with hyper-threading technology, and 64 GB of RAM. Our evaluation application runs in parallel using the 12 threads provided by the CPU.

5   Conclusion

In this paper, we present a cascade tree of sparse regressors to solve the problem of head pose estimation. This work is built upon the work in [15]. We use the face appearance as the only input, and generate normalized pixel features for training a cascade of MVRVMs. The simple features used are inexpensive to compute on a CPU.

We significantly improve the work in [15] by building a more complex structure that can handle more input data and improve the accuracy of the head pose estimation. Moreover, we make further analysis of the MVRVMs kernels. We also, use a more challenging dataset for training the cascade and validating our method. Finally, we generalize our method where we train using different subjects and not only one subject. Our new proposed approach works on unseen faces.

We tested our approach on two challenging datasets, the YouTube faces dataset and the PaSC dataset. Although, if the values provided by the datasets or the Chehra library [2] are not the absolute correct head rotation angles, we show that we can learn these numbers without the need of model fitting or complex landmark localization. Besides our extensive cross validation experiments which we ran on hundreds of thousands of images from the datasets, we compared our approach to another model-free one by [6], and we show that we significantly reduce the average error on the HPEG dataset [1].

Acknowledgments. This work has been partially funded by the University project Zentrums für Nutzfahrzeugtechnologie (ZNT), and the European project Eyes of Things (EoT) under contract number GA643924.

References

1. Asteriadis, S., Soufleros, D., Karpouzis, K., Kollias, S.: A natural head pose and eye gaze dataset. In: Proceedings of the International Workshop on Affective-Aware Virtual Agents and Social Robots. ACM (2009)
2. Asthana, A., Zafeiriou, S., Cheng, S., Pantic, M.: Incremental face alignment in the wild. In: 2014 IEEE Conference on Computer Vision and Pattern Recognition (CVPR), pp. 1859–1866. IEEE (2014)
3. Beveridge, J.R., Phillips, P.J., Bolme, D.S., Draper, B.A., Givens, G.H., Lui, Y.M., Teli, M.N., Zhang, H., Scruggs, W.T., Bowyer, K.W., Flynn, P.J., Cheng, S.: The challenge of face recognition from digital point-and-shoot cameras. In: 2013 IEEE Sixth International Conference on Biometrics: Theory, Applications and Systems (BTAS), pp. 1–8, September 2013
4. Blanz, V., Vetter, T.: Face recognition based on fitting a 3d morphable model. IEEE Trans. Pattern Anal. Mach. Intell. **25**(9), 1063–1074 (2003)
5. Cootes, T.F., Wheeler, G.V., Walker, K.N., Taylor, C.J.: View-based active appearance models. Image Vis. Comput. **20**(9), 657–664 (2002)
6. Cristina, S., Camilleri, K.P.: Model-free head pose estimation based on shape factorisation and particle filtering. In: Azzopardi, G., Petkov, N. (eds.) CAIP 2015. LNCS, vol. 9257, pp. 628–639. Springer, Cham (2015). doi:10.1007/978-3-319-23117-4_54
7. Fanelli, G., Gall, J., Van Gool, L.: Real time head pose estimation with random regression forests. In: 2011 IEEE Conference on Computer Vision and Pattern Recognition (CVPR), pp. 617–624. IEEE (2011)
8. Gu, L., Kanade, T.: 3D alignment of face in a single image. In: 2006 IEEE Computer Society Conference on Computer Vision and Pattern Recognition, vol. 1, pp. 1305–1312. IEEE (2006)
9. Huang, G.B., Ramesh, M., Berg, T., Learned-Miller, E.: Labeled faces in the wild: A database for studying face recognition in unconstrained environments. Technical report 07–49, University of Massachusetts, Amherst, October 2007
10. Jones, M., Viola, P.: Fast multi-view face detection. Mitsubishi Electric Research Lab TR-20003-96, 3:14 (2003)
11. Mukherjee, S.S., Robertson, N.M.: Deep head pose: Gaze-direction estimation in multimodal video. IEEE Trans. Multimedia **17**(11), 2094–2107 (2015)
12. Murphy-Chutorian, E., Trivedi, M.M.: Head pose estimation in computer vision: A survey. IEEE Trans. Pattern Anal. Mach. Intell. **31**(4), 607–626 (2009)

13. Pentland, A., Moghaddam, B., Starner, T.: View-based and modular eigenspaces for face recognition. In 1994 IEEE Computer Society Conference on Computer Vision and Pattern Recognition, 1994. Proceedings CVPR'94, pp. 84–91. IEEE (1994)
14. Jonathon Phillips, P., Moon, H., Rizvi, S.A., Rauss, P.J.: The feret evaluation methodology for face-recognition algorithms. IEEE Trans. Pattern Anal. Mach. Intell. **22**(10), 1090–1104 (2000)
15. Selim, M., Pagani, A., Stricker, D.: Real-time head pose estimation using multivariate rvm on faces in the wild. In: Computer Analysis of Images and Patterns (2015)
16. Thayananthan, A., Navaratnam, R., Stenger, B., Torr, P.H.S., Cipolla, R.: Multivariate relevance vector machines for tracking. In: Leonardis, A., Bischof, H., Pinz, A. (eds.) ECCV 2006. LNCS, vol. 3953, pp. 124–138. Springer, Heidelberg (2006). doi:10.1007/11744078_10
17. Valenti, R., Sebe, N., Gevers, T.: Combining head pose and eye location information for gaze estimation. IEEE Trans. Image Process. **21**(2), 802–815 (2012)
18. Wolf, L., Hassner, T., Maoz, I.: Face recognition in unconstrained videos with matched background similarity. In: 2011 IEEE Conference on Computer Vision and Pattern Recognition (CVPR), pp. 529–534. IEEE (2011)
19. Zhu, X., Ramanan, D.: Face detection, pose estimation, and landmark localization in the wild. In: 2012 IEEE Conference on Computer Vision and Pattern Recognition (CVPR), pp. 2879–2886. IEEE (2012)

On the Use of the Tree Structure of Depth Levels for Comparing 3D Object Views

Fabio Bracci[1(✉)], Ulrich Hillenbrand[1], Zoltan-Csaba Marton[1], and Michael H.F. Wilkinson[2]

[1] Institute of Robotics and Mechatronics,
German Aerospace Center (DLR), Weßling, Germany
{fabio.bracci,ulrich.hillenbrand,zoltan.marton}@dlr.de,
fabio.bracci@freenet.de, michael.h.f.wilkinson@rug.nl
[2] Johann Bernoulli Institute for Mathematics and Computer Science,
University of Groningen (RuG), Groningen, The Netherlands

Abstract. Today the simple availability of 3D sensory data, the evolution of 3D representations, and their application to object recognition and scene analysis tasks promise to improve autonomy and flexibility of robots in several domains. However, there has been little research into what can be gained through the explicit inclusion of the structural relations between parts of objects when quantifying similarity of their shape, and hence for shape-based object category recognition. We propose a Mathematical Morphology inspired hierarchical decomposition of 3D object views into peak components at evenly spaced depth levels, casting the 3D shape similarity problem to a tree of more elementary similarity problems. The matching of these trees of peak components is here compared to matching the individual components through optimal and greedy assignment in a simple feature space, trying to find the maximum-weight-maximal-match assignments. The matching thus achieved provides a metric of total shape similarity between object views. The three matching strategies are evaluated and compared through the category recognition accuracy on objects from a public set of 3D models. It turns out that all three methods yield similar accuracy on the simple features we used, while the greedy method is fastest.

Keywords: 3D shape similarity · Tree matching · Mathematical Morphology · Object recognition · Scene analysis

1 Introduction

Nowadays 3D sensing technologies provide an ever growing number of low-cost and reliable sensors like the popular Kinect® and the Xtion® in both first and second versions. The depth data produced by these devices come at a high frame rate and provide a dense representation of 3D surface geometry. Speed of processing and representational geometric quality are important aspects in robotics, which often makes 3D sensing and processing the procedure of choice

© Springer International Publishing AG 2017
M. Felsberg et al. (Eds.): CAIP 2017, Part I, LNCS 10424, pp. 251–263, 2017.
DOI: 10.1007/978-3-319-64689-3_21

for enhancing robot autonomy and adaptivity. In particular, object recognition and scene analysis are increasingly based upon analysis of 3D data, and both imply matching and comparing objects.

Naive pixel-based or 3D point-based global matching is intractable for the combinatorial explosion of computations. Hence, several methods were developed to compare object shape through an abstract representation, e.g. histograms of local or regional descriptors on keypoints, global descriptors, or graphs of parts and their relations. Apart from the latter approach, the vast majority of representations has not explicitly regarded the relations of parts or components of objects. By contrast, we are here specifically interested in the effects of those relations when included in a *structural* object representation.

One particular kind of structure is the object's topology; comparing objects then means comparing topologies, which can be formalized as a graph matching problem. However, graph matching in general can be costly, even unfeasible for practical problems, therefore approaches like [14] take approximations and simplifications. Efficient matching algorithms are known for the graph subclass of trees, which are often produced as a problem approximation, e.g., by minimum spanning trees. Here we propose a representation where the favorable tree structure comes out naturally in the exact formulation.

The shape of 3D objects is here represented in a view-based fashion. Range images of object views are decomposed into *peak components*, which are the height profiles over 2D cross-sections taken at evenly spaced depth levels from the sensor. Considering depth levels is very suitable for depth data and this decomposition yields the inclusion relationship between each pair of such peaks: the inclusion hierarchy. A tree structure thus naturally arises from the proposed decomposition, the representation by a 3D *shape tree*.

Individual peak components, that is, the vertices of the 3D shape tree, need to be described in some feature space. For the sake of this study, we have here used a very simple feature space with limited descriptive capacity, thus relying on the aggregation effect of simple components. A 3D shape metric can be obtained by computing and quantifying a match between trees or simply between the sets of components.

We quantify the contribution of the tree structure among the components to the matching accuracy achieved in shape-based category recognition by a nearest-neighbor classifier: matching is done using the tree-structural constraints and also with structureless assignment of the components. Elaborating and comparing the results is the main contribution made by this work.

While intuitively one might expect that an explicit regard for structural relations among object components should generally improve matching accuracy, it here turns out that it does not improve similarity-based category recognition in this particular setup. It seems hence worthwhile to always compare a structural matching against a structureless baseline, and the latter may suffice. The analysis of the results is presented in Sect. 6.

2 Related Work

A taxonomy of shape matching methods is proposed in [22]; there shape match-ing methods are subdivided into *feature* based, *graph* based and *geometry* based. Our method uses the ability of a tree to describe the relations among the ele-ments, which qualifies it as a graph-based method. At the same time the target to be described is a range image of an object taken from a given viewpoint, which also puts it among the geometry-based methods. The proposed method also relies on features of peak components, which makes it feature-based too.

An inspiration to our work has been the depth decomposition proposed in [8]. They slice 3D CAD models in three orthogonal directions aligned with the object's bounding box, and the cross-sections produced by scanning through the object along those directions are collected from the front to the back. The 3D descriptor is made by a histogram computed by binning simple 2D slice features.

A popular histogram-based shape descriptor is the Viewpoint Feature His-togram (VFH) of [19]. For any patch of points a histogram of three angular values between all the couples of point normals is computed and extended with the angular values between the point normals and the central viewpoint direc-tion. A fast approximated kNN is then used to classify the objects. The methods based on histograms as global object descriptors do not take into account any structure among the constituent elements whose features are accumulated.

As discussed in [7], perceptual organization should be captured using models that take account of the part structure of objects and capture the properties of 3D shapes. As argued for example by Huber [6], part-based detection has the advantage of generalizing to unknown instances of object types. While in [6] and for the part-based VFH (called CVFH) feature [1] objects need to be singulated first, other part-based approaches like [13] can efficiently detect objects in clutter. Recently, Richtsfeld *et al.* [18] presented a multi-level approach to fit planar or curved surfaces to over-segment parts, and then define inter-segment relations to decide if they should be merged or not, but the method performs best for merging touching segments and for convex shapes.

A graph-based method is proposed in [21] where objects are matched through their skeletons. The skeletonization is made from a voxelized object and the skele-tal voxels are connected with a minimum spanning tree algorithm. The actual object matching is restricted to the nearest graphs by indexing. The matching is modeled as a modified maximum-cardinality-minimum-weight matching, and is computed with a recursive depth-first coarse-to-fine search.

In the VRML community [27] also proposed a graph-based method. They decompose 3D objects in concave patches by watershed, and formulate the object matching as a graph isomorphism between attributed graphs, a computation-ally difficult problem without polynomial time solutions. The authors tackle the problem by merging patches such that their number stays small.

Trees for object recognition have been used in [4,5] for human airways recog-nition. The authors restrict the modeled trees to planted trees with bifurcations and trifurcations; edge weights represent the distance among vertices. The node similarity is computed from the ratio of the length of the airway to the vertex, the branch point type, and structural similarity. The matching between their structures minimizes a tree-edit distance, where a set of allowed matching combinations is previously determined to limit the $\mathcal{NP}$-complexity.

An example of a histogram-based region descriptor method is the shape context proposed in [2]. The authors first align the objects, and then sample a number of points along the object contour with uniform spacing, and in each contour point they take a distribution of the relative position of the other contour points. Correspondences between two objects are found by solving an optimal assignment problem; the total weight is used to estimate the similarity between pairs of objects. This method was extended to 3D in [11] to a histogram of relative 3D point positions of all the other shape points of a patch. The object matching goes through the optimal assignment formulation as well.

Deep learning is a class of methods of growing popularity, see [15]. The methods proposed so far are not able to process megapixel-size images and require large amount of training data. An example of such methods is [26] where a global object representation is learned, while in [3] a patches-representation is learned across deformed shapes.

The structural-representation-based method here investigated exploits the descriptiveness of geometric features of the object's parts at multiple scales, used along with a tree's ability to represent the relationship among those parts, while gearing on efficient tree matchings. This method was chosen as a promising structured matching procedure for depth data, as it is different from generic graph methods that are mainly applied rather on a small scale and on abstract decompositions/labellings of the objects due to the computationally costly matching.

This approach is similar to the work done in [10] where attributed graphs are built on a triangulation of Harris corners; they match image graphs by optimal assignment and locally preserve structure through the heterogeneous Euclidean overlap metric, a metric jointly considering vertex attributes, vertex degree, and attributes of the incident vertices. We also draw on the work of [24], where they measure the total distance between two trees with the maximum-cardinality-maximum-weight tree isomorphism through a recursive descent which maximizes the total accumulated similarity for different root vertex choices. They evaluate their work on matching skeletal graphs of 2D images.

In our application of these ideas we borrow the image decomposition from the work done in [25]. This method is based on 2D Shape-Size Pattern Spectra, which are 2D histograms of shape (moment of inertia divided by squared area) and size (area) of the peak components of an image. They classify images trough decision trees of the image's pattern spectra. Connected components and peak components are Mathematical Morphology's concepts which have been proven as a valuable decomposition; in [20] we find an efficient method to extract and store these components.

3   Method

The structural representation we investigate in this study aims at capturing shape details as well as the coarse outline. The representation is built from range images of objects taken from any viewpoint, either by a range sensor or through rendering of a 3D model. The core idea is to represent a 3D shape by a hierarchy of range image peaks, split into their connected components, and hence yielding a tree structure.

We now explain in turn the decomposition of a range image into its peak components and their descriptor; the construction of the tree structure; the matching algorithm for the trees; and how a 3D shape metric is derived from a match between two trees. The baseline of a structureless matching we compare to is a straightforward assignment between individual peak components from two object views.

Peak Components and Descriptors. As input we have a range image, which is a raster of depth values $D(x, y)$ for pixel coordinates (x, y) of an object from a given viewpoint. This raster describes a height profile with isolines shrinking towards the observer. Occluded parts and sides of the object are not visible to the observer and hence no information about those is available.

Slicing, or thresholding at any depth level h produces one or more peak components P_i^h, which is the i-th set of connected image component of depth h and lower:

$$P_i^h \{(x, y) \,|\, D(x, y) \leq h\} \tag{1}$$

In other words, as a sequence of decreasing depth levels are used for slicing, the generated peak components are the connected regions in the remaining depth image. These are nested into each other like the isolines at the different values for h; the largest component lays at the maximal depth and the smallest lies close to the observer.

We use the same features as in [25] to describe the peak component shape, namely area, elongation and entropy. Additionally we consider relative area (the ratio of the peak component's area and the root peak component's area) and normalized entropy (the ratio of entropy and the logarithm of the number of pixels per peak component); see Sect. 4 for further motivation.

The individual peak components are compared by means of Euclidean distances in a metric space. In a space of descriptive features, similar elements are expected to be close to each other while dissimilar elements are expected to be far apart. We map the point-wise distance to a similarity measure

$$s(u, v) = \max(1 - \|u - v\|/r, 0) \tag{2}$$

where $\|\,\|$ denotes the Euclidean norm, u and v are descriptor vectors of peak components, and r is a cutoff radius. In particular, the similarity is set to zero for a distance larger than r.

3D Shape Tree. So far we have an unorganized collection of peak components; now we represent their mutual inclusion by building the hierarchy. This is achieved through the computation of the Max-Tree proposed in [20], a well known data structure in the Mathematical Morphology. The Max-Tree is an inclusion tree, which is a tree of nested connected components, where each component includes its children. The Max-Tree represents morphological changes across the depth in the object through branches of arbitrary degree and depth, and through the vertex descriptors. The Max-Tree is a special level-set method.

The Max-Tree computation is linear in the number of pixels. This is achieved through a recursive flood-fill procedure starting at the highest depth; each time a pixel of lower depth is found, the flooding continues at the new depth level. This way the tree vertices correspond to varying discrete depth changes, and we have a non-uniform sampling of depth levels. We overcome this with an expansion step where all the vertices with depth difference to the parent $\triangle h > 1$ are substituted with a small series of $\triangle h$ non-branching vertices of unit depth increase.

Structural and Structureless Matching. Once object views are transformed into trees of peak component descriptors, we need a matching and comparison scheme defining quantitatively a 3D shape similarity. General graph matching is costly and when done by edit-distance minimization is $\mathcal{NP}$ complete; tree matching instead is solved efficiently in polynomial time.

We define a tree as $T = (V, E)$ with vertices V and edges E. A matching of the vertices of two trees without regarding the edges can be obtained straightforwardly through optimal assignment of the vertices. Specifically, we look for a matching $\varphi \subseteq V_1 \times V_2$ that maximizes the total similarity

$$W(\varphi) = \sum_{(u,v)\in\varphi} s(u, v). \tag{3}$$

where φ represents the one-to-one vertex mapping between the two trees.

This optimal assignment is one baseline of a structureless matching we compare to. The problem was originally solved by Kuhn [12] and Munkres [16]. The original algorithm has $\mathcal{O}(n^4)$ complexity for n nodes to match; instead we use the algorithm of Jonker and Volgenant proposed in [9] which has $\mathcal{O}(n^3)$ complexity.

In the literature greedy algorithms are known which iteratively couple labeled vertices. We extend this approach with similarity (2), so that the correspondence with the largest weight is selected and added to the matching; the involved vertices are removed from the trees afterwards; this assignment proccess is iterated over the complete set. The procedure is shown in Algorithm 1: it requires a distance matrix between all the vertex pairs, which has quadratic time complexity, and a sorting step on it, which gives $\mathcal{O}(n \log(n))$ total time complexity.

The above matching methods are generic and flexible, but they disregard the tree hierarchies and should be considered as *structureless* matching. For instance, if $u \in T_1$ is matched to $v \in T_2$, the child u' of u might be matched to the parent v' of v, as there is no constraint modeled in the assignment. We consider also a tree matching algorithm proposed in [24], which is proven to have $\mathcal{O}(bn^3)$ complexity, where b is the maximum degree of the tree vertices. For any given pair of vertices, u in T_1 and v in T_2, the algorithm performs a recursive descent and accumulates the similarity between u and v plus the maximal similarity among all the pairs of children of u and v in the matching: $(u', v') \in Ch(u) \times Ch(v)$ where $Ch()$ gives the vertex children. The algorithm finds the maximal tree matching by iterating on $(u, v) \in \{root(T_1)\} \times T_2$ and $(u, v) \in T_1 \times \{root(T_2)\}$ where $root()$ is the tree root.

Algorithm 1. Sequential Greedy Match for MWMM

$(w, W) = \texttt{SequentialGreedyMWMM}(V_1, V_2)$ $(V_1, V_2$ vertex sets$)$
1. **if** $|V_1| = 0 \vee |V_2| = 0$ **then**
2.   **return** $(0, \emptyset)$
3. **else**
4. $(u, v) = \text{argmin}\{s(u, v) \,|\, u \in V_1, \;\; v \in V_2\}$
5. $V_1' = V_1 \setminus \{u\}$; $V_2' = V_2 \setminus \{v\}$
6. $(w', W') = \texttt{SequentialGreedyMWMM}(V_1', V_2')$
7.   **return** $(s(u, v) + w', \{(u, v)\} \cup W')$

It needs a pairwise similarity $s(u, v)$. This procedure resembles counting the similar vertices found in a parallel descent of two trees. Such matching enforces inclusion relationships by construction and is therefore a *structural* matching.

3D Shape Similarity Metric. Now a (structural or structureless) tree matching is computed in order to find the corresponding parts of two trees representing two object views. Let φ_{12} be the optimal mapping found through optimal assignment, greedy assignment or structural matching, hence maximizing (3). The total similarity $W(\varphi_{12})$ accumulated through φ_{12} is not meaningful as such: generally matching two small trees produces a small total similarity while matching two large trees produces a large one because of the different number of involved vertices. In order to avoid this size bias effect, we weight $W(\varphi_{12})$ relative to the original tree sizes. For this purpose we apply the four metrics proposed in [23], with $|V_i|$ the number of vertices in V_i:

$$d_1(T_1, T_2) = \max(|V_1|, |V_2|) - W(\varphi_{12}) \tag{4}$$

$$d_2(T_1, T_2) = |V_1| + |V_2| - 2W(\varphi_{12}) \tag{5}$$

$$d_3(T_1, T_2) = 1 - W(\varphi_{12}) / \max(|V_1|, |V_2|) \tag{6}$$

$$d_4(T_1, T_2) = 1 - W(\varphi_{12}) / (|V_1| + |V_2| - W(\varphi_{12})) \tag{7}$$

4 Evaluation

We evaluate the structural and structureless matching variants of our trees in an experiment on shape-based category recognition. We are interested in the accuracy of 1NN classification using the shape metrics (4) through (7) for finding the nearest neighbor.

Our experimental setting is similar to the one used in [2]. We take 10 objects each from 10 classes from the SHREC-2010 database [17], take four specific views of each object (front, top, elevated back diagonal and elevated left diagonal view) and render a total of 400 range images of 900×1200 pixels. Each image is rendered with the focal length of the Kinect® and with an object-to-viewpoint distance such that the object fits in and fills the image. This set of arbitrary views is meant to contain different and therefore discriminable representations of the chosen objects. Example images, one for each class, are shown in Fig. 1.

| bird | fish | non flying insect | flying insect | biped | quadru- ped | apart- ment | sky- scraper | bottle | mug |

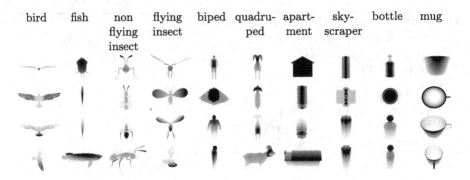

Fig. 1. Example range images from our evaluation set; the four views of an object from each class. Object models taken from [17].

In this study we focus on the aspect of shape similarity and disregard the absolute object size. Our 3D shape tree disregards absolute size by choice of the component descriptors. Moreover, to make the representation independent of the image size of an object as well, we investigate a version where the image area of a peak component is quantified as the fraction of the object silhouette area. Likewise, the entropy of a peak component is normalized by the logarithm of the component area. Finally, a depth normalization step is applied before the peak component decomposition in order to make the depth slicing comparable across objects. Depending on the actual application, when object size matters real physical units of the measures can be used for the descriptor vectors. Note that the used features are rotationally invariant which avoids the need for sampling of different angles around the viewing axis.

We compute the accuracy of a 1NN classifier of object category on three subsets of our range images set: a set of low-detail objects (fish, skyscraper, bottle, mug), a set of mixed-detail objects (bird, fish, biped, single house, mug), and a set of high-detail objects (bird, non flying insect, flying insect, biped, quadruped). This way we evaluate the quality of the proposed descriptor and relation between the different match procedures for different shape detail levels. Classification accuracies are computed in a leave-one-out scheme, where each queried object instance is left out from the database of category samples to match against. The evaluations cover the four shape metrics (4) through (7). The cutoff radius on similarity of peak components (cf. Eq. 2) is set to a range of percentile values (10%, 20%, ..., 90%, 99%) from the full sample of distances between all peak components. This way we adapt the parameter to the actual statistics of occurring feature distances.

5   Results

The evaluation was performed with four different features combinations; here we present results only for the feature set leading to the best classification accuracy,

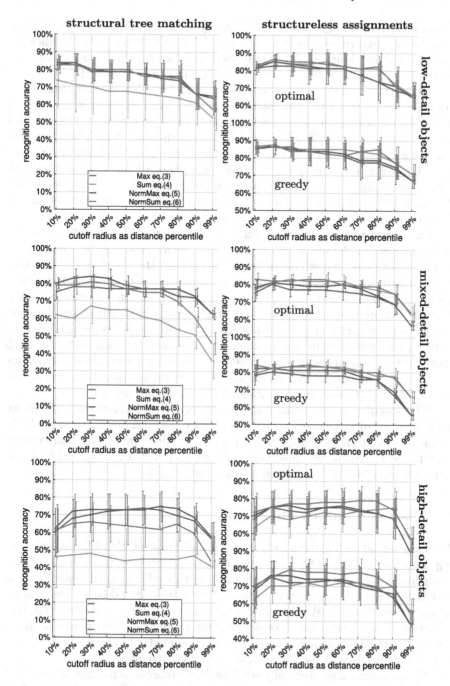

Fig. 2. Classification experiments results: mean classification accuracies and standard errors (bars) depending on similarity cutoff radius (cf. Eq. 2). Shown is the performance separately on low-, mixed-, and high-detail shape classes for the structural and structureless matching methods and for the resulting shape metrics according to Eqs. (4) through (7), as indicated.

which is the relative area, normalized entropy and elongation. The results of the classification experiments are shown in Fig. 2. There we can see the classification accuracy for matching the shape trees (structural match), and through optimal and greedy assignment (structureless matches) of peak components.

All the three matching methods show a similar best classification accuracy on the same data sets: on the low-detail shapes the greedy assignment is best with a 87.5% accuracy, on the mixed-details shapes both the greedy and the optimal assignments reach 84% accuracy, while on the high-detail shapes both the tree matching and the greedy assignment matchings show an 83% classification accuracy. The differences of best achievable accuracy between matching methods fall within the standard error of the mean in all these cases, are hence not significant. The overall best classification accuracy is achieved for the low-detail classes; it is overall worst for the high-detail classes.

The accuracy differences between best achievable accuracies fall within the standard error of the mean also across the different data sets. Nonetheless it's worth to note that the largest difference is between the accuracies on the low-detail and the high-detail shape classes, showing that low-detail shapes are somewhat easier to discriminate properly in the simple descriptor space considered. Very similar are the accuracies between different metrices within each data set: in most cases those are very close to each other with similar dependence on the different similarity cutoffs; only for the tree matching the Sum metric of Eq. 5 does have a lower accuracy. The general trend is that accuracies degrade for the highest similarity cutoffs, which indicates that there should be an indifferent zero similarity set from some moderate feature distance upward.

When comparing computation times for the different levels of shape detail, we found that low-detail classes are less computationally demanding than the high-detail classes for all the techniques. Also we observe an increasing computation time for both structureless matching variants when increasing the cutoff radius, while the structural matching remains insensitive to it.

6   Discussion and Conclusions

We have introduced a tree-structured representation of 3D object views that captures the hierarchy of peak components at different depth levels. The peak components have been described in a simple three-dimensional space of area, elongation, and entropy. In this setting, shape is represented by an aggregation of simple components. We have investigated the contribution made by the tree structure to the performance of a shape metric derived from matching these components between objects: matchings are computed with and without explicit regard for the edges in the tree as a structural constraint.

We have found no statistically significant difference in achievable accuracy of shape-based category recognition for the structural and structureless matches. When comparing the behavior for shape classes of different levels of geometric detail, it turnes out that low-detail shapes are somewhat easier to discriminate than high-detail shapes in our simple descriptor space. A very general conclusion to be drawn from this study is that it is worthwhile to compare any structural matching against a structureless baseline, such as the naive assignment. The latter may give similar results, be easier to implement, and may even run faster with off-the-shelf libraries.

It is interesting to understand the reason behind the somewhat counter-intuitive result that the tree structure of the shape components doesn't need to be considered during matching to find a reasonable correspondence. Our hypothesis is that this is due to the effect of the spatial arrangement of the peak components in the used feature space. On visual inspection it appears that their descriptors, as a point set in feature space, are generally laid out in the shape of the tree they are extracted from. Hence, two structurally and vertex-wise similar trees align quite well in feature space, such that the naive strategy of assigning components of the one tree to the nearest component of the other tree turns out effective and mostly does not disrupt the inclusion relationships.

As the found equivalence of structureless and structural assignments seems tightly linked to the nature of our feature space, it seems likely that results may turn out differently in different feature spaces. In particular, the role of structure in shape representations based on more descriptive features will be investigated in our research. Moreover, the structural aspect may be enhanced through, e.g., taking orientation relations between peak components into account.

References

1. Aldoma, A., Blodow, N., Gossow, D., Gedikli, S., Rusu, R., Vincze, M., Bradski, G.: CAD-model recognition and 6DOF pose estimation using 3D cues. In: ICCV Workshop on 3D Representation and Recognition (3dRR 2011), Barcelona (2011)
2. Belongie, S., Malik, J., Puzicha, J.: Shape context: a new descriptor for shape matching and object recognition. In: Proceedings of the 13th International Conference on Neural Information Processing Systems (NIPS 2000), pp. 798–804. MIT Press, Cambridge (2000). http://portal.acm.org/citation.cfm?id=3008867
3. Boscaini, D., Masci, J., Melzi, S., Bronstein, M.M., Castellani, U., Vandergheynst, P.: Learning class-specific descriptors for deformable shapes using localized spectral convolutional networks. Comput. Graph. Forum 34(5), 13–23 (2015). http://dx.doi.org/10.1111/cgf.12693
4. Graham, M.W., Higgins, W.E.: Optimal graph-theoretic approach to 3D anatomical tree matching. In: 3rd IEEE International Symposium on Biomedical Imaging: Macro to Nano, pp. 109–112. IEEE, April 2006. http://dx.doi.org/10.1109/isbi.2006.1624864
5. Graham, M.W., Higgins, W.E.: Globally optimal model-based matching of anatomical trees. In: Medical Imaging, vol. 6144, pp. 614415-1–614415-15 (2006). http://dx.doi.org/10.1117/12.651719
6. Huber, D., Kapuria, A., Donamukkala, R.R., Hebert, M.: Parts-based 3D object classification. In: Proceedings of the IEEE Conference on Computer Vision and Pattern Recognition (CVPR 04), July 2004
7. Jacobs, D.W.: Perceptual organization as generic object recognition, North-Holland, vol. 130, pp. 295–329, December 2001. http://www.sciencedirect.com/science/article/pii/S0166411501800303

8. Jiantao, P., Yi, L., Guyu, X., Hongbin, Z., Weibin, L., Uehara, Y.: 3D model retrieval based on 2D slice similarity measurements. In: Proceedings of the 2nd International Symposium on 3D Data Processing, Visualization, and Transmission (3DPVT 2004), pp. 95–101 (2004). http://dx.doi.org/10.1109/3dpvt.2004.3

9. Jonker, R., Volgenant, A.: A shortest augmenting path algorithm for dense and sparse linear assignment problems. Computing **38**(4), 325–340 (1987). http://dx.doi.org/10.1007/bf02278710

10. Jouili, S., Mili, I., Tabbone, S.: Attributed graph matching using local descriptions. In: Blanc-Talon, J., Philips, W., Popescu, D., Scheunders, P. (eds.) ACIVS 2009. LNCS, vol. 5807, pp. 89–99. Springer, Heidelberg (2009). doi:10.1007/978-3-642-04697-1_9

11. Körtgen, M., Novotni, M., Klein, R.: 3D shape matching with 3D shape contexts. In: The 7th Central European Seminar on Computer Graphics (2003). http://citeseerx.ist.psu.edu/viewdoc/summary?doi=10.1.1.67.9266

12. Kuhn, H.W.: The Hungarian method for the assignment problem. Nav. Res. Logist. **2**(1–2), 83–97 (1955). http://dx.doi.org/10.1002/nav.3800020109

13. Lai, K., Fox, D.: Object recognition in 3D point clouds using web data and domain adaptation. Int. J. Robot. Res. **29**(8), 1019–1037 (2010). http://ijr.sagepub.com/cgi/doi/10.1177/0278364910369190

14. Marton, Z.C., Balint-Benczedi, F., Mozos, O., Blodow, N., Kanezaki, A., Goron, L., Pangercic, D., Beetz, M.: Part-based geometric categorization and object reconstruction in cluttered table-top scenes. J. Intell. Robot. Syst. **76**(1), 35–56 (2014). http://dx.doi.org/10.1007/s10846-013-0011-8

15. Masci, J., Rodolà, E., Boscaini, D., Bronstein, M.M., Li, H.: Geometric deep learning. In: SIGGRAPH ASIA 2016 Courses (SA 2016), NY, USA (2016). http://dx.doi.org/10.1145/2988458.2988485

16. Munkres, J.: Algorithms for the assignment and transportation problems. J. Soc. Ind. Appl. Math. **5**(1), 32–38 (1957)

17. Pratikakis, I., Spagnuolo, M., Theoharis, T., Editors, R.V., Dutagaci, H., Godil, A., Cheung, C.P., Furuya, T., Hillenbrand, U., Ohbuchi, R.: SHREC 2010 - shape retrieval contest of range scans. In: Proceedings of Eurographics (2010). http://citeseerx.ist.psu.edu/viewdoc/summary?doi=10.1.1.361.8068

18. Richtsfeld, A., Morwald, T., Prankl, J., Zillich, M., Vincze, M.: Segmentation of unknown objects in indoor environments. In: 2012 IEEE/RSJ International Conference on Intelligent Robots and Systems (IROS), pp. 4791–4796 (2012)

19. Rusu, R.B., Bradski, G., Thibaux, R., Hsu, J.: Fast 3D recognition and pose using the viewpoint feature histogram. In: 2010 IEEE/RSJ International Conference on Intelligent Robots and Systems (IROS), pp. 2155–2162. IEEE (2004). http://dx.doi.org/10.1109/iros.2010.5651280

20. Salembier, P., Serra, J.: Flat zones filtering, connected operators, and filters by reconstruction. IEEE Trans. Image Process. **4**(8), 1153–1160 (1995). http://dx.doi.org/10.1109/83.403422

21. Sundar, H., Silver, D., Gagvani, N., Dickinson, S.: Skeleton based shape matching and retrieval. In: Shape Modeling International, pp. 130–139. IEEE (2003)

22. Tangelder, J.W.H., Veltkamp, R.C.: A survey of content based 3D shape retrieval methods. In: Shape Modeling International, pp. 145–156 (2004). http://citeseerx.ist.psu.edu/viewdoc/summary?doi=10.1.1.97.8133

23. Torsello, A., Hidovic, D., Pelillo, M.: Four metrics for efficiently comparing attributed trees. In: Proceedings of the 17th International Conference on Pattern Recognition (ICPR 2004), vol. 2, pp. 467–470. IEEE, August 2004. http://dx.doi.org/10.1109/icpr.2004.1334263

24. Torsello, A., Hidovic-Rowe, D., Pelillo, M.: Polynomial-time metrics for attributed trees. IEEE Trans. Pattern Anal. Mach. Intell. **27**(7), 1087–1099 (2005). http://dx.doi.org/10.1109/tpami.2005.146

25. Urbach, E.R., Roerdink, J.B.T.M., Wilkinson, M.H.F.: Connected shape-size pattern spectra for rotation and scale-invariant classification of gray-scale images. IEEE Trans. Pattern Anal. Mach. Intell. **29**(2), 272–285 (2007). http://dx.doi.org/10.1109/tpami.2007.28

26. Wu, Z., Song, S., Khosla, A., Yu, F., Zhang, L., Tang, X., Xiao, J.: 3D ShapeNets: a deep representation for volumetric shapes - IEEE Xplore Document. In: The IEEE Conference on Computer Vision and Pattern Recognition (CVPR), June 2015. http://ieeexplore.ieee.org/abstract/document/7298801/

27. Zuckerberger, E., Tal, A., Shlafman, S.: Polyhedral surface decomposition with applications. Comput. Graph. **26**(5), 733–743 (2002). http://dx.doi.org/10.1016/s0097-8493(02)00128--0

The Classic Wave Equation Can Do Shape Correspondence

Robert Dachsel[✉], Michael Breuß, and Laurent Hoeltgen

Institute for Mathematics, Brandenburg Technical University,
Platz der Deutschen Einheit 1, 03046 Cottbus, Germany
{dachsel,breuss,hoeltgen}@b-tu.de

Abstract. A major task in non-rigid shape analysis is to retrieve correspondences between two almost isometric 3D objects. An important tool for this task are geometric feature descriptors. Ideally, a feature descriptor should be invariant under isometric transformations and robust to small elastic deformations. A successful class of feature descriptors employs the spectral decomposition of the Laplace-Beltrami operator. Important examples are the heat kernel signature using the heat equation and the more recent wave kernel signature applying the Schrödinger equation from quantum mechanics.

In this work we propose a novel feature descriptor which is based on the classic wave equation that describes e.g. sound wave propagation. We explore this new model by discretizing the underlying partial differential equation. Thereby we consider two different time integration methods. By a detailed evaluation at hand of a standard shape data set we demonstrate that our approach may yield significant improvements over state of the art methods for finding correct shape correspondences.

Keywords: Feature descriptor · Shape analysis · Wave equation

1 Introduction

For the purpose of shape analysis applications, it is useful to describe the shape of a three dimensional geometric object by its bounding surface $\mathcal{M}$. In this setting, two shapes may be considered similar if there exists an almost isometric transformation between them. Such a transformation allows small elastic deformations such as stretching and contractions. In order to investigate the similarity of shapes, its geometry has to be analyzed. To this end, often a simplified shape representation, called feature descriptor, is employed. Ideally, a feature descriptor should be invariant under almost isometric transformations which is challenging to achieve.

Over the last decade many feature descriptors have been presented. Classic feature descriptors are mostly invariant under rigid transformation only. As examples for this class of descriptors let us mention here spin images [7] and integral volume descriptors [12]. Some more recent approaches, such as the one in [10] which relies on the Möbius transform, are also invariant under isometric transformations.

© Springer International Publishing AG 2017
M. Felsberg et al. (Eds.): CAIP 2017, Part I, LNCS 10424, pp. 264–275, 2017.
DOI: 10.1007/978-3-319-64689-3_22

A modern class of feature descriptors that can handle almost isometric transformations is based on the spectral decomposition of the Laplace-Beltrami operator. In this context, shapes can be thought as a vibrating membrane and the eigenfunctions can be interpreted as its vibration modes and the eigenvalues have the meaning of the corresponding vibration frequencies. Unfortunately, this eigenspectrum can not fully determine the shape of the domain as captured in [8]. Nevertheless, for shape analysis these spectral methods were first proposed in [11]. Based on developments in [16], the heat kernel signature (HKS) has been introduced [17]. It assigns each point on an object surface a unique signature based on the fundamental solution of the heat equation. This amounts effectively to the computation of a series expansion of the heat kernel. A scale invariant extension of this approach was developed in [4]. In [1] another feature descriptor inspired by the Schrödinger equation was proposed. This feature descriptor is called the Wave Kernel Signature (WKS) and represents the average probability of measuring a quantum mechanical particle at a specific location. Let us note that in order to make the series expansion techniques efficient, it is advocated to employ heuristics such as the scaling of time [17] or energy domain [1]. This avoids the computation of the full spectrum. As an alternative approach to the construction of the feature descriptor, the authors of [5] proposed to solve the corresponding partial differential equations (PDEs) numerically, i.e. the heat equation and the Schrödinger equation are discretized directly in space and time. From the constructed solution of the corresponding PDE the feature descriptor can be extracted.

Our Contribution. In this paper we introduce a novel feature descriptor based on the classic wave equation. To our best knowledge, this rather fundamental model has not been considered before for that purpose. Our motivation for proposing the wave equation is that the wave propagation described by this PDE may yield by the arising complex wave interaction phenomena a more unique signature as in previous models. Let us also note that the physical basis of the wave equation is inherently very different to the quantum mechanical approach based on the Schrödinger equation. Similarly to the proceeding in [5] we solve the wave equation numerically by direct discretization of the PDE. As a consequence of the results of that work, we perform time integration by implicit schemes. Since the suitable numerical solution can have a significant influence on the quality of the feature descriptor, we opt to give here a detailed study of important numerical baseline methods at hand of one standard shape data set. We demonstrate experimentally that our novel feature descriptor may give superior matching results compared to the state of the art WKS based on the Schrödinger equation. Especially, we observe a substantially higher accuracy in detecting correct one-to-one correspondences at the first match.

2   The Models

In this section we first briefly sketch the mechanism of the spectral eigenfunction expansion methods that are employed for comparison purposes. Afterwards, we present the framework of the classical wave equation.

2.1   Existing Spectral Methods

The key mechanism of the eigenfunction expansion methods is the spectral theorem which is a result stating when a linear and self-adjoint operator can be diagonalized. Fortunately, the Laplace-Beltrami operator is such an operator on the space $L_2(\mathcal{M})$. We assume the existence of a discrete spectral decomposition of eigenvalues $\lambda_0 < \lambda_1 \leq \ldots$ and corresponding orthogonal eigenfunctions $\varphi_0, \varphi_1, \ldots$ satisfying the Helmholtz equation $\Delta_\mathcal{M}\, \varphi_i = -\lambda_i\varphi_i$. Here, λ_i is the ith eigenvalue and φ_i denotes the corresponding i^{th} eigenfunction of the Laplace-Beltrami operator $\Delta_\mathcal{M}$, respectively.

Heat Kernel Signature. For $p \in \mathcal{M}$ the heat equation $\partial_t u(p,t) = \Delta_\mathcal{M}\, u(p,t)$ describes how a heat distribution $u(p,t)$ would propagate along a surface $\mathcal{M}$. Its solution at time t can be expressed as convolution of the heat kernel $K(p,p',t)$ with the initial heat distribution at $p' \in \mathcal{M}$:

$$u(p,t) = \int_\mathcal{M} K(p,p',t)\, u(p',0)\, \mathrm{d}\mu(p'), \tag{1}$$

where $K(p,p',t)$ describes the volume of heat transmitted from p' to p after time t. According to the spectral decomposition of the Laplace-Beltrami operator, the heat kernel can be expressed as

$$K(p,p',t) = \sum_{i=1}^{\infty} \exp\left(-\lambda_i t\right) \varphi_i(p)\varphi_i(p') \tag{2}$$

where λ_i are the (ordered) eigenvalues and φ_i the corresponding eigenfunctions. The quantity [17]

$$\mathrm{HKS}(p,t) = K(p,p,t) = \sum_{i=1}^{\infty} \exp\left(-\lambda_i t\right) \varphi_i^2(p) \tag{3}$$

describes the amount of heat present at point p at time t. In [17], the authors proposed to associate each point p on the surface with a vector sampled at a finite set of times $t_1, \ldots, t_M$. This feature descriptor is called the Heat Kernel Signature.

Wave Kernel Signature. The Schrödinger equation $\partial_t u(p,t) = i\Delta_\mathcal{M} u(p,t)$, where $i^2 = -1$ allows to study how a free and massive quantum particle would move on the surface $\mathcal{M}$. In quantum mechanics, the dynamics of a particle is described by its complex-valued function $u(p,t)$ which can be expressed in terms of eigenfunctions and eigenvalues of the Laplace-Beltrami operator. In [1], the authors introduced the WKS to be defined as the average probability over time to measure a particle at position p:

$$\mathrm{WKS}(p,e) = \lim_{T \to \infty} \frac{1}{T} \int_0^T |u(p,t)|^2 \mathrm{d}t = \sum_{i=1}^{\infty} c_e^2(\lambda_i)\phi_i^2(p). \tag{4}$$

where c_e is an appropriate log-normal energy distribution and a logarithmic energy scale e was chosen. Finally, the Wave Kernel Signature can be written as:

$$\text{WKS}(p, e) = C_e \sum_{i=1}^{\infty} \phi_i^2(p) \exp\left(\frac{-(e - \log \lambda_i)^2}{2\sigma^2}\right) \tag{5}$$

where the variance of the energy distribution is denoted by σ and C_e represents a normalization factor.

2.2 The Geometric Wave Equation

Let us elaborate a bit on our new model with the aim to convey a physical intuition of the underlying process. To this end we observe in the following the motion of disturbances originating from a localized point in all directions over a surface $\mathcal{M}$. The resulting phenomena are called waves and their dynamics will be described by $u(p, t)$. The surface may then be considered as a mesh of point masses connected by elastic strings. Applying an initial displacement of a point mass at a fixed position $u(p, 0) = \phi$, this displacement takes place in outward normal direction with respect to $\mathcal{M}$, i.e. $\phi > 0$. The motion of the displaced point mass is governed by Newton's second law of motion with the acting tension force of the elastic strings.

The resulting dynamic process is a wave which propagates as a result of that initial displacement. The mathematical formulation reads as:

$$\frac{1}{c^2}\partial_{tt}u(p, t) = \Delta_{\mathcal{M}}\, u(p, t) \tag{6}$$

where the constant c is the speed of the wave's propagation, assumed here to be at unit velocity $c = 1$. The wave equation is a hyperbolic second order linear partial differential equation and requires therefore a further initial value $\partial_t u(p, 0) = \psi$ which has the interpretation of the initial velocity of ϕ.

3 Discretization Aspects

As indicated, in contrast to the framework of the spectral descriptors, we will construct a feature descriptor by direct discretization of the underlying wave equation.

In order to approximate the wave equation on a shape we have to take care of three things. Firstly, a discrete approximation of our continuous and closed surface as well as of the time domain is needed. Secondly, a suitable discrete Laplace-Beltrami operator has to be defined. Thirdly, a discrete approximation for the second order time derivative has to be found.

Discrete Space and Time Domain. A suitable surface representation is given by a triangular mesh. In more detail, a triangulated surface is given by the tuple $\mathcal{M}_d = (P, E)$. The point cloud $P := \{p_1, ..., p_N\}$ contains the finite number of coordinate points a shape consists of. The edges E contain the neighborhood relations between the coordinate points. The entire mesh can be formed by connecting the coordinate points p_i so that one obtains two-dimensional triangular cells.

The time axis $t \in [0, T]$ is sampled uniformly by $M + 1$ grid points $0 =: t_0 < ... < t_k < ... < t_M := T$ and τ denotes the (uniform) time increment. Thus, a function u can be approximated via $u(p, t) \approx (u(p_1, t_k), ..., u(p_N, t_k))^\top = (u_1^k, ..., u_N^k)^\top =: u^k \in \mathbb{R}^N$, where the vector contains the spatial components from the kth time layer.

Discrete Laplace-Beltrami Operator. Many schemes have been proposed to estimate the Laplace-Beltrami operator for a triangular meshed surface [2,14,15]. A commonly used method is the cotangent weight scheme introduced in [13]. The authors present a formal derivation using the mixed finite-element/finite-volume paradigm. The arising formulae can be transferred into matrix notation:

$$(L_\mathcal{M})_{ij} = \begin{cases} -\frac{1}{2A_j} \sum_{i \in N_j} (\cot \alpha_{ij} + \cot \beta_{ij}), & \text{if } i = j \\ \frac{1}{2A_j} (\cot \alpha_{ij} + \cot \beta_{ij}), & \text{if } i \neq j \text{ and } i \in N_j \\ 0, & \text{else} \end{cases} \tag{7}$$

where $L_\mathcal{M} \in \mathbb{R}^{N \times N}$ and $\Delta_\mathcal{M} u(p, t) \approx L_\mathcal{M} u$ is a linear approximation. With N_j we denote the set of points adjacent to p_j, and A_j represents the barycentric area of the cell that corresponds to p_j. Furthermore, α_{ij} and β_{ij} denote the two angles opposite to the edge (p_i, p_j) as shown in Fig. 1.

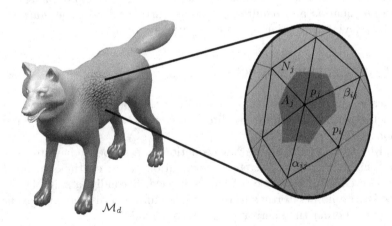

Fig. 1. Illustration of the cotangent weight scheme.

Discrete Time Operator. Finally, the discrete representation of the second order time derivative $\partial_{tt}u(p,t)$ has to be defined. In this paper we use the finite difference method to obtain the following baseline methods for time integration:

$$\text{centered:}\qquad \partial_{tt}u(p,t) \approx \frac{u^{k+1} + 2u^k - u^{k-1}}{\tau^2} \tag{8}$$

$$\text{backward:}\qquad \partial_{tt}u(p,t) \approx \frac{u^k + 2u^{k-1} - u^{k-2}}{\tau^2} \tag{9}$$

3.1   Some Details on Our Approach

Let us comment that the definition of a time integration method not only refers to the use of a discrete time operator, but also on the time level used for evaluating the expressions of the spatial discretization. To this end the data sets u^k and u^{k-1} from time levels k and $k-1$ are considered as given.

Since explicit finite differences schemes are known to be just conditionally stable, leading to limitations for the time increment τ in dependence of the cell area A_j, they are not usable without restriction for shape analysis tasks. Therefore we study here two possible implicit time integration schemes as described below.

Backward Time Integration (BT Method). The second order time derivative $\partial_{tt}u(p,t)$ is replaced by the second order backward differences quotient to obtain the backward time integration method for the wave equation

$$\frac{u^k - 2u^{k-1} + u^{k-2}}{\tau^2} = L_\mathcal{M}u^k \quad \Leftrightarrow \quad \left(I - \tau^2 L_\mathcal{M}\right)u^{k+1} = 2u^k - u^{k-1} \quad \text{(BT-W)}$$

where $k = 1, ..., M-1$ and an index shift $k \to k+1$ is applied. Computing values at time $k+1$ requires solving a system of linear equations at each time step.

Centered in Time Integration (CT Method). The wave equation is approximated by the second order central difference quotient to obtain the centered time integration method for the wave equation

$$\frac{u^{k+1} - 2u^k + u^{k-1}}{\tau^2} = \frac{1}{2}\left(L_\mathcal{M}u^{k-1} + L_\mathcal{M}u^{k+1}\right) \tag{10}$$

$$\Leftrightarrow \quad \left(I - \frac{\tau^2}{2}L_\mathcal{M}\right)u^{k+1} = 2u^k - \left(I - \frac{\tau^2}{2}L_\mathcal{M}\right)u^{k-1} \quad \text{(CT-W)}$$

where $k = 1, ..., M-1$ and u^k was linearly interpolated using u^{k+1} and u^{k-1}. The centered in time scheme is an implicit method. To obtain values at time $k+1$ it requires both solving a system of linear equations as well as an additional matrix-vector multiplication compared to (BT-W) for each time step.

Approximation of the Initial Condition. The wave equation requires two initial conditions. For $u(p, 0)$, the initial point displacement ϕ is approximated by the Kronecker peak $\phi = \delta_{p_i}(p)$. In vector notation the initial condition reads

$$u^0 = \delta_{p_i}(p) = (u_1^0, ..., u_i^0, ..., u_N^0)^\top = (0, ..., 1, ..., 0)^\top \tag{11}$$

and we give it the interpretation of a sampled Gaussian distribution with a small variance parameter. Also the initial velocity of the point displacement $\partial_t(p, 0) = \varphi$ has to be approximated. This second initial condition is used in turn to obtain an expression for u^2. Therefore $\partial_t(p, 0) = \psi$ becomes approximated by the central difference quotient at the time-layer $k = 0$, to result in

$$\partial_t(p, 0) \approx \frac{u^1 - u^{-1}}{2\tau} = \psi \quad \Leftrightarrow \quad u^{-1} = -2\tau\psi + u^1 \tag{12}$$

where the point u^{-1} is a virtual point. To obtain u^1 we simply use (BT-W) and (CT-W) for the time layer $k = 0$ and replace the virtual term u^{-1} by (12). For the backward integration method we get

$$\left(I - \tau^2 L_\mathcal{M}\right) u^1 = 2\phi + 2\tau\psi - u^1 \quad \Leftrightarrow \quad \left(I - \frac{\tau^2}{2} L_\mathcal{M}\right) u^1 = \phi + \tau\psi \tag{13}$$

and finally, for the centered integration method one obtains

$$\left(I - \frac{\tau^2}{2} L_\mathcal{M}\right) u^1 = 2\phi - \left(I - \frac{\tau^2}{2} L_\mathcal{M}\right) u^1 + \left(2\tau I - \tau^3 L_\mathcal{M}\right)\psi \tag{14}$$

$$\Leftrightarrow \quad \left(I - \frac{\tau^2}{2} L_\mathcal{M}\right) u^1 = \phi + \left(\tau I - \frac{\tau^3}{2} L_\mathcal{M}\right)\psi. \tag{15}$$

For the following framework, the initial velocity of the displacement is assumed to be zero $\psi = 0$.

3.2   The Wave Equation Based Feature Descriptor

In order to obtain the feature descriptor from the wave equation at the location $p = p_i$ we follow the two key steps:

- Solve the wave equation $\partial_{tt} u(p, t) = \Delta_\mathcal{M} u(p, t)$ with $u(p, 0) = \delta_{p_i}(p)$ and $\partial_t u(p, 0) = 0$ to obtain $u(p, t)$ as shown for example in Fig. 2.
- From the solution $u(p, t)$ the feature descriptors can be extracted by restricting its spatial component $f_i(t) := u(p, t)|_{p=p_i} = \left(u_i^0, ..., u_i^M\right)^\top \in \mathbb{R}^{M+1}$, such it is considered at the point p_i, i.e. where the initial condition is triggered.

Since the underlying PDE describes physical phenomena it is possible to assign those to the feature descriptor. The wave based feature descriptor describes the motion amplitudes of an emitted wave front over the time at the considered point p_i. Therefore, the feature descriptor catches partially reflected waves, influenced by the intrinsic geometry of the surface. Over time the waves are spread throughout the surface.

Fig. 2. The dynamics described by the wave equation on a triangular meshed surface. In this case, the solution was modelled with the BT scheme. The initial point displacement $u(p,0) = \delta_{blue}(p)$ takes place at the blue point on the man's chest. The time evolution of $|u|$ is shown from left to the right. (Color figure online)

4   The Correspondence Problem

Let us consider two almost isometric and triangular meshed surfaces $\mathcal{M}_d$ and $\tilde{\mathcal{M}}_d$. Then, the almost isometric transformation $Q_{aiso} : \mathcal{M}_d \to \tilde{\mathcal{M}}_d$ unfolds one surface onto the other. This transformation $Q_{aiso} := Q_{iso} \circ Q_\varepsilon$ can be decomposed into a purely isometric part Q_{iso} and a small distortion Q_ε. We call $\mathcal{M}_d$ the reference shape because we are able to allocate a fixed labelling to its points $P = \{p_1, ..., p_N\}$. For a fixed labeling $p_i \in P$ the assignment map $match : P \to \tilde{P}$, $p_i \mapsto \tilde{p}_j$ gives $\tilde{p}_j \in \tilde{P}$ a label on the deformed shape, and the points with matching indices belong together. Therefore we write $p_i \leftrightarrow \tilde{p}_j \Leftrightarrow match(p_i) = \tilde{p}_j$ if p_i and $\tilde{p}_j$ describe the same point on $\mathcal{M}_d$ and $\tilde{\mathcal{M}}_d$. Now, the feature descriptor is used as a local shape representation that is invariant to isometric transformations $f_i = Q_{iso}(f_i)$ and robust to almost isometric transformations $f_i \approx Q_{iso}(f_i)$. Let us apply this condition to the correspondence problem: $p_i \leftrightarrow \tilde{p}_j \Leftrightarrow f_i \approx Q_{aiso}(f_j) := \tilde{f}_j$. The feature descriptors $f_i(t)$ and $\tilde{f}_j(t)$ are real valued functions, sampled point-wise on the time axis. Therefore, the distance between two discrete feature descriptors is simply measured by using the normalized l_1-norm:

$$d(f_i, \tilde{f}_j)(t) := \frac{1}{M} \sum_{k=1}^{M} |f_i(t_k) - \tilde{f}_j(t_k)|. \tag{16}$$

The points with the smallest feature distance should belong together. This condition can be written as a minimisation problem for all points:

$$p_i \leftrightarrow \tilde{p}_j \quad \Leftrightarrow \quad match(p_i) = \underset{\tilde{p}_j \in \tilde{\mathcal{M}}_d}{\arg\min} \left\{ d(f_i, \tilde{f}_j)(t) \right\}. \tag{17}$$

An example for a correspondence task is shown in Fig. 3.

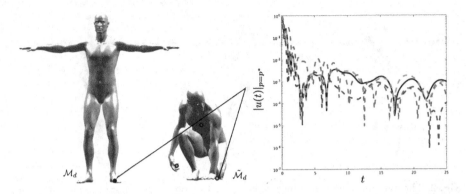

Fig. 3. An almost isometric correspondence task: The black point on the man's toe has to be retrieved on the almost isometric counterpart (green point). The spectrum of the feature descriptor at corresponding locations is similar. The problem leads to a minimisation problem, finding the feature descriptors with the smallest distance. (Color figure online)

5   Experiments

In the following we give a detailed quantitative evaluation showing that the proposed new model based on the wave equation may give in some interesting aspects results that are superior to previous kernel based descriptors.

The Data Set. For our illustrative experiment six *centaur shapes* are used, see Fig. 4, taken from the TOSCA data set [3], available in the public domain. For those shapes we evaluate a dense shape correspondence by investigating the performance to find correct corresponding pairs for all N points on the shapes.

Parameters and Reference Model. For the experiment, the time interval size is evaluated for $T = 25$ and $T = 100$. The time increment is set to $\tau = 1$. For experimental comparison we considered the state-of-the-art wave kernel descriptor. Its parameters were set as described in [1]. Further, we evaluate the time integration feature descriptor obtained from the heat (BT-H and CT-H) and Schrödinger equation (BT-S and CT-S) [5].

Numerical Implementation Details. The implementation of our methods was done in Matlab and the sparse linear system was LU-decomposed with the SuiteSparse package [6] to reduce the cost of solving a system of linear equations to approximately $\mathcal{O}(N^2)$. To obtain feature descriptors for all points of one centaur shape it requires $M \cdot N$ times a $\mathcal{O}(N^2)$ action. For example for $T = 100$ and $\tau = 1$ it needs 1.58 million times solving a system of 15768 equations.

Cumulative Match Characteristic (CMC). The CMC curve evaluates the probability of finding the correct match within the first k best matches. The value $k = 1, ..., 100/N$ is scaled such we evaluate the correct corresponding points among the first 1% of the best matches.

Fig. 4. The transformed centaur shapes are almost isometric modifications of the reference shape (left).

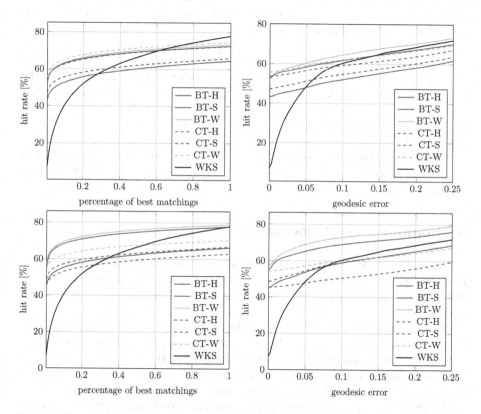

Fig. 5. The evaluation of the centaur data set. For reference, the WKS descriptor with constant 100 evaluation steps was evaluated. **Top:** The small interval size $T = 25$ was evaluated. **Bottom:** The larger interval size $T = 100$ was investigated. **Left:** The CMC curves. For $T = 25$ and for less then 0.6% of best matchings, the CT-W, BT-W, CT-S and BT-W descriptors gain better results in this experiment than the WKS descriptor. For $T = 100$, the BT-W and BT-S descriptors outperform the WKS descriptor for 1% of best matches. **Right:** The geodesic error. The CT-W method outperforms the WKS descriptor on both time sizes.

The Geodesic Error. For the evaluation of the correspondence quality, we followed the Princeton benchmark protocol [9]. This procedure evaluates the precision of the matching performance by determining how far are the computed matches away from the actual ground-truth correspondence. For a correct correspondence, this inaccuracy is zero.

Results of Quantitative Comparison. On the left-hand side of Fig. 5 the hit rate for increasing values of $100 \cdot k/N$ is shown. By increasing the size of time interval from $T = 25$ to $T = 100$ the corresponding feature descriptor contains geometric information of a larger neighborhood. This in fact leads to a better hit rate for the BT-S and BT-W descriptors. For the setting $T = 100$ the BT-W descriptors clearly beat the WKS descriptor. However, the CT-S and CT-W descriptors lose their performance at a certain time size. Yet, the combination of the heat equation with the centered time method CT-H is not affected by a performance collapse.

The evaluation of the geodesic error on the right-hand side of Fig. 5 shows the superiority of the BT-W methods over the WKS descriptor. Especially, the experimental results indicate that the time integration descriptors have a higher accuracy to find correct correspondences at the first match $k = 1$. This holds in particular for the BT-W descriptor. By comparing the wave and Schrödinger equation alone, the descriptor based on the wave equation gains better results.

Concluding the above discussion, a feature descriptor based on backward integration of the wave equation leads to the best results ($T = 100$) on the centaur data set.

6   Summary and Conclusion

We introduced a novel feature descriptor based on the classic wave equation. Further, two different time integration methods for the direct numerical solution of the wave equation PDE were investigated for the purpose to obtain feature descriptors.

Experimental results performed on the centaur data set illustrate the properties of numerical solvers and the usefulness of the approach for tackling the correspondence problem. They confirm exemplarily that our novel wave approach may have a higher precision to find correct correspondences compared to the state of the art WKS, especially for small percentages of best matchings.

In future work we aim to perform more exhaustive benchmark tests in order to evaluate on a broader basis the properties of our method. The results documented here are, to our impression, very promising and may show that our new method can at least be an alternative to kernel based methods. We also aim to consider other numerical solvers to make the research of our direct integration approach more complete.

References

1. Aubry, M., Schlickewei, U., Cremers, D.: The wave kernel signature: a quantum mechanical approach to shape analysis. In: IEEE Computer Vision Workshops (ICCV Workshops), pp. 1626–1633 (2011)
2. Belkin, M., Sun, J., Wang, Y.: Discrete Laplace operator on meshed surfaces. In: Proceedings of SOCG, pp. 278–287 (2008)
3. Bronstein, A.M., Bronstein, M.M., Kimmel, R.: Numerical Geometry of Non-rigid Shapes. Springer, New York (2009). doi:10.1007/978-0-387-73301-2
4. Bronstein, M.M., Kokkinos, I.: Scale-invariant heat kernel signatures for non-rigid shape recognition. In: Proceedings of the International Conference on Computer Vision and Pattern Recognition, pp. 1704–1711 (2010)
5. Dachsel, R., Breuß, M., Hoeltgen, L.: Shape matching by time integration of partial differential equations. In: Lauze, F., Dong, Y., Dahl, A.B. (eds.) SSVM 2017. LNCS, vol. 10302, pp. 669–680. Springer, Cham (2017). doi:10.1007/978-3-319-58771-4_53
6. Davis, T.A.: Algorithm 930: factorize: an object-oriented linear system solver for Matlab. ACM Trans. Math. Softw. **39**(4), 1–18 (2013)
7. Johnson, A., Herbert, M.: Using spin images for efficient object recognition in cluttered 3D scenes. IEEE Trans. Pattern Anal. Mach. Intell. **21**(5), 433–449 (1999)
8. Kac, M.: Can one hear the shape of a drum? Am. Math. Mon. **73**(4), 1–23 (1966)
9. Kim, V.G., Lipman, Y., Funkhouser, T.A.: Blended intrinsic maps. ACM Trans. Graph. **30**, 79 (2011)
10. Lipman, Y., Funkhouser, T.A.: Möbius voting for surface correspondence. ACM Trans. Graph. **28**(3), 1–12 (2009)
11. Lévy, B.: Laplace-Beltrami eigenfunctions towards an algorithm that 'understands' geometry. In: International Conference on Shape Modeling and Applications (2006)
12. Manay, S., Hong, B.W., Yezzi, A.J., Soatto, S.: Integral invariant signatures. In: Proceedings of ECCV, pp. 87–99 (2004)
13. Meyer, M., Desbrun, M., Schröder, P., Barr, A.H.: Discrete differential-geometry operators for triangulated 2-manifolds. In: Hege, H.C., Polthier, K. (eds.) Visualization and Mathematics III, pp. 35–57. Springer, Heidelberg (2002). doi:10.1007/978-3-662-05105-4_2
14. Pinkall, U., Polthier, K.: Computing discrete minimal surfaces and their conjugates. Exp. Math. **2**(1), 15–36 (1993)
15. Reuter, M., Wolter, F.E., Peinecke, N.: Laplace-Beltrami spectra as shape-DNA of surfaces and solids. Comput. Aided Des. **38**(4), 342–366 (2006)
16. Rustamov, R.: Laplace-Beltrami eigenfunctions for deformation invariant shape representation. In: Symposium on Geometry Processing, pp. 225–233 (2007)
17. Sun, J., Ovsjanikov, M., Guibas, L.: A concise and provably informative multi-scale signature based on heat diffusion. Comput. Graph. Forum **28**(5), 1383–1392 (2009)

Image/Video Indexing and Retrieval

A Hypergraph-Based Reranking Model for Retrieving Diverse Social Images

Noura Bouhlel[(✉)], Ghada Feki, Anis Ben Ammar, and Chokri Ben Amar

REGIM: Research Groups in Intelligent Machines,
National Engineering School of Sfax (ENIS),
University of Sfax, BP 1173, 3038 Sfax, Tunisia
noura.bouhlel.tn@ieee.org

Abstract. With the proliferation of social networks and photo-sharing websites, the need for an effective image retrieval system has become crucial. To match the users' intents, retrieval results are expected to be not only relevant to the query but also diverse. In this way, they depict a comprehensive summarization of the user query. Motivated by this observation, we propose a hypergraph-based reranking model for retrieving diverse social images. Indeed, a visual hypergraph is constructed to capture high-order relationships among images. Different from exiting hypergraph ranking that usually ranks images according to their relevance to a given query, our approach emphasizes diversity by integrating absorbing nodes into the ranking process. This way, redundant images are prevented from getting high ranking scores, thereby ensuring diversity. Extensive experiments conducted on the MediaEval 2016 dataset demonstrate that our approach can achieve competitive performance to the existing diversification approaches.

Keywords: Image retrieval · Visual reranking · Diversity · Hypergraph

1 Introduction

During the past few years, we have witnessed an explosive growth of social networks and photo-sharing websites (e.g. Facebook[1] and Flickr[2]). In view of this, the amount of community-contributed photos has been exponentially increased. Accordingly, exploiting the huge amount of photos for multimedia applications is of great importance [2].

Among multimedia applications, image retrieval has attracted increasing research attention [3,8,16,17]. Until recently, exiting image retrieval engines (e.g. Google and Flickr) commonly relied on the textual descriptions associated with images for indexing and retrieval. Nevertheless, text-based image retrieval presents some critical limitations i.e. noisy, language-dependent and irrelevant textual information. Moreover, the rich content of images is not appropriately

---

[1] www.facebook.com.
[2] www.flickr.com.

© Springer International Publishing AG 2017
M. Felsberg et al. (Eds.): CAIP 2017, Part I, LNCS 10424, pp. 279–291, 2017.
DOI: 10.1007/978-3-319-64689-3_23

described through textual information. Accordingly, retrieval results may contain some irrelevant images.

To tackle this issue, visual reranking has been proposed [5,22]. It aims at refining the text-based retrieval results by incorporating the visual information. Most visual reranking models in earlier years focused primarily on providing relevant results to the query. However, the retrieved results may contain a large number of redundant information (e.g. duplicate or visually similar images). Therefore, an effective visual reranking model should make a trade-off between relevance and diversity. In fact, diversity has been considered to be a key criteria of image retrieval results expected by users [10,11,16,20,24]. It provides a summarization of retrieval results which constitutes a complete and comprehensive representation of the query and enables faster and better access to the desired information. Recently, diversity has been the focus of several international challenges (ImageCLEF [19] and MediaEval Retrieving Diverse Social Images Task [15]). Particularly, the MediaEval task focuses on enhancing both relevance and diversity of image retrieval results within the social context [15].

In this paper, we tackle the aforementioned issue by proposing an approach for retrieving diverse social images using an hypergraph ranking with absorbing nodes. In the first place, we model the high-order relationships among social images through a visual hypergraph. Next, we perform an iterative hypergraph ranking to learn the visual ranking scores of different images. In order to achieve diversity, the hypergraph ranking algorithm is extended with the concept of absorbing nodes. The role of absorbing nodes is to prevent visually similar images in the dataset from having higher ranking scores. The proposed approach is evaluated on the MediaEval 2016 "Retrieving Diverse Social Images" dataset. Experimental results demonstrate that our approach achieve competitive performance compared to the existing diversification approaches.

The reminder of this paper is organized as follows. Section 2 presents the related work. The proposed approach is described in details in Sect. 3. Experimental results are shown in Sect. 4. Finally, Sect. 5 concludes the paper.

2   Related Work

With the proliferation of social networks and photo-sharing websites, the need for an effective image retrieval system has become crucial. In view of this, visual reranking has been broadly investigated for boosting retrieval effectiveness [5,22,28]. The prior objective of early visual reranking approaches is the relevance of retrieval results. However, in recent literature, both relevance and diversity are considered as two key criteria of effective image retrieval system [16,20]. In this way, retrieval results depict a complete and comprehensive representation of the query. For instance, Tollari et al. [23] propose an approach that uses textual reranking to improve the relevance. After reranking, an Agglomerative Hierarchical Clustering is performed to ensure the diversity. Boteanu et al. [1] introduce a novel approach based on pseudo-relevance feedback in which an automatic selection of image is applied instead of human feedback.

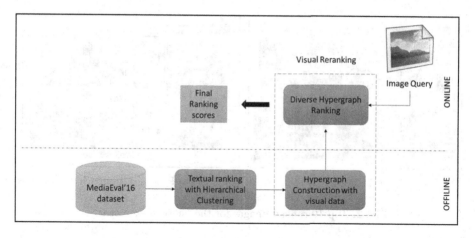

Fig. 1. Schematic illustration of the proposed diversification approach.

Diversification is achieved by applying (1) a Hierarchical Clustering and (2) diversification strategy based on a round robin approach. Zaharieva et al. [27] select the most appropriate combinations of features and clustering method for each query. The selection is performed using clustering internal validation measures. The clustering methods investigated in this method are: Affinity Propagation (AP), expectation maximization (EM), k-means (KM) and X-means (XM). Diversification is also achieved using a Round-Robin approach. Feki et al. [9] score each image by the mean average of a visual score and a textual score. The visual score is obtained by clustering the visual information using the EM and Make Density-based clustering algorithm. The textual score is obtained by an Hierarchical Clustering using the textual information. Ferreira et al. [12] propose a reranking approach that consists of 3 steps: (1) re-ranking the initial list provided by Flickr using the textual information, (2) aggregating re-ranked lists by several text-based descriptors using Genetic Programming (GP) and finally (3) employing Agglomerative and Birch methods to achieve diversification. Castellanos et al. [6] compute the relevance of each image to the query based on the visual information and a relevance feedback algorithm. Next, a textual-based FCA clustering is performed. Finally, the image at the top of each textual FCA cluster is selected to generate the final list of images. Overall, the aforementioned techniques try to improve the relevance and diversity separately. In general, they employ clustering algorithm followed by a method to select the most representative images from each cluster. In this paper, we propose a hypergraph-based reranking model that aims to jointly consider the relevance and diversity.

3   A Hypergraph-Based Reranking Model

An effective image retrieval system should generate results that are not only relevant to the query but also diverse. In this section, we propose an approach for

(a) Flickr   (b) Proposed approach

Fig. 2. Top 10 ranked images for the query *sailing_boat*

retrieving diverse social images using an hypergraph-based ranking with absorbing nodes. A schematic illustration of the proposed approach is presented in Fig. 1. After ranking the initial list provided by Flickr using the textual information and Hierarchical Clustering, we construct a visual hypergraph where the vertices are the images initially retrieved from Flickr. Each image is represented by the convolutional neural network based descriptor (CNN). Thereafter, we perform an iterative ranking over the constructed hypergraph in order to learn the ranking scores of different images. To ensure diversity, we extend the hypergraph ranking algorithm with the concept of absorbing nodes. By the way, we are able to not only rank the images based on their visual relevance but also to consider the diversity of retrieval results.

Next, we emphasize the different steps of the hypergraph-based ranking with absorbing nodes.

3.1 Visual Hypergraph Model

Graphs are widely used to represent relationships among different objects [10, 26,28]. Particularly, a hypergraph is a graph in which an edge can link more than two vertices [2]. Regarding its effectiveness in modeling high-order relationships among objects, the hypergraph has been successfully employed in numerous application such as 3D object recognition [25], image retrieval [14,18], person re-identification and tracking [21], music recommendation [4] and so on.

In our approach, the hypergraph is used for retrieving diverse social images and high-order relationships among social images are represented using hyperedges.

More formally, a hypergraph $G = (V, E, \omega)$ is consisted of a finite vertex set V, an hyperedge set E which is composed by a family of subsets e of V such that $\bigcup_{e \in E} = V$. Each hyperedge e is weighted with a positive scalar $\omega(e)$ [2]. The hypergraph G is generally represented with an incidence matrix $H \in \mathcal{M}^{|V| \times |E|}$

where

$$h(e,v) = \begin{cases} S(u,v), & v \in e \\ 0, & otherwise \end{cases} \tag{1}$$

and $S(i,j)$ is the similarity between u the centroid of hyperedge e and v. The degree of hyperedge $e \in E$ is defined as follows:

$$\delta(e) = \sum_{v \in V} h(e,v) \tag{2}$$

Each hyperedge is weighted with a positive scalar expressing its importance in the hypergraph. The weight of the hyperedge $e \in E$ is defined as follows:

$$\omega(e) = \sum_{v \in e} S(e,v) \tag{3}$$

The degree of vertex $v \in V$ is defined as follows:

$$d(v) = \sum_{e \in E} \omega(e)h(e,v) \tag{4}$$

Let D_v, D_e and W be diagonal matrices containing the vertex degrees, the hyperedge degrees and the hyperedge weights respectively.

In our image retrieval framework, we construct a visual hypergraph in which vertices are the visual descriptors of social images.

Let $\chi = \{x_1, ...x_N\}$ be a collection of social images and $F = \{f_1, ...f_N\}$ a set of visual descriptors where f_i is the adapted convolutional neural network based descriptor (CNN) of image x_i. CNN descriptors are based on the reference convolutional neural network (CNN) model provided along with the Caffe framework [15].

In order to capture high-order relationships of visually similar images, we build k-nearest neighbors graphs based on the similarity between different CNN descriptors measured with Euclidean distance. Indeed, we consider each descriptor f_i as a 'centroid' vertex and compose a hyperedge by connecting each centroid with its k-nearest neighbors.

3.2 Hypergraph-Based Diversified Ranking

Given the constructed visual hypergraph and a user query, exiting image retrieval approaches perform an hypergraph ranking that rank all vertices in the hypergraph with respect to their relevance to the query [13,14]. Nevertheless, they do not consider the diversity of visual search results. In fact, beside relevance, diversity has also been considered as critical criterion for ranking [7]. Diversified retrieval results provide users with a complete and comprehensive representation of the query. To tackle this issue, we extend the hypergraph ranking algorithm with the concept of absorbing nodes which improve the diversity of retrieval results [21].

Let $q \in \Re^{|V| \times 1}$ be the query vector containing the initial ranking scores:

$$q_i = \begin{cases} 1 & \text{if i is the query index} \\ 0 & \text{if } otherwise \end{cases} \tag{5}$$

Let y be the to-be-learned ranking function that assigns a ranking score $y(v)$ to vertex $v \in V$ such that,

$$y : |V| \to \mathbb{R}, v \to y(v) \tag{6}$$

In our framework, we formulate the hypergraph ranking as a regularization framework similarly to [29]:

$$\arg \min_{y} \{\Omega(y) + \mu R_{emp}(y)\} \tag{7}$$

where $\Omega(y)$ is a regularization term which ensures that vertices sharing many hyperedges will probably have similar ranking scores.

$$\Omega(y) = \frac{1}{2} \sum_{e \in E} \sum_{v,u \in V} \frac{\omega(e)h(e,v)h(e,u)}{\delta(e)} \\ (\frac{y(v)}{\sqrt{d(v)}} - \frac{y(u)}{\sqrt{d(u)}})^2 \tag{8}$$

and $R_{emp}(y)$ denotes the empirical loss which imposes that the final ranking scores are not far away from the initial ones.

$$\mu R_{emp}(y) = \mu \|y - q\|^2 = \sum_{v \in V} (y(v) - q)^2 \tag{9}$$

where μ is a positive weighting parameter. To solve the cost function in Eq. 7, let $\Delta = I - \Theta = I - D_v^{-1/2} H W D_e^{-1} H^T D_v^{-1/2}$ be the Laplacian of the hypergrpah. Then,

$$\Omega(f) = y^T (I - \Theta)y \tag{10}$$

By following the step described in [29], the final ranking scores is obtained by solving the following linear equation:

$$((1 + \mu)I - \Theta)y = \mu q \tag{11}$$

Let $\gamma = \frac{1}{1+\mu}$, the final ranking scores can be obtained as

$$y = (1 - \gamma)(I - \gamma\Theta)^{-1} q \tag{12}$$

The aforementioned hypergraph ranking algorithm enables to perform relevant-based image retrieval. However, an effective visual search should return not only relevant results to the query but also diverse results. To enhance the diversity, redundant items in the ranked images list should be avoided. Therefore, we extend the aforementioned hypergraph ranking with the concept of absorbing nodes. Actually, to the best of our knowledge, this concept has not been applied to the context of image retrieval yet. Different from the hypergraph ranking, the hypergraph ranking with absorbing nodes computes the ranking scores in an iterative manner. At each iteration, previously-ranked items are turned into absorbing nodes, i.e. their ranking scores are set equal to zero. In this way, any ranking score is propagated from the absorbing nodes to their neighbors. Accordingly, the redundant candidates are prevented from having high ranking scores. Hence, the diversity of visual search is enhanced.

More formally, the ranking scores are computed iteratively as follows.

$$y^{(t+1)} = \gamma \Theta I_f y^{(t)} + (1 - \gamma)q \tag{13}$$

Where I_f is an identity matrix such that

$$I_f(i, i) = \begin{cases} 0, \text{if the (i,i)-entry is an absorbing nodes} \\ 1, otherwise \end{cases} \tag{14}$$

The sequence $\{y^{(t)}\}$ converges to the ranking scores obtained previously in the analytic analysis, i.e. $y = (1 - \gamma)(I - \gamma \Theta)^{-1}q$ [7].

Proof. Starting from (13) used in the iterative hypergraph ranking, we have

$$y^{(t)} = (\gamma \Theta I_f)^{(t)} f^{(0)} + (1 - \gamma) \sum_{i=0}^{t-1} (\gamma \Theta I_f)^i q$$

Let $\hat{P} = D_v^{-1} H W D_e^{-1} H^T I_f$ be the similarity transformation of ΘI_f as follows:

$$\begin{aligned} \Theta I_f &= D_v^{-1/2} H W D_e^{-1} H^T D_v^{-1/2} I_f \\ &= D_v^{1/2} D_v^{-1} H W D_e^{-1} H^T D_v^{-1/2} I_f \\ &= D_v^{1/2} \hat{P} D_v^{-1/2} \end{aligned}$$

Therefore, $\hat{P}$ and ΘI_f are similar and have the same eigenvalues. By Gershgorin circle theorem, we have

$$|\rho| \le \sum_{i \ne j} |\hat{P}_{ii}| \le 1$$

where ρ is the largest eigenvalue of $\hat{P}$. Then eigenvalues of ΘI_f are not greater than one.

Since $0 \le \gamma \le 1$ and $|\rho| \le 1$, we have

$$\lim_{t \to \infty} (\gamma \Theta I_f)^t = 0$$

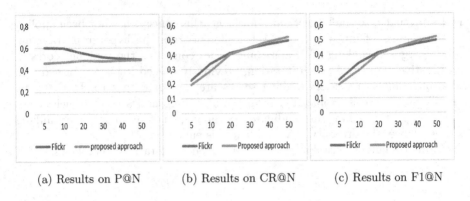

(a) Results on P@N   (b) Results on CR@N   (c) Results on F1@N

Fig. 3. Comparison with Flickr baseline in terms of relevance and diversity

and

$$\lim_{t\to\infty}\sum_{i=0}^{t-1}(\gamma\Theta I_f)^i = (I - \gamma\Theta I_f)^{-1}$$

Hence, the sequence $\{y^{(t)}\}$ converges to

$$y^* = (1 - \gamma)(I - \gamma\Theta I_f)^{-1}q$$

4   Experimental Results

4.1   Data and Evaluation Metrics

In order to evaluate the performance of the proposed approach, we conducted the experiments on the test set of the public dataset MediaEval 2016 "Retrieving Diverse social images Task" [15]. The test set consists of 65 general-purpose multi-topic queries. Each query is represented with a list of around 300 Flickr-provided images which are typically ranked using the default 'relevance' algorithm of Flickr. Furthermore, each image is represented with several content descriptors (e.g. CNN visual descriptors, text information) [15].

Performance in terms of both relevance and diversity is measured with the following metrics: Precision, Cluster recall and the F1-measure [15].

- Precision at X (P@X): this measure is used to assess the relevance. It denotes the number of relevant photos among the top X ranked results.
- Cluster Recall at X (CR@X): this measure is used to assess the diversity. It denotes the number of different clusters from the ground truth that are represented among the first X results.
- F1-measure (F1@X). The F1 measure is the harmonic mean of P@X and CR@X. It is used to measure both relevance and diversity.

We conduct the following experiments: Sect. 4.2 compares the proposed approach with Flickr Baseline; Sect. 4.3 deals with performance stability of the proposed approach; finally, Sect. 4.4 presents a comparison to diversification approaches presented in the MediaEval 2016. In these experiments, we randomly choose an image from the Flickr-provided list of images as a visual query in our proposed approach.

4.2 Refinement of Flickr Retrieval Results

Given a Flickr-provided list of images, our focus is to evaluate the performance of proposed approach in refinement of Flickr retrieval results. Accordingly, in this experimentation, we consider the default 'relevance' algorithm of Flickr as baseline.

As shown in the Fig. 2, the proposed approach achieves a diversity improvement over the baseline approach.

Figure 3 presents the comparison results with Flickr baseline in terms of relevance and diversity. Indeed, starting from the official ranking metric of the MediaEval 2016 reranking task, which was fixed to a cutoff of 20 images, we achieve 0,3738 of CR@20 compared to 0,3609 for Flickr baseline. Interestingly, the proposed approach achieves a steady diversity improvement over the baseline from a cutoff at 20 images (see Fig. 3b). However, it is necessary to note that the Flickr baseline relevance is maintained for a cutoff at 30 images and above (see Fig. 3a).

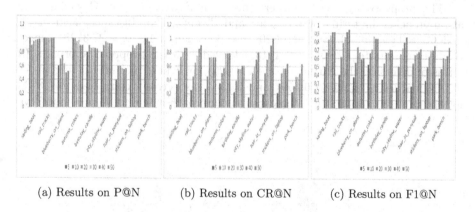

(a) Results on P@N   (b) Results on CR@N   (c) Results on F1@N

Fig. 4. Evolution curve of relevance and diversity for different query topics

Overall, as revealed in Fig. 3c, our approach outperforms the Flickr baseline in terms of both relevance and diversity from a cutoff at 30 images.

4.3 Refinement Stability Analysis

In this section, we aim to investigate the performance stability of our approach at different cutoff points for different query topics.

Figure 4 illustrates the performance per query of the proposed approach in terms of precision, cluster recall and F1-measure for various cutoff points. We rank the queries based on the F1-measure and select the top ten queries for the evaluation.

In terms of relevance, Fig. 4a show that, for most queries, the precision decreases when the number of cutoff points is increased. For instance, for the query *"blueberry_on_plant"*, the precision P@20 = 0,75 compared to 0,52 for P@50. This results reveals the fact that the more number of images are returned, the more irrelevant images are arisen.

Contrary to the precision, the cluster recall increases proportionally with the number of returned images. For instance, for the topic class *"rail_tracks"*, we reach the best cluster recall at a cutoff at 50 images (e.g., CR@50 = 0,9 whereas CR@20 = 0,65).

Overall, the above observations demonstrate the fact that our approach is stable in terms of precision. However, the diversity of our retrieval system is more dependent on the number of cutoff points that the more the number of evaluated images rises, the more our approach is effective (see Fig. 4c).

4.4   Comparison to Diversification Approaches

In this section, we compare our approach to diversification approach presented during the MediaEval'16 in terms of Precision (P@20), Cluster Recall (CR@20) and F1-score (F1@20) as illustrated in Fig. 5.

The proposed approach outperforms both Feki et al. and Castellanos et al. approaches in terms of relevance and diversity. For instance, Feki et al. and Castellanos et al. have $F1@20 = 0,3964$ and $F1@20 = 0,3745$ respectively compared to $F1@20 = 0,4013$ achieved with the proposed approach. In terms of diversity, the cluster recall (CR@20) for Feki et al. and Castellanos et al. approaches is equal to 0,3501 and 0,3544 respectively compared to 0,3738 for the proposed approach. Moreover, our approach achieved very competitive

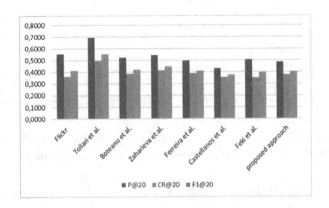

Fig. 5. Comparison to diversification approaches

performance with both Ferreira et al. and Boteanu et al. approaches. As shown in Fig. 5, our proposed approach have $F1@20 = 0,4013$ compared to $F@20 = 0.4107$ and $F1 = 0.4208$ for Ferreira et al. and Boteanu et al. respectively.

It is interesting to note that the best performance in terms of both relevance and diversity is achieved by the approach of Tollari et al. One explanation is the fact that authors have proceeded with a re-ranking step based on textual features to enhance relevance. Furthermore, different from other approaches, they use an effective clustering method based on ScalableColor features [23] as visual information and also textual information.

5   Conclusion

In this paper, a novel approach is proposed to retrieve diverse social images based on a modified hypergraph ranking. A visual hypergraph is first constructed with database images. Then, an iterative hypergraph ranking is performed. During the ranking process, ranked items is turned into absorbing nodes to prevent redundant items from having higher scores. In this way, the diversity of visual search is enhanced. Experimental results, conducted on the MediaEval 2016 "Retrieving Diverse Social Images Task" dataset, depict that the proposed method has achieved a significant enhancement of search results diversity. Future work will focus on integrating more social information to further enhance both relevance and diversity.

References

1. Boteanu, B., Constantin, M., Ionescu, B.: LAPI @ 2016 retrieving diverse social images task: a pseudo-relevance feedback diversification perspective. In: MediaEval 2016 (2016)
2. Bouhlel, N., Ksibi, A., Ammar, A.B., Amar, C.B.: Semantic-aware framework for mobile image search. In: ISDA 2015, pp. 479–484 (2015)
3. Bouhlel, N., Ben Ammar, A., Ksibi, A., Ben Amar, C.: Soff: scalable and oriented fast-based local features. In: Proceedings of the SPIE, Ninth International Conference on Machine Vision (ICMV 2016), vol. 10341, pp. 1034102-1–1034102-6 (2017)
4. Bu, J., Tan, S., Chen, C., Wang, C., Wu, H., Zhang, L., He, X.: Music recommendation by unified hypergraph: combining social media information and music content. In: Proceedings of the 18th ACM International Conference on Multimedia (MM 2010), pp. 391–400. ACM, New York (2010)
5. Cai, J., Zha, Z.J., Wang, M., Zhang, S., Tian, Q.: An attribute-assisted reranking model for web image search. IEEE Trans. Image Process. **24**(1), 261–272 (2015)
6. Castellanos, A., Benavent, X., García-Serrano, A., de Ves Cuenca, E.: UNED-UV @ retrieving diverse social images task. In: MediaEval 2016 (2016)
7. Cheng, X.Q., Du, P., Guo, J., Zhu, X., Chen, Y.: Ranking on data manifold with sink points. IEEE Trans. Knowl. and Data Eng. **25**(1), 177–191 (2013)

8. Fakhfakh, R., Feki, G., Ammar, A.B., Amar, C.B.: Personalizing information retrieval: a new model for user preferences elicitation. In: 2016 IEEE International Conference on Systems, Man, and Cybernetics (SMC), pp. 002091–002096, October 2016

9. Feki, G., Fakhfakh, R., Bouhlel, N., Ammar, A.B., Amar, C.B.: REGIM @ 2016 retrieving diverse social images task. In: MediaEval 2016 (2016)

10. Feki, G., Ksibi, A., Ammar, A.B., Amar, C.B.: Improving image search effectiveness by integrating contextual information. In: CBMI 2013, pp. 149–154 (2013)

11. Feki, G., Ammar, A.B., Amar, C.B.: Adaptive semantic construction for diversity-based image retrieval. In: Proceedings of the International Conference on Knowledge Discovery and Information Retrieval (IC3K 2014), pp. 444–449 (2014)

12. Ferreira, C.D., Calumby, R.T., do C. Araujo, I.B.A., Dourado, Í.C., Muñoz, J.A.V., Penatti, O.A.B., Li, L.T., Almeida, J., Torres, R.: Recod @ mediaeval 2016: diverse social images retrieval. In: MediaEval 2016 (2016)

13. Gao, Y., Wang, M., Luan, H., Shen, J., Yan, S., Tao, D.: Tag-based social image search with visual-text joint hypergraph learning. In: Proceedings of the ACM Conference on Multimedia, pp. 1517–1520 (2011)

14. Huang, Y., Liu, Q., Zhang, S., Metaxas, D.N.: Image retrieval via probabilistic hypergraph ranking. In: 2010 IEEE Computer Society Conference on Computer Vision and Pattern Recognition. pp. 3376–3383, June 2010

15. Ionescu, B., Gînsca, A., M., Boteanu, B., Lupu, M., Müller, H.: Retrieving diverse social images at mediaeval 2016: challenge, dataset and evaluation. In: MediaEval 2016 (2016)

16. Ionescu, B., Popescu, A., Radu, A.L., Müller, H.: Result diversification in social image retrieval: a benchmarking framework. Multimed. Tools Appl. 75(2), 1301–1331 (2016)

17. Ksibi, A., Ben Ammar, A., Ben Amar, C.: Adaptive diversification for tag-based social image retrieval. Int. J. Multimed. Inf. Retr. 3(1), 29–39 (2014). http://dx.doi.org/10.1007/s13735-013-0045-5

18. Liu, Y., Shao, J., Xiao, J., Wu, F., Zhuang, Y.: Hypergraph spectral hashing for image retrieval with heterogeneous social contexts. Neurocomputing 119, 49–58 (2013). Intelligent Processing Techniques for Semantic-based Image and Video Retrieval

19. Paramita, M.L., Sanderson, M., Clough, P.: Diversity in photo retrieval: overview of the ImageCLEFPhoto task 2009. In: Peters, C., Caputo, B., Gonzalo, J., Jones, G.J.F., Kalpathy-Cramer, J., Müller, H., Tsikrika, T. (eds.) CLEF 2009. LNCS, vol. 6242, pp. 45–59. Springer, Heidelberg (2010). doi:10.1007/978-3-642-15751-6_6

20. Spyromitros-Xioufis, E., Papadopoulos, S., Ginsca, A.L., Popescu, A., Kompatsiaris, Y., Vlahavas, I.: Improving diversity in image search via supervised relevance scoring. In: Proceedings of the 5th ACM on International Conference on Multimedia Retrieval (ICMR 2015), pp. 323–330. ACM, New York (2015)

21. Sunderrajan, S., Manjunath, B.S.: Context-aware hypergraph modeling for re-identification and summarization. IEEE Trans. Multimed. 18(1), 51–63 (2016)

22. Tian, X., Yang, L., Lu, Y., Tian, Q., Tao, D.: Image search reranking with hierarchical topic awareness. IEEE Trans. Cybern. 45(10), 2177–2189 (2015)

23. Tollari, S.: UPMC at mediaeval 2016 retrieving diverse social images task. In: MediaEval 2016 (2016)

24. Wang, M., Yang, K., Hua, X.S., Zhang, H.J.: Towards a relevant and diverse search of social images. IEEE Trans. Multimed. 12(8), 829–842 (2010)

25. Xia, S., Hancock, E.R.: 3D object recognition using hyper-graphs and ranked. In: da Vitoria Lobo, N., et al. (eds.) SSPR/SPR 2008. LNCS, vol. 5342, pp. 117–126. Springer, Heidelberg (2008). doi:10.1007/978-3-540-89689-0_16
26. Xu, B., Bu, J., Chen, C., Wang, C., Cai, D., He, X.: EMR: a scalable graph-based ranking model for content-based image retrieval. IEEE Trans. Knowl. Data Eng. **27**(1), 102–114 (2015)
27. Zaharieva, M.: An adaptive clustering approach for the diversification of image retrieval results. In: MediaEval 2016 (2016)
28. Zhang, S., Yang, M., Cour, T., Yu, K., Metaxas, D.N.: Query specific rank fusion for image retrieval. IEEE Trans. Pattern Mach. Intell. **37**(4), 803–815 (2015)
29. Zhou, D., Huang, J., Schólkopf, B.: Learning with hypergraphs: clustering, classification, and embedding. In: Advances in Neural Information Processing Systems (NIPS), p. 19. MIT Press (2006)

Learning a Limited Text Space for Cross-Media Retrieval

Zheng Yu, Wenmin Wang$^{(\boxtimes)}$, and Mengdi Fan

School of Electronic and Computer Engineering, Shenzhen Graduate School,
Peking University, Lishui Road 2199, Nanshan District, Shenzhen 518055, China
yuzheng@pku.edu.cn, wangwm@ece.pku.edu.cn, fanmengdi@sz.pku.edu.cn

Abstract. In this paper, we propose a novel model for cross-media retrieval which relies on a limited text space rather than a common space or an image space. More specifically, the model consists of three parts: A visual part that consists of a convolutional neural network and an image understanding network; A language model part that achieves sentence understanding by recurrent neural network; An embedding part that contains a fusion layer to capture both visual label information and semantic correlations between images and sentences, as well as learn the final limited text space by optimizing pairwise ranking loss. Experimental results on three benchmark datasets show that our proposed model gains promising improvement in accuracy for cross-media retrieval especially on sentence retrieval compared with the related state-of-the-art methods.

Keywords: Cross-media retrieval · Limited text space · Fusion layer · Image understanding network · Recurrent neural network

1 Introduction

Along with the popularization of the Internet, there has been a rapid growth of multimedia data such as images, texts, videos and audios which always appear together. As a result, single-media retrieval can not meet people's daily needs since some people want to search sentences that can best describe a given image or show images that can best depict a given sentence. Therefore, cross-media retrieval has been proposed which comprises two problems. The first problem is how to efficiently represent multimedia data. Traditional methods such as the bag-of-words for sentences and SIFT for images transformed multimedia data into low-level features. In order to learn more abstractive representations, deep neural network was proposed to represent data at a higher level which was proven to be more efficient than low-level features. However, the semantic gap among heterogeneous data features still exists.

W. Wang—This project was supported by Shenzhen Peacock Plan (20130408-183003656).

© Springer International Publishing AG 2017
M. Felsberg et al. (Eds.): CAIP 2017, Part I, LNCS 10424, pp. 292–303, 2017.
DOI: 10.1007/978-3-319-64689-3_24

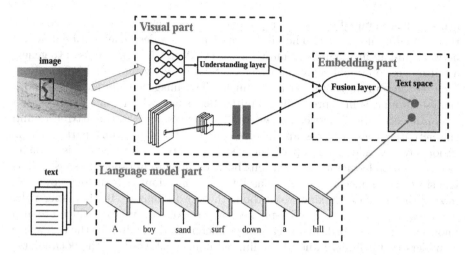

Fig. 1. A detailed illustration of our proposed model which consists of three parts: Visual part, Language model part and Embedding part.

Hence, the second problem is how to embed heterogeneous multimedia data such as images and sentences into a homogeneous space so that their similarity can be measured directly. Since we only focus on the retrieval between images and sentences, cross-media retrieval can be achieved based on a common space [4, 7, 8, 19, 21], a text space [9, 18], or an image space [10]. When performing cross-media retrieval, from the human point of view, we always try to understand the images and sentences sufficiently before retrieval. It is simple and intuitive for brains to understand the sentences but a little bit complicated to understand the images. Given an image, we first caption the image subconsciously by nature language and then understand it. In order to make the model behave as similar as human, we hope it is able to understand images and sentences sufficiently before retrieval. Therefore, we aim to perform cross-media retrieval in a text space. Current methods based on a text space mostly employed the Word2Vec space. Image understanding was achieved by a convex combination of the word embedding vectors of the visual labels predicted to be the most relevant to the image. However, the visual labels only reflect the objects contained in an image but ignore how these objects relate to each other as well as their attributes and the activities they are involved in. Thus, the Word2Vec space is not an effective text space for cross-media retrieval.

Accordingly, we propose a novel model to learn the text space effectively which is capable of understanding images like human. For the first problem mentioned above, we propose a visual part to learn deep representations for images. Meanwhile, recurrent neural network is used to learn dense representations for sentences. For the second problem, we propose an embedding part to learn a limited text space. More specifically, the whole model contains a language model part, a visual part and an embedding part. The language model part learns a dense limited text space representation for each sentence which contains rich semantic information. The visual part contains a deep convolutional neural network and an image understanding network. Deep CNN can be used to generate deep convolutional representations containing rich visual label information, such as the objects contained in an image. The image understanding network represents images in a pre-trained limited text space which can capture strong semantic correlations between images and sentences, such as the attributes and the activities these objects are involved in. For the embedding part, a single fusion layer is added on top of the visual part to capture both visual label information and semantic correlations between images and sentences, as well as transform the image representations into the final limited text space by optimizing the pairwise ranking loss. About the word "limited" in the paper title, it means that the text space is spanned by a set of base vectors which are also known as different words in a vocabulary. Therefore, the ability for the text space to understand is limited due to the limited number of words in the vocabulary.

Our core contributions are: (1) We propose a novel model to do the retrieval of images and sentences humanly in a limited text space. Pairwise ranking loss function is exploited as the objective function to be optimized. (2) The image understanding network is capable of modeling strong semantic correlations between images and sentences. (3) A single fusion layer is added on top of the visual part in order to capture both visual label information and semantic correlations between images and sentences, as well as transform the image representations into a limited text space.

The rest of the paper is organized as follows. Section 2 reviews the related work for cross-media retrieval. Section 3 describes details of our proposed model. Section 4 presents the experimental results on three datasets. Finally, we make a summary of the paper in Sect. 5.

2   Related Work

There are a lot of methods that have been proposed to handle the aforementioned two problems. For the first problem of learning efficient image and sentence representations, Sharif et al. [1] argued that a pre-trained deep convolutional neural network (CNN) was an effective image feature extractor which had achieved the state-of-the-art performances on many image processing tasks. Simonyan et al. [2] investigated how the depth of convolutional neural networks affected their performance and proposed VGG which had won the first and the second places in the localization and classification tasks respectively. For sentence, traditional methods including Word2Vec [3], LDA [17], or FV [5] were used to learn low-level representations for sentences without concerning the rich contextual information. Recently, with the great progress on machine translation [6], recurrent neural network is found to be a more powerful tool for language modeling which is able to take advantage of the contextual information of the whole sentence. Wang et al. [23] proposed a deep alternative neural network (DANN) to extract contextual information for action recognition in video.

For the second problem of learning a homogeneous space, the mainstream approach is to learn a common space by affine or deep transformation on both sentence and image sides. Canonical correlation analysis [4] learned a common space by maximizing the correlations between relevant sentences and images. Karpathy et al. [19] broke down both images and sentences into fragments and embedded them into a common multi-modal space. Fan et al. [21] performed coupled feature mapping and correlation mining successively for cross-media retrieval. Coupled feature mapping learned two projection matrices to map the multimodal features into a common category space and then correlation mining was used to take advantage of the semantic category information. Yan et al. [7] stacked fully connected layers together to represent the sentences and used deep canonical correlation analysis for matching images and sentences. Ma et al. [8] proposed multimodal convolutional neural networks (m-CNNs) which captured relations between images and sentences at different level. In addition to a common space, in the DeViSE model developed by Frome et al. [18], the text space was formed by a pre-trained Word2Vec model. In a follow-up work, Norouzi et al. [9] employed Word2Vec for both sentence and image embeddings. The text space vector of an image was obtained by a convex combination of the word embedding vectors of the visual labels predicted to be the most relevant to the image. Recently, a distributional visual embedding space provided by Word2VisualVec [10] was found to be an effective space to perform cross-media retrieval by embedding sentences into a visual feature space.

Apart from those models designed for cross-media retrieval, image caption models can also be used to learn an appropriate homogeneous space. For example, multimodal recurrent neural network (m-RNN) [11], neural image caption (NIC) [13], deep visual-semantic alignments (DVSA) [14], Unifying Visual-Semantic Embeddings (VSE) [15] were used to learn the relations between images and sentences and generate the captions for a given image. Before translating the image representations to descriptive sentences, those models first transformed the image representations into a limited text space. Thus, the limited text space vectors for images contain rich semantic correlations between images and sentences.

3 Proposed Method

The architecture of our proposed model is shown in Fig. 1 which contains a language model part, a visual part and an embedding part.

3.1 Language Model Part

We first review GRU which is used for learning dense limited text space representations for sentences. Cho et al. [16] proposed Gated Recurrent Unit as a simpler alternative to the LSTM. The GRU uses reset gates r and update gates z to control the flow of information inside the unit. h represents the activation of the GRU and $\hat{h}$ is the previous computed activation.

Let $S = (s_0, s_1, \cdots, s_t), t \in \{0 \cdots T\}$ be an input sentence, where we represent each word as a one-hot vector s_t of dimension equals to the size of the dictionary. Note that we denote by s_T a special end word which designates the end of the sentence. Before fed into the GRU, s_t should be embedded into a more dense space such as the Word2Vec space:

$$x_t = W_e s_t, t \in \{0 \cdots T\}, \tag{1}$$

The word embedding matrix W_e maps the one-hot vectors to a more dense text space. As mentioned in [16], the GRU takes the form:

$$
\begin{aligned}
h_t^j &= (1 - z_t^j)h_{t-1}^j + z_t^j \hat{h}_t^j \\
z_t^j &= \sigma(W_z x_t + U_z h_{t-1})^j \\
\hat{h}_t^j &= tanh(W x_t + U(r_t \odot h_{t-1}))^j \\
r_t^j &= \sigma(W_r x_t + U_r h_{t-1})^j
\end{aligned}
\tag{2}
$$

As shown in Eq. (2), the activation h_t^j of the GRU at time t is a linear interpolation between the previous activation h_{t-1}^j and the candidate activation $\hat{h}_t^j$. An update gate z_t^j decides how much the unit updates its activation. The reset gate r_t^j controls the unit whether to forget the previous computed states or not. Finally, the representation of a sentence S is the hidden state of the GRU at time T.

3.2 Visual Part

The visual part contains a deep convolutional neural network and an image understanding network to achieve image understanding. For convolutional neural network, we employ VGG [2] to extract 4096-dimensional image representations X_{vgg} which contain rich visual label information.

For the image understanding network, inspired by the idea of automatic image captioning, we propose a novel method to map image pixels to pre-trained limited text space representations which contain strong semantic correlations between images and sentences similar to NIC [13]. The understanding process can be divided into two sub-processes:

(1) **Learning image representations.** We choose *Inception v3* image recognition model pre-trained on the ILSVRC-2012-CLS image classification dataset as an image feature extractor:

$$X_{img} = Inception\_v3(Image) \tag{3}$$

where X_{img} is a 1024-dimensional vector in an image space.

(2) **Embedding image representations into a pre-trained limited text space.** A single linear embedding layer is added on top of the *Inception v3* model to transform the image representations into a pre-trained limited text space:

$$X_{txt} = X_{img}W_{img} \tag{4}$$

where W_{img} maps X_{img} to a 512-dimensional vector.

3.3   Embedding Part

As mentioned in the previous subsection, X_{vgg} is able to capture rich visual label information but ignores semantic correlations between images and sentences. X_{txt} is particularly good at modeling semantic correlations between images and sentences which is complementary to X_{vgg}. According to it, we add a linear fusion layer on top of the visual part to combine X_{txt} with X_{vgg} as well as embed them into a limited text space:

$$X_{final} = X_{vgg}W_{vgg\_fuse} + X_{txt}W_{txt\_fuse}, \qquad (5)$$

where W_{vgg_fuse} and W_{txt_fuse} are embedding matrices for X_{vgg} and X_{txt} respectively.

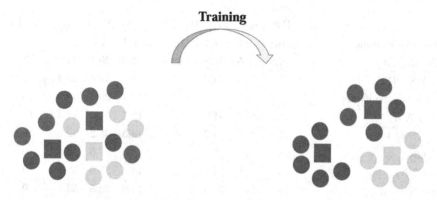

Fig. 2. Illustration of the pairwise ranking loss for learning the limited text space. Rectangles represent images and circles represent sentences. Matching image-sentence pairs are denoted in the same color.

In order to optimize the model parameters, pairwise ranking loss function is exploited to be the objective function. That is, as shown in Fig. 2, given an image query x, we want the distance between x and its matching sentences to be smaller than the distance between x and its non-matching sentences by a margin, and vice versa for a sentence query. Thus, we optimize the following loss function:

$$L = \min_{\Theta} \sum_{x} \sum_{k} max\left\{0, margin - d(x,s) + d(x,s_k)\right\}$$
$$+ \sum_{v} \sum_{k} max\left\{0, margin - d(s,x) + d(s,x_k)\right\} \qquad (6)$$

where s_k is a negative sentence for a given image x and x_k is a negative image for a given sentence s. In order to obtain the non-matching terms, we choose them randomly from the training set and re-sampled every epoch.

4   Experiments

In order to evaluate the effectiveness of our proposed model on cross-media retrieval, we have performed an extensive set of experiments on three benchmark datasets. We follow the evaluation metrics adopted in [15] for a fair comparison using Recall@K and Med r. The R@K (with K = 1, 5, 10) computes the mean number of images for which the correct caption is ranked within the top-K retrieved results and vice versa for sentences. Med r is the median rank of the first correct result in the ranking list. Higher R@K and lower Med r thus mean better performance.

Table 1. Bidirectional image and sentence retrieval results on Flickr8K

	Sentence retrieval				Image retrieval			
	R@1	R@5	R@10	Med r	R@1	R@5	R@10	Med r
Random ranking	0.1	0.6	1.1	631	0.1	0.5	1.0	500
DeViSE [18]	4.8	16.5	27.3	28.0	5.9	20.1	29.6	29
m-RNN [11]	14.5	37.2	48.5	11	11.5	31.0	42.4	15
Deep fragment [19]	12.6	32.9	44.0	14	9.7	29.6	42.5	15
DCCA [7]	17.9	40.3	51.9	9	12.7	31.2	44.4	13
m-CNN$_{wd}$ [8]	15.6	40.1	55.7	8	14.5	38.2	52.6	9
m-CNN$_{phs}$ [8]	18.0	43.5	57.2	8	14.6	39.5	53.8	9
m-CNN$_{phl}$ [8]	16.7	43.0	56.7	7	14.4	38.6	52.2	9
m-CNN$_{st}$ [8]	18.1	44.1	57.9	7	14.6	38.5	53.5	9
m-CNN$_{ENS}$ [8]	24.8	53.7	67.1	5	20.3	47.6	61.7	5
FV (GMM + HGLMM) [5]	**31.0**	**59.3**	**73.7**	**4**	**21.3**	**50.0**	**64.8**	**5**
VSE [15]	18.0	40.9	55.0	8	12.5	37.0	51.5	10
Ours_single	21.8	50.2	64.5	5	13.8	37.5	52.2	10
Ours_fusion	21.5	50.3	66.2	5	15.4	40.5	54	9

4.1   Datasets

For evaluation we use three benchmark datasets consisting of images and their corresponding descriptive sentences. The statistics of the datasets are as follows:

Flickr8K [24]: This dataset consists of 8,000 images. Each image is annotated with 5 sentences describing the content. We use 6,000 images for training, 1,000 images for validation and 1,000 images for testing.

Flickr30K [22]: This dataset consists of 31,783 images. Each image is also annotated with 5 sentences describing the content of the image. We use 29,000 images for training, 1,000 images for validation and 1,000 images for testing.

MSCOCO [12]: This dataset consists of 82,783 training, 40,504 validation, and 40,775 testing images. Each image is also annotated with 5 sentences describing the content of the image. We reserve 1,000 random images from the MSCOCO validation set as test and use it to report results.

4.2 Experimental Configurations

For the visual part, we adopt the similar architecture as NIC [13] to embed image pixels into a pre-trained limited text space which was pre-trained on MSCOCO. More specifically, we use 512 dimensions for the word embeddings. For the language model part, we randomly initialize the word embeddings W_e to be 1024-dimensional vectors. Similar to [15], our GRU uses one layer with 1024 units and weights initialized uniformly from [−0.08, 0.08]. For the embedding part, Xavier initialization [20] is used to initialize the embedding matrices W_{vgg_fuse} and W_{txt_fuse}. In order to choose an appropriate $margin$ that achieves the best performance on all datasets, we set $margin = [0.1, 0.2, ..., 0.9]$ for validation and finally select $margin = 0.7$. The whole model is implemented in TensorFlow and Theano based on NVIDIA Tesla K80 GPU. We use minibatches of 40 on Flickr8K, 100 on Flickr30K and MSCOCO in the training procedure.

Table 2. Bidirectional image and sentence retrieval results on Flickr30K

	Sentence retrieval				Image retrieval			
	R@1	R@5	R@10	Med r	R@1	R@5	R@10	Med r
Random ranking	0.1	0.6	1.1	631	0.1	0.5	1.0	500
DeViSE [18]	4.5	18.1	29.2	26	6.7	21.9	32.7	25
Deep fragment [19]	14.2	37.7	51.3	10	10.2	30.8	44.2	14
m-RNN-vgg [11]	**35.4**	63.8	73.7	3	22.8	50.7	63.1	5
DCCA [7]	16.7	39.3	52.9	8	12.6	31.0	43.0	15
m-CNN$_{wd}$ [8]	21.3	53.2	66.1	5	18.2	47.2	60.9	6
m-CNN$_{phs}$ [8]	25.0	54.8	66.8	4.5	19.7	48.2	62.2	6
m-CNN$_{phl}$ [8]	23.9	54.2	66.0	5	19.4	49.3	62.4	6
m-CNN$_{st}$ [8]	27.0	56.4	70.1	4	19.7	48.4	62.3	6
m-CNN$_{ENS}$ [8]	33.6	**64.1**	74.9	3	**26.2**	**56.3**	**69.6**	4
FV (GMM + HGLMM) [5]	35.4	62.0	73.8	3	25.0	52.7	66.0	5
VSE [15]	23.0	50.7	62.9	5	16.8	42.0	56.5	8
Ours_single	24.5	54.2	69.3	5	17.7	43.6	55.9	8
Ours_fusion	31.2	62.5	**75.8**	3	21.5	48.9	61.5	6

4.3 Experimental Results

We aim to show the experimental results on two aspects. Firstly, in order to emphasize the importance of the fusion layer, we design three contrast models:

Table 3. Bidirectional image and sentence retrieval results on MSCOCO

	Sentence retrieval				Image retrieval			
	R@1	R@5	R@10	Med r	R@1	R@5	R@10	Med r
Random ranking	0.1	0.6	1.1	631	0.1	0.5	1.0	500
m-RNN-vgg [11]	41.0	73.0	83.5	2	29.0	42.2	77.0	3
DVSA [14]	38.4	69.9	80.5	1	27.4	60.2	74.8	3
m-CNN$_{wd}$ [8]	34.1	66.9	79.7	3	27.9	64.7	80.4	3
m-CNN$_{phs}$ [8]	34.6	67.5	81.4	3	27.6	64.4	79.5	3
m-CNN$_{phl}$ [8]	35.1	67.3	81.6	5	27.1	62.8	79.3	3
m-CNN$_{st}$ [8]	38.3	69.6	81.0	2	27.4	63.4	79.5	3
m-CNN$_{ENS}$ [8]	42.8	73.1	84.1	2	**32.6**	**68.6**	**82.8**	3
FV (GMM + HGLMM) [5]	39.4	67.9	80.9	2	25.1	59.8	76.6	4
VSE [15]	43.4	75.7	85.8	2	31	66.7	79.9	3
Ours_single	34.6	68.5	82.9	3	17.8	49.9	66.9	6
Ours_fusion	**45.5**	**78.7**	**88.8**	2	30.2	66	80.5	3

(1) VSE is the baseline model which uses VGG features to represent images; (2) Ours_single removes the fusion layer and uses the image understanding network to learn image representations; (3) Ours_fusion reserves the fusion layer. Secondly, we compare the three models with the related state-of-the-art methods so as to verify the effectiveness of the limited text space. The experimental results on Flickr8K, Flickr30K and MSCOCO are illustrated in Tables 1, 2, and 3 where the best performance of each evaluation metric has been highlighted.

The experimental results among the three contrast models show that our proposed fusion model Ours_fusion outperforms VSE and Ours_single on all datasets. It demonstrates that the fusion layer is able to capture both visual label information and semantic correlations between images and sentences which is beneficial to the performance of cross-media retrieval compared with VGG features and the pre-trained limited text space representations. Moreover, Ours_single outperforms VSE as well especially on sentence retrieval task due to the sufficient understanding of images by the visual part.

The contrast experiments between the fusion model Ours_fusion and the related state-of-the-art methods are shown as follows. On Flickr8K, FV achieves the best performance. Ours_fusion performs inferiorly to FV and m-CNN$_{ENS}$. Due to the lack of training samples, Fisher vector is proven to be the most powerful method on modeling sentences. However, recurrent neural network is essentially a kind of temporally deep neural network and thus needs sufficient data to tune the parameters adequately. Except FV, m-CNN$_{ENS}$ performs better than us due to the integration of four separate models. It is worth mentioning that Ours_fusion outperforms the four models on sentence retrieval and matches their results on image retrieval.

On Flickr30K, with more training samples than Flickr8K, $Ours\_fusion$ gains a significant improvement on sentence retrieval task which shows competitive experimental results compared with FV and m-CNN$_{ENS}$. However, the model performs poor on image retrieval task. The most probable cause may be the insufficient understanding of sentences by RNN which may lead to the ambiguity during the retrieval.

On MSCOCO, with the largest number of training samples, the performance of $Ours\_fusion$ on sentence retrieval has been significantly improved, compared with all the other methods. Only DVSA outperforms $Ours\_fusion$ in terms of Med r. It demonstrates that with enough training samples, the parameters of GRU and the embedding matrices can be more adequately tuned. On image retrieval task, $Ours\_fusion$ performs inferiorly to m-CNN$_{ENS}$ but still superior to the other methods.

Table 4 shows three examples of sentence retrieval. It can be observed that our model finds the closest results for a given image query. For example, the groundtruth descriptive sentences for the first image query are: (1) A wooden

Table 4. Three examples of sentence retrieval. The first column contains the image queries and the second column shows the top five retrieved sentences on MSCOCO dataset. The correctly retrieved sentences for each image query are denoted in blue. The incorrectly identified objects are marked in red.

Image queries	Top five retrieved sentences
	1. Wooden spoons are lined up on a table.
	2. Wooden spoons and forks are all over a table.
	3. The table is full of wooden spoons and utensils.
	4. Multiple wooden spoons are shown on a table top.
	5. A wood table holding an assortment of wood cooking utensils.
	1. A person is riding a bicycle but there is a train in the background.
	2. A man on a bicycle riding next to a train.
	3. A guy that is riding his bike next to a train.
	4. A red and white train and a man riding a bicycle.
	5. A man riding a bike past a train traveling along tracks.
	1. A tall woman is standing in a small kitchen.
	2. A woman observing something on a kitchen stove.
	3. A kitchen with two windows and two metal sinks.
	4. A kitchen has the windows open and plaid curtains.
	5. The view of a kitchen from across the room.

ball on top of a wooden stick. (2) The table is full of wooden spoons and utensils. (3) A wood table holding an assortment of wood cooking utensils. (4) A selection of wooden kitchen tools on a counter. (5) Wooden spoons are lined up on a table. Although retrieved sentences "Wooden spoons and forks are all over a table" and "Multiple wooden spoons are shown on a table top" are regarded as irrelevant to the image, they can describe the content more accurately than "A wooden ball on top of a wooden stick" and "A selection of wooden kitchen tools on a counter" from a subjective perspective of human. However, there are some unreasonable results for the third image query. As shown by the words marked in red, our model identifies visual concept "woman" incorrectly which is nonexistent in the image.

5   Conclusion

In this paper, we propose a novel model to perform cross-media retrieval in a limited text space which aims to learn limited text space representations for both images and sentences. Firstly, the visual part learns image representations including deep convolutional features and pre-trained limited text space representations. The language model part learns a dense limited text space representation for each sentence. Secondly, the embedding part captures both visual label information and semantic correlations between images and sentences, as well as learns the final limited text space. Experimental results on three benchmark datasets demonstrate the importance of the fusion layer and the effectiveness of the limited text space. Our proposed fusion model achieves promising improvement compared with the related state-of-the-art methods. In the future, we will pay more attention to the image retrieval task and further improve the accuracy of it.

References

1. Razavian, A.S., Azizpour, H., Sullivan, J., Carlsson, S.: CNN features off-the-shelf: an astounding baseline for recognition. In: Proceedings of the IEEE Conference on Computer Vision and Pattern Recognition Workshops, pp. 806–813 (2014)
2. Simonyan, K., Zisserman, A.: Very deep convolutional networks for large-scale image recognition. CoRR, vol. abs/1409.1556 (2014)
3. Mikolov, T., Sutskever, I., Chen, K., Corrado, G.S., Dean, J.: Distributed representations of words and phrases and their compositionality. In: Advances in Neural Information Processing Systems, pp. 3111–3119 (2013)
4. Thompson, B.: Canonical correlation analysis. In: Encyclopedia of Statistics in Behavioral Science (2005)
5. Klein, B., Lev, G., Sadeh, G., Wolf, L.: Fisher vectors derived from hybrid Gaussian-Laplacian mixture models for image annotation. arXiv preprint arXiv:1411.7399 (2014)
6. Sutskever, I., Vinyals, O., Le, Q.V.: Sequence to sequence learning with neural networks. In: Advances in Neural Information Processing Systems, pp. 3104–3112 (2014)

7. Yan, F., Mikolajczyk, K.: Deep correlation for matching images and text. In: Proceedings of the IEEE Conference on Computer Vision and Pattern Recognition, pp. 3441–3450 (2015)
8. Ma, L., Lu, Z., Shang, L., Li, H.: Multimodal convolutional neural networks for matching image and sentence. In: Proceedings of the IEEE International Conference on Computer Vision, pp. 2623–2631 (2015)
9. Norouzi, M., Mikolov, T., Bengio, S., Singer, Y., Shlens, J., Frome, A., Corrado, G.S., Dean, J.: Zero-shot learning by convex combination of semantic embeddings. arXiv preprint arXiv:1312.5650 (2013)
10. Dong, J., Li, X., Snoek, C.G.: Word2visualvec: cross-media retrieval by visual feature prediction. arXiv preprint arXiv:1604.06838 (2016)
11. Mao, J., Xu, W., Yang, Y., Wang, J., Huang, Z., Yuille, A.: Deep captioning with multimodal recurrent neural networks (m-RNN). arXiv preprint arXiv:1412.6632 (2014)
12. Lin, T.-Y., et al.: Microsoft COCO: common objects in context. In: Fleet, D., Pajdla, T., Schiele, B., Tuytelaars, T. (eds.) ECCV 2014. LNCS, vol. 8693, pp. 740–755. Springer, Cham (2014). doi:10.1007/978-3-319-10602-1_48
13. Vinyals, O., Toshev, A., Bengio, S., et al.: Show and tell: Lessons learned from the 2015 MSCOCO image captioning challenge. IEEE Trans. Pattern Anal. Mach. Intell. 39(4), 652–663 (2017)
14. Karpathy, A., Fei-Fei, L.: Deep visual-semantic alignments for generating image descriptions. In: Proceedings of the IEEE Conference on Computer Vision and Pattern Recognition, pp. 3128–3137 (2015)
15. Kiros, R., Salakhutdinov, R., Zemel, R.S.: Unifying visual-semantic embeddings with multimodal neural language models. arXiv preprint arXiv:1411.2539 (2014)
16. Cho, K., Van Merriënboer, B., Bahdanau, D., Bengio, Y.: On the properties of neural machine translation: encoder-decoder approaches. arXiv preprint arXiv:1409.1259 (2014)
17. Blei, D.M., Ng, A.Y., Jordan, M.I.: Latent Dirichlet allocation. J. Mach. Learn. Res. 3, 993–1022 (2003)
18. Frome, A., Corrado, G.S., Shlens, J., Bengio, S., Dean, J., Mikolov, T., et al.: Devise: a deep visual-semantic embedding model. In: Advances in Neural Information Processing Systems, pp. 2121–2129 (2013)
19. Karpathy, A., Joulin, A., Li, F.F.F.: Deep fragment embeddings for bidirectional image sentence mapping. In: Advances in Neural Information Processing Systems, pp. 1889–1897 (2014)
20. Glorot, X., Bengio, Y.: Understanding the difficulty of training deep feedforward neural networks. In: Aistats, vol. 9, pp. 249–256 (2010)
21. Fan, M., Wang, W., Wang, R.: Coupled feature mapping and correlation mining for cross-media retrieval. In: 2016 IEEE International Conference on Multimedia & Expo Workshops (ICMEW), pp. 1–6. IEEE (2016)
22. Young, P., Lai, A., Hodosh, M., Hockenmaier, J.: From image descriptions to visual denotations: new similarity metrics for semantic inference over event descriptions. Trans. Assoc. Comput. Linguist. 2, 67–78 (2014)
23. Wang, J., Wang, W., Wang, R., Gao, W., et al.: Deep alternative neural network: exploring contexts as early as possible for action recognition. In: Advances in Neural Information Processing Systems, pp. 811–819 (2016)
24. Hodosh, M., Young, P., Hockenmaier, J.: Framing image description as a ranking task: data, models and evaluation metrics. J. Artif. Intell. Res. 47, 853–899 (2013)

Speeding-Up Graph-Based Keyword Spotting by Quadtree Segmentations

Michael Stauffer[1,4(✉)], Andreas Fischer[2,3], and Kaspar Riesen[1]

[1] Institute for Information Systems,
University of Applied Sciences and Arts Northwestern Switzerland,
Riggenbachstr. 16, 4600 Olten, Switzerland
{michael.stauffer,kaspar.riesen}@fhnw.ch
[2] Department of Informatics, University of Fribourg, 1700 Fribourg, Switzerland
andreas.fischer@unifr.ch
[3] Institute for Complex Systems,
University of Applied Sciences and Arts Western Switzerland,
1705 Fribourg, Switzerland
[4] Department of Informatics, University of Pretoria, Pretoria, South Africa

Abstract. Keyword Spotting (KWS) improves the accessibility to handwritten historical documents by unconstrained retrievals of keywords. The proposed KWS framework operates on segmented words that are in turn represented as graphs. The actual KWS process is based on matching graphs by means of a cubic-time graph matching algorithm. Although this matching algorithm is quite efficient, the polynomial time complexity might still be a limiting factor (especially in case of large documents). The present paper introduces a novel approach that aims at speeding up the retrieval process. The basic idea is to first segment individual graphs into smaller subgraphs by means of a quadtree procedure. Eventually, the graph matching procedure can be conducted on the resulting pairs of smaller subgraphs. In an experimental evaluation on two benchmark datasets we empirically confirm substantial speed-ups while the KWS accuracy is nearly not affected.

Keywords: Handwritten keyword spotting · Bipartite graph matching · Quadtree graph segmentation

1 Introduction

In the last decades, handwritten historical documents have been made increasingly digitally available around the world. Example documents are the *Barcelona marriage registry* [1], the *Saint Gall manuscript* [2] or the *George Washington letters* [3], to mention just a few. Yet, the accessibility to these documents with respect to browsing and searching is still an issue since automatic full transcriptions are often not feasible. Thus, *Keyword Spotting (KWS)* as an alternative to transcriptions has been proposed for this type of document [3–6]. KWS allows to retrieve any instances of a given keyword in a certain document. In case of

© Springer International Publishing AG 2017
M. Felsberg et al. (Eds.): CAIP 2017, Part I, LNCS 10424, pp. 304–315, 2017.
DOI: 10.1007/978-3-319-64689-3_25

historical documents, KWS is limited to document images only and is thus an *offline* task. Generally, offline KWS is regarded as the more difficult case when compared to *online* KWS, where temporal information about the writing process is available as well.

KWS approaches can be roughly distinguished into *template-based* or *learning-based* algorithms. In the case of template-based KWS, handwritten words are often represented by sequences of feature vectors used to store certain characteristics of the handwriting. A query word can then be retrieved in a set of document words by matching sequences of features vectors, for example by means of *dynamic time warping* [5,7,8]. Learning-based KWS on the other hand is based on a statistical model that is trained *a priori* on a relatively large set of training words [3,6,9]. Comparing both approaches with each other, we observe that learning-based approaches lead to higher accuracies in general, while template-based approaches are characterised by a higher flexibility (as no training is required). In the present paper, we focus on template-based KWS using graphs for the formal representation of words.

In various fields of pattern recognition, graphs have been employed as a versatile representation formalism [10–12]. Yet, for applications based on KWS, graphs are rarely used [13–18]. This is somehow surprising as graphs offer a natural and comprehensive representation for handwritten words. In particular, graphs are able to adapt both their size and structure to the complexity of the underlying handwritten words. Moreover, graphs are able to represent binary relationships that might exists between parts of the handwritten words.

In case of graph-based KWS, the actual spotting process includes matching a query graph with a set of document graphs (using some graph matching algorithm [11]). When large amounts of graph matchings are necessary, the complete KWS process might take too much time, even with fast approximate matching algorithms (e.g. [16]).

In the present paper, we focus on speeding up the graph-based KWS process by adapting the actual graph matching procedure. Graphs are first iteratively segmented into smaller subgraphs by means of a *quadtree* segmentation. Second, the graph matching is conducted on corresponding small subgraphs rather than on the large graphs representing the complete word. The rationale for this procedure is to substantially speed up the graph matching, as the matching time depends on the number of nodes of the involved graphs.

The remainder of this paper is organised as follows. In Sect. 2, the basic KWS framework for graph-based word representations is reviewed. The actual speed-up procedure for graph-based KWS is introduced in Sect. 3 and evaluated in Sect. 4. Finally, Sect. 5 concludes the paper and outlines possible future research activities.

2   Graph-Based Keyword Spotting

The present paper proposes a novel method for speeding up a framework for graph-based KWS [16]. The basic KWS framework consists of four different

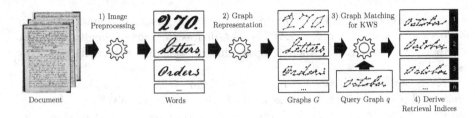

Fig. 1. Process of Graph-based keyword spotting of the word "October"

processing steps as illustrated in Fig. 1. In the following four subsections these four steps are briefly reviewed.

2.1 Image Preprocessing

For the purpose of evaluation, we employ two historical documents, viz. the *George Washington letters (GW)* and the *Parzival manuscript (PAR)*. GW is based on twenty pages with a total of 4,894 handwritten words[1]. The letters are written in English by George Washington and his associates during the American Revolutionary War in 1755. Variations caused by both degradation and writing style are low. PAR consists of 45 pages with a total of 23,478 handwritten words[2]. The manuscript is written in Middle High German and originates in the 13th century. Variations caused by degradation are markable, while variations caused by writing style are low. Two exemplary words are given for both documents in the first row of Fig. 2.

	GW		PAR	
Original				
Preprocessed				
Keypoint				
Projection				

Fig. 2. Different representations of two sample words from both datasets.

---

[1] George Washington Papers at the Library of Congress, 1741–1799: Series 2, Letterbook 1, pp. 270–279 & pp. 300–309, http://memory.loc.gov/ammem/gwhtml/gwseries2.html.

[2] Parzival at IAM historical document database, http://www.fki.inf.unibe.ch/databases/iam-historical-document-database/parzival-database.

The original document images are first preprocessed to reduce variations caused, for instance, by skew scanning, noisy background, and document degradation. On the basis of preprocessed document images, single word images are automatically segmented by means of their projection profile (and if necessary manually corrected). In the second row of Fig. 2 we show the result of our preprocessing on some examples. For details regarding both image preprocessing and word segmentation we refer to [16].

2.2 Graph Representation

A graph g is generally defined as a four-tuple $g = (V, E, \mu, \nu)$ where V and E are finite sets of nodes and edges, and $\mu : V \rightarrow L_V$ as well as $\nu : E \rightarrow L_E$ are labelling functions for nodes and edges, respectively. Graphs can be divided into *undirected* and *directed* graphs, where pairs of nodes are either connected by undirected or directed edges, respectively. Additionally, graphs are often distinguished into *unlabelled* and *labelled* graphs. In the latter case, both nodes and edges can be labelled with an arbitrary numerical, vectorial, or symbolic label from L_v or L_e, respectively. In the former case we assume empty label alphabets, i.e. $L_v = L_e = \{\}$.

The following two graph extraction algorithms (originally presented in [17]) result in graphs where nodes are labelled with two-dimensional numerical labels, while the undirected edges remain unlabelled, i.e. $L_V = \mathbb{R}^2$ and $L_E = \{\}$.

- **Keypoint**: The first graph extraction algorithm makes use of keypoints in the word images such as start, end, and junction points. These keypoints are represented as nodes and labelled with the (x, y)-coordinates of the corresponding keypoint. Between pairs of keypoints further intermediate points (in equidistant intervals) are converted to nodes and added to the graph. Finally, undirected edges are inserted between pairs of nodes that are directly connected by a stroke.
- **Projection**: The second graph extraction algorithm is based on an adaptive segmentation of word images by means of horizontal and vertical projection profiles. A node is inserted into the graph for every segment and labelled by the (x, y)-coordinates of the centre of mass of the corresponding segment. Undirected edges are inserted into the graph for each pair of nodes that is directly connected by a stroke in the original word image.

In the third and fourth row of Fig. 2 we show the resulting graphs of Keypoint and Projection, respectively.

For both graph representations, the dynamic range of the (x, y)-coordinates of each node label $\mu(v)$ is normalised with a z-score. Formally,

$$\hat{x} = \frac{x - \mu_x}{\sigma_x} \quad \text{and} \quad \hat{y} = \frac{y - \mu_y}{\sigma_y},$$

where (μ_x, μ_y) and (σ_x, σ_y) represent the mean and standard deviation of all (x, y)-coordinates in the graph under consideration.

2.3 Graph Matching for Keyword Spotting

In our general KWS approach, a query graph q is individually matched with every graph g from a set of document graphs $G = \{g_1, \ldots, g_N\}$. For this particular task, we focus on inexact graph matching and employ the concept of *Graph Edit Distance (GED)* [19]. Note that any other graph matching algorithm could be used as well. Yet, GED is particularly interesting as it allows matchings of arbitrary graphs.

Given a query graph q and document graph $g \in G$, the basic principle of graph edit distance is to transform q into g using some edit operations (i.e. *insertions*, *deletions*, and *substitutions*) for both nodes and edges. A set $\{e_1, \ldots, e_k\}$ of k edit operations e_i that transform q completely into g is called an *edit path* $\lambda(q, g)$ between q and g.

To find the most suitable edit path, one commonly introduces a domain-specific cost function $c(e)$ for every edit operation e. This cost function is used to measure the degree of deformation of a given edit operation. Given an adequate cost model, the graph edit distance $d_{\mathrm{GED}}(q, g)$, or d_{GED} for short, between q and g is defined by

$$d_{\mathrm{GED}}(q, g) = \min_{\lambda \in \Upsilon(q,g)} \sum_{e_i \in \lambda} c(e_i),$$

where $\Upsilon(q, g)$ denotes the set of all edit paths between q and g.

For the exact computation of d_{GED}, it is common to employ A*-based search techniques using some heuristics [20, 21]. However, these exhaustive search procedures are exponential with respect to the number of nodes of the involved graphs. Formally, GED belongs to the family of *Quadratic Assignment Problems (QAPs)* [22], which in turn belong to the class of $\mathcal{NP}$-complete problems.

In order to overcome this limitation, we make use of an approximation algorithm for the computation of GED [23]. This method basically reduces the problem of GED computation to an instance of the *Linear Sum Assignment Problem (LSAP)*. Both QAPs and LSAPs deal with the optimal alignment of entities of two sets. Yet, by encoding the GED problem as an LSAP, we have to neglect the global edge structures of the graphs. This actually leads to a general overestimation of the true GED. However, with this transformation, we benefit from the polynomial complexity of LSAPs (see [24] for an exhaustive survey on LSAP solving algorithms). For the remainder of this paper, we make use of this graph matching algorithm and name the corresponding suboptimal graph edit distance $d_{\mathrm{BP}}(q, g)$, or d_{BP} for short[3].

2.4 Derive Retrieval Indices

Our approach for keyword spotting relies on retrieval indices which are based on the suboptimal graph edit distance d_{BP}. We define retrieval indices for both *local* and *global* threshold scenarios. In case of local thresholds, the KWS accuracy is

---

[3] BP stand for bipartite (LSAPs are also termed *bipartite matching problem*).

independently measured for every keyword, while in case of global thresholds, the KWS accuracy is measured for every keyword with one single threshold.

In both scenarios, d_{BP} is first normalised by the sum of the maximum cost edit path between q and g, i.e. the sum of the edit path that results from deleting all nodes and edges of q and inserting all nodes and edges in g. Formally,

$$\hat{d}_{BP}(q, g) = \frac{d_{BP}(q, g)}{(|V_q| + |V_k|)\, \tau_v + (|E_q| + |E_g|)\, \tau_e},$$

where τ_v and τ_e denote the node and edge insertion/deletion costs. In case a query consists of a set of graphs $\{q_1, \ldots, q_t\}$ that represents the same keyword, the normalised graph edit distance $\hat{d}_{BP}$ is given by the minimal distance achieved on all t query graphs. This normalised graph edit distance is used to derive a first retrieval index for local thresholds by

$$r_1(q, g) = -\hat{d}_{BP}(q, g).$$

To derive a retrieval index for global thresholds, $\hat{d}_{BP}$ is further normalised by using the average distance of a query graph q to its k nearest document graphs, i.e. the document graphs $\{g_{(1)}, \ldots, g_{(k)}\}$ with smallest distance values to q. Formally, we use

$$\bar{d}_k(q) = \frac{1}{k} \sum_{i=1}^{k} \hat{d}_{BP}(q, g_{(i)}),$$

to derive

$$\hat{\hat{d}}_{BP}(q, g) = \frac{\hat{d}_{BP}(q, g)}{\bar{d}_k(q)}.$$

Finally, the distance $\hat{\hat{d}}_{BP}$ is used to derive the retrieval index for global thresholds by

$$r_2(q, g) = -\hat{\hat{d}}_{BP}(q, g).$$

Rather than defining k as a constant, we dynamically adapt k for every query graph q. In particular, k is defined such that the distance $d_{BP}(q, g_{(k)})$ of q to its k-th nearest document graph $g_{(k)}$ is equal to

$$\bar{d}_m(q) + \theta\, (\bar{d}_N(q) - \bar{d}_m(q)),$$

where m and θ are user defined parameters and N refers to the number of document graphs. The value of $\bar{d}_m(q)$ refers to the mean distance of q to its m nearest neighbours and $\bar{d}_N(q)$ refers to the mean distance to all document graphs available. This sum reflects the level of the dissimilarities of q to the graphs in its direct neighbourhood. If the sum is large, k is automatically defined large, too. This in turn increases $\bar{d}_k(q)$, which ultimately increases the scaling to $\hat{\hat{d}}_{BP}$.

3 Speeding-Up the Graph Matching

The contribution of the present paper is a novel method that aims at faster computations of pairwise graph dissimilarities. That is, we focus on reducing the time for computing d_{BP}. Basically, rather than matching complete graphs, we first apply a *quadtree* segmentation to individual graphs. Next, we match the small subgraphs (corresponding to each other w.r.t. the segmentation) and sum up the individual matching costs. This procedure might substantially speed up the graph matching procedure as the complexity of the graph matching algorithm is a cubic function of the number of nodes of the involved graphs.

The graph segmentation is carried out as follows. First, the bounding box surrounding a graph g is segmented at the *Centre of Mass* (x_m, y_m) into four segments as illustrated in Fig. 3a. To make this segmentation more robust against variations in the underlying graphs, we overlap each segment depending on a user defined factor $\alpha \in \,]0, 1[$. Parameter α defines the overlap of a segment to its neighbouring segments with respect to width and height of the corresponding segment. That is, for $\alpha = 0.10$, for instance, the overlapping region is 10% of the size of the corresponding segment. For each of the resulting segments one subgraph is created that includes all nodes (and edges) of the corresponding segment. Hence, we obtain four (not necessarily disjoint) subgraphs. These subgraphs are iteratively segmented at their corresponding centre of mass into further subgraphs until the recursion level l is equal to a maximum recursion depth $r > 0$ (defined by the user). This procedure is illustrated in Fig. 3b and c.

The actual procedure for computing a dissimilarity between two graphs g and g' using the proposed segmentation is formalised in Algorithm 1 (termed *Quadtree Graph Matching*). The proposed procedure is initialised by an external call with recursion level $l = 1$, i.e. $BPQ(1, g, g')$. First, both graphs g and g' are segmented into four subgraphs with the procedure described above. Each of these subgraphs represent the nodes and edges in one of the four segments under consideration (with a relative overlap of α) (see line 2 of Algorithm 1). Eventually, the sum of the four bipartite graph edit distances computed on the corresponding four subgraphs is built (see line 3). Finally, the subgraph pairs are further segmented by means of a recursive function call of BPQ (see line 6). This procedure is repeated until the current recursion level l is equal to the user-defined maximum depth r (see line 4 and 5).

4 Experimental Evaluation

The proposed speed-up procedure of quadtree segmentation (termed BP-Q from now on) is compared with two reference systems. First, we use the original KWS framework presented in [16] (termed BP from now on). Second, we use BP in conjunction with a fast rejection procedure recently proposed in [18] (termed BP-FR from now on). BP-FR also aims at speeding up the KWS process. Yet, this approach is based on a reduction of the number of graph matchings. In particular, pairs of query and document graphs are first compared with respect

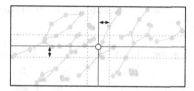

(a) Recursion Level $l = 1$ with Centre of Mass (x_m, y_m) and Overlap Factor α

(b) Recursion Level $l = 1$ with Subgraphs g_1 (highlighted), g_2, g_3, and g_4

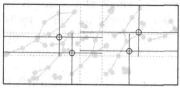

(c) Recursion Level $l = 2$ with Centres of Mass (x_{m_1}, y_{m_1}), $\ldots$, (x_{m_4}, y_{m_4})

Fig. 3. Quadtree graph segmentation.

Algorithm 1. Quadtree Graph Matching

Input: Graphs g and g', overlap factor α, maximum recursion depth $r > 0$
Output: Graph distance d_{BP_Q} between graph g and g'

1: **function** BPQ(l, g, g')
2:     Quadtree segment g and g' to g_1, g_2, g_3, g_4 and g'_1, g'_2, g'_3, g'_4
3:     $d_{BP_Q} = \sum\limits_{i=1}^{4} d_{BP(g_i, g'_i)}$
4:     **if** l equal r **then**
5:         **return** d_{BP_Q}
6:     **return** $d_{BP_Q} + \left(\sum\limits_{i=1}^{4} BPQ(l+1, g_i, g'_i) \right)$

---

to their node distributions in a polar coordinate system. If these distributions are similar enough, the graph matching is actually carried out (otherwise the document graph is rejected without further computations).

In the following subsection, the optimisation of the proposed KWS system is described. Eventually, the results are presented and discussed in Subsect. 4.2.

4.1 Optimisation of the Parameters

For the optimisation of the KWS framework, we manually select ten different keywords (with different word lengths) on both datasets (GW and PAR). Moreover, a validation set is defined consisting of 1,000 different random words including at least ten instances of all ten keyword instances. The KWS experiments

Table 1. Number of keywords as well as the size of the training and test set for both benchmark datasets.

Dataset	Keywords	Train	Test
GW	105	2,447	1,224
PAR	1,217	11,468	6,869

are finally conducted with optimised parameter settings on the same training and test sets as proposed in [3]. In Table 1, the number of keywords, as well as the size of the training- and test set are shown for both datasets.

In case of global thresholds, the accuracy of KWS systems is often measured by the *Average Precision (AP)*, which is the area under the *Recall-Precision (RP)* curve for all keywords given one single global threshold. In case of local thresholds, the KWS accuracy is commonly measured by the *Mean Average Precision (MAP)*, that is the mean over the AP of each individual keyword query. In a real-world scenario, global thresholds are regarded as the more realistic but also more difficult case.

The optimal parameters for the KWS system BP, the fast rejection method BP-FR, as well as the two graph extraction methods `Keypoint` and `Projection` are adopted from previous works [16–18]. The parameters of the quadtree segmentation, i.e. the maximum recursion depth l and the overlap factor α, are optimised as follows. We evaluate five maximum recursion depths $r \in \{1, 2, 3, 4, 5\}$ in combination with 20 overlap factors $\alpha \in \{0.01, 0.02, \ldots, 0.20\}$. On both datasets and for both extraction methods a maximum recursion depth of 1 turns out to be optimal. On PAR an overlap factor α of 0.01 is optimal for both graph representations, while on GW $\alpha = 0.01$ and $\alpha = 0.02$ is optimal for `Keypoint` and `Projection`, respectively.

The retrieval index r_2 is optimised for the scenario with global thresholds. In particular, parameter m and threshold scaling factor θ are optimised. To this end, we evaluate 2,000 parameter pairs (m, θ) with $m = \{10, 20, \ldots, 990, 1000\}$ and $\theta = \{0.01, 0.02, \ldots, 0.19, 0.20\}$. In Table 2, the optimal parameter settings for r_2 are given for both graph extraction methods and benchmark datasets.

Table 2. Optimal parameters m and θ for retrieval index r_2 for both graph extraction methods and benchmark datasets.

Method	GW		PAR	
	m	θ	m	θ
`Keypoint`	70	0.01	950	0.20
`Projection`	60	0.02	1,000	0.20

4.2 Results and Discussion

In Table 3, the MAP and AP for local and global threshold scenarios are given for all three KWS systems, i.e. the original framework BP [16,17], the BP framework with fast rejection BP-FR [18], as well as our novel procedure BP-Q. Additionally, we indicate the speed-up factor[4] as well as the relative gain or loss of the KWS accuracy of both speed-up approaches when compared with the original system BP.

Table 3. Mean average precision (MAP) using local thresholds, average precision (AP) using a global threshold, and speed-up factor (SF) for KWS using the original bipartite graph matching without rejection (BP), with fast rejection (BP-FR), and with quadtree segmentation (BP-Q). With ± we indicate the relative percental gain or loss in the accuracy of BP-FR and BP-Q when compared with BP.

Method		GW					PAR				
		MAP	±	AP	±	SF	MAP	±	AP	±	SF
BP	Keypoint	66.08		55.22			62.04		60.76		
	Projection	61.43		49.34			66.23		62.38		
BP-FR	Keypoint	68.81	+4.1	54.10	−2.0	3.2	67.70	+9.1	63.01	+3.7	2.4
	Projection	64.65	+5.2	48.94	−0.8	2.6	72.02	+8.7	63.49	+1.8	2.3
BP-Q	Keypoint	65.92	−0.2	54.91	−0.6	17.1	56.83	−8.4	54.66	−10.0	21.2
	Projection	59.57	−3.0	48.13	−2.5	15.0	64.62	−2.4	61.72	−1.1	21.5

When compared to BP, the proposed method BP-Q achieves speed-up factors of about 15–17 and 21 on GW and PAR, respectively. This refers to a substantial improvement of the performance, especially as the previous method for speeding up the KWS process (BP-FR) leads to speed-up factors of about 2 to 3 only. However, for both datasets and both threshold scenarios an accuracy loss has to be taken into account with BP-Q, while BP-FR outperforms BP in three out of four cases. Yet, this deterioration of BP-Q in the KWS accuracy is negligible. In particular, when we consider the results of Keypoint on GW and Projection on PAR (where the relative loss of accuracy is lower than 1% and 2.5%, respectively). Hence, we can summarise that BP-Q achieves comparable results as BP but needs about 20 times less computation time for KWS.

5   Conclusion and Outlook

In the present paper a procedure for speeding up graph-based keyword spotting is presented. The basic idea is to iteratively segment graphs into smaller subgraphs by means of a quadtree segmentation. These small subgraphs, rather than

---

[4] We carry out our experiments on a high performance computing cluster with dozens of 2.2 GHz CPU nodes. Hence, these readings refer to the average matching time per keyword measured in a sequential scenario.

complete graphs, are eventually matched during the KWS process. The motivation for this procedure is to decrease the runtime of the KWS process. This is actually reasonable as the time complexity of the employed graph matching algorithm is a cubic function of the number of nodes of the involved graphs.

We compare the proposed speed-up procedure BP-Q with the original framework BP and a recent fast rejection method BP-FR on two different benchmark datasets. On both datasets, BP-Q achieves remarkable speed-up factors of 15 to 21 when compared with BP (BP-FR leads to substantially smaller speed-up factors of 2 to 3). However, these performance improvements are accomplished with a marginal loss in accuracy when compared with BP.

In future work we aim at combining both speed-up approaches BP-Q and BP-FR to further speed up the KWS process. That is, graphs might be first filtered by the fast rejection method [18] and eventually segmented and matched by means of the quadtree graph matching procedure. Moreover, we see great potential in applying our fast matching procedure in other fields of graph-based pattern recognition. Last but not least, it would be interesting to employ our general method in a parallelised computation scenario.

Acknowledgments. This work has been supported by the Hasler Foundation Switzerland.

References

1. Fernandez-Mota, D., Almazan, J., Cirera, N., Fornes, A., Llados, J.: BH2M: The Barcelona historical, handwritten marriages database. In: International Conference on Pattern Recognition, pp. 256–261 (2014)
2. Fischer, A., Frinken, V., Fornés, A., Bunke, H.: Transcription alignment of Latin manuscripts using hidden Markov models. In: Workshop on Historical Document Imaging and Processing, New York, p. 29 (2011)
3. Fischer, A., Keller, A., Frinken, V., Bunke, H.: Lexicon-free handwritten word spotting using character HMMs. Pattern Recognit. Lett. **33**(7), 934–942 (2012)
4. Manmatha, R., Han, C., Riseman, E.: Word spotting: a new approach to indexing handwriting. In: Computer Vision and Pattern Recognition, pp. 631–637 (1996)
5. Rath, T., Manmatha, R.: Word image matching using dynamic time warping. In: Computer Vision and Pattern Recognition, vol. 2, pp. II-521–II-527 (2003)
6. Rodríguez-Serrano, J.A., Perronnin, F.: Handwritten word-spotting using hidden Markov models and universal vocabularies. Pattern Recognit. **42**(9), 2106–2116 (2009)
7. Rodriguez, J.A., Perronnin, F.: Local gradient histogram features for word spotting in unconstrained handwritten documents. In: International Conference on Frontiers in Handwriting Recognition, pp. 7–12 (2008)
8. Rodríguez-Serrano, J.A., Perronnin, F.: A model-based sequence similarity with application to handwritten word spotting. IEEE Trans. Pattern Anal. Mach. Intell. **34**(11), 2108–2120 (2012)
9. Perronnin, F., Rodriguez-Serrano, J.A.: Fisher kernels for handwritten word-spotting. In: International Conference on Document Analysis and Recognition, pp. 106–110 (2009)

10. Conte, D., Foggia, P., Sansone, C., Vento, M.: Thirty years of graph matching in pattern recognition. Int. J. Pattern Recognit. Artif. Intell. **18**(03), 265–298 (2004)
11. Riesen, K.: Structural Pattern Recognition with Graph Edit Distance. Advances in Computer Vision and Pattern Recognition. Springer, Cham (2015). doi:10.1007/978-3-319-27252-8
12. Stauffer, M., Tschachtli, T., Fischer, A., Riesen, K.: A survey on applications of bipartite graph edit distance. In: Foggia, P., Liu, C.-L., Vento, M. (eds.) GbRPR 2017. LNCS, vol. 10310, pp. 242–252. Springer, Cham (2017). doi:10.1007/978-3-319-58961-9_22
13. Wang, P., Eglin, V., Garcia, C., Largeron, C., Llados, J., Fornes, A.: A novel learning-free word spotting approach based on graph representation. In: International Workshop on Document Analysis Systems, pp. 207–211 (2014)
14. Bui, Q.A., Visani, M., Mullot, R.: Unsupervised word spotting using a graph representation based on invariants. In: International Conference on Document Analysis and Recognition, pp. 616–620 (2015)
15. Riba, P., Llados, J., Fornes, A.: Handwritten word spotting by inexact matching of grapheme graphs. In: International Conference on Document Analysis and Recognition, pp. 781–785 (2015)
16. Stauffer, M., Fischer, A., Riesen, K.: Graph-based keyword spotting in historical handwritten documents. In: International Workshop on Structural, Syntactic, and Statistical Pattern Recognition (2016)
17. Stauffer, M., Fischer, A., Riesen, K.: A novel graph database for handwritten word images. In: International Workshop on Structural, Syntactic, and Statistical Pattern Recognition (2016)
18. Stauffer, M., Fischer, A., Riesen, K.: Speeding-up graph-based keyword spotting in historical handwritten documents. In: Foggia, P., Liu, C.-L., Vento, M. (eds.) GbRPR 2017. LNCS, vol. 10310, pp. 83–93. Springer, Cham (2017). doi:10.1007/978-3-319-58961-9_8
19. Bunke, H., Allermann, G.: Inexact graph matching for structural pattern recognition. Pattern Recognit. Lett. **1**(4), 245–253 (1983)
20. Berretti, S., Del Bimbo, A., Vicario, E.: Efficient matching and indexing of graph models in content-based retrieval. IEEE Trans. Pattern Anal. Mach. Intell. **23**(10), 1089–1105 (2001)
21. Fankhauser, S., Riesen, K., Bunke, H.: Speeding up graph edit distance computation through fast bipartite matching. In: Jiang, X., Ferrer, M., Torsello, A. (eds.) GbRPR 2011. LNCS, vol. 6658, pp. 102–111. Springer, Heidelberg (2011). doi:10.1007/978-3-642-20844-7_11
22. Koopmans, T.C., Beckmann, M.: Assignment problems and the location of economic activities. Econometrica **25**(1), 53 (1957)
23. Riesen, K., Bunke, H.: Approximate graph edit distance computation by means of bipartite graph matching. Image Vis. Comput. **27**(7), 950–959 (2009)
24. Burkard, R., Dell'Amico, M., Martello, S.: Assignment Problems (2009)

Shape Representation and Analysis

Ellipse Detection for Visual Cyclists Analysis "In the Wild"

Abdelrahman Eldesokey$^{(\boxtimes)}$, Michael Felsberg, and Fahad Shahbaz Khan

Computer Vision Laboratory, Linköping University, Linköping, Sweden
{abdelrahman.eldesokey,michael.felsberg,fahad.khan}@liu.se

Abstract. Autonomous driving safety is becoming a paramount issue due to the emergence of many autonomous vehicle prototypes. The safety measures ensure that autonomous vehicles are safe to operate among pedestrians, cyclists and conventional vehicles. While safety measures for pedestrians have been widely studied in literature, little attention has been paid to safety measures for cyclists. Visual cyclists analysis is a challenging problem due to the complex structure and dynamic nature of the cyclists. The dynamic model used for cyclists analysis heavily relies on the wheels. In this paper, we investigate the problem of ellipse detection for visual cyclists analysis in the wild. Our first contribution is the introduction of a new challenging annotated dataset for bicycle wheels, collected in real-world urban environment. Our second contribution is a method that combines reliable arcs selection and grouping strategies for ellipse detection. The reliable selection and grouping mechanism leads to robust ellipse detections when combined with the standard least square ellipse fitting approach. Our experiments clearly demonstrate that our method provides improved results, both in terms of accuracy and robustness in challenging urban environment settings.

1 Introduction

Visual cyclists analysis is gaining considerable attention, especially due to the growing demand for autonomous driving safety. The analysis mainly involves understanding cyclists' behavior and their intentions. A cyclist has a complex structure composed of a bicycle and a pedestrian. Therefore, it cannot be processed as a pedestrian nor a bicycle. The work of [35] introduced a dynamic model for cyclists with nine state parameters that define the cyclist pose in the global coordinate system. Among these parameters are the wheel base, the steering angle, and the normal vector to the rear wheel, as well as the normal to the front wheel, which can be estimated from the steering angle. Such sophisticated dynamic model requires a robust and accurate ellipse estimation for the front and the back wheel to facilitate the state estimation. The assumption is that by tracking these states, the behavior of cyclists can be analyzed and their intentions can be predicted. This would have a great impact on autonomous driving safety, allowing vehicles to interact with cyclists efficiently by knowing their current state and their intentions.

© Springer International Publishing AG 2017
M. Felsberg et al. (Eds.): CAIP 2017, Part I, LNCS 10424, pp. 319–331, 2017.
DOI: 10.1007/978-3-319-64689-3_26

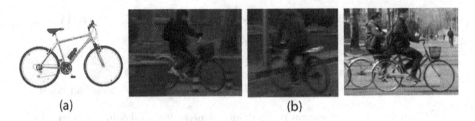

Fig. 1. (a) An example from the Caltech 256 dataset [14], (b) Examples for challenges in the TDCB dataset [20].

Thus, ellipse detection plays a crucial rule for visual cyclists analysis. There are plethora of good ellipse detectors in literature, which have been extensively evaluated [2,11,26]. However, they have been evaluated on either synthetic data or clean images taken in controlled environments. To our knowledge, there exist no dataset for ellipse detection in realistic imagery acquired in uncontrolled environment. Hence, we introduce a new dataset, E-TDCB, with annotated ellipses of wheels for visual cyclists analysis in urban environment. The images in the E-TDCB dataset are taken from the *Tsinghua-Daimler Cyclist Detection Benchmark Dataset (TDCB)* [20], and we provide rich annotations of the bicycle wheels. Our motivation to generate this dataset is to provide an evaluation benchmark for ellipse fitting in real-world urban imagery that is more challenging than the standard datasets and to produce a baseline for visual cyclists analysis methods that apply a dynamic cyclist model relying on ellipse estimates. Figure 1 shows a comparison between a bicycle image from the Caltech 256 dataset [14], taken in a controlled environment, and some challenging examples from the TDCB dataset. We also introduce a novel ellipse detector which combines several ellipse fitting approaches into a light-weight detector with real-time performance, high accuracy, and robustness. We perform comprehensive experiments by evaluating our method and state-of-the-art ellipse detectors on the E-TDCB dataset. The results clearly demonstrate that our method outperforms existing state-of-the-art detectors, while providing an exceptional balance between accuracy and robustness. In summary, our contributions are:

- A new dataset with wheels annotations for visual cyclists analysis in the wild.
- A robust ellipse-based wheel detection method facilitating cyclists analysis.
- Comparison to existing state-of-the-art ellipse detectors on the new dataset.

2   Related Work

One of the few existing works on visual cyclists analysis is the recent work by Zernetsch et al. [34] that predicts the trajectory of cyclists and their intentions such as "Starting", "Stopping", "Waiting", or "Passing". The approach is based on Artificial Neural Networks that are trained on annotated tracks captured using a stereo camera and a laser scanner. Another work by Ardeshiri et al. [1]

estimates and tracks the cyclist's state based on measurements from a conventional monocular camera instead of special hardware. It applies an advanced dynamic cyclist model [35] and particle filters to predict the future state of cyclists. For the ellipse extraction, a simple method fits ellipses to bicycles wheels based on the assumption that tires have reflective rings on them, thus unsuitable for uncontrolled imagery. A robust ellipse detector that works with uncontrolled bicycle imagery would be essential to make this method applicable in practice.

Shape matching, as a generalization to ellipse extraction, has been a frequently studied computer vision approach to object recognition tasks. Several methods have been proposed based on geometric context [3,4], shape descriptors in the spatial domain [12], and in the frequency domain [19]. These methods have been applied to numerous challenging problems such as object recognition [24,25], character recognition [29], traffic sign recognition [19], pedestrian detection, and motion analysis [21]. *Ellipses* are one of the most frequently observed shapes in digital imagery since they are projections of circles commonly available in real-world objects. This has prompted ellipse fitting to be one of the prerequisites for several shape matching methods employed in many applications, including facial gesture analysis [15,30], medical image analysis [27,31], vehicle wheels detection [7], visual cyclists analysis [1], and traffic sign detection [22].

An ellipse is defined by five parameters and a common approach was to use the Hough Transform (HT) to estimate these parameters. The approach is similar to HT line detection, but in case of ellipses, the accumulator has five dimensions instead of two. This imposes high computational and memory demands to explore the space and several attempts have been made to reduce its size. Mclaughlin [23] eliminated two parameters by geometrically finding the center of the ellipse and then performing Randomized HT (RHT) [33] to obtain the other three parameters. In [32], the dimensionality of the accumulator was reduced to one dimension by randomly selecting pairs of pixels that match certain geometric constraints and estimate the center, major axis, and the angle for candidate ellipses from these pairs. The minor axis is then calculated using 1D HT. Similarly, two optimized versions of [32] were proposed in [2,6] which require less memory, fewer computations, and are more robust against noise and false detections. All these HT-based methods require a proper selection of several control parameters that define geometric constraints.

Another approach is based on the canonical representation of ellipses and formulating the problem as a least-squares minimization. Fitzgibbon [10] introduced direct least squares fitting of ellipses by minimizing the algebraic distance between some scattered points and an ellipse hypothesis that is represented in canonical form and constrained to produce only ellipses, not parabolas or hyperbolas. A more stable version [28] addresses the problem of singularities in the design matrix [10] and produces more stable solutions for the least-squares problem. RANSAC has also been used to randomly sample a subset of the data, to fit an ellipse using least-squares minimization, and to iterate until the best ellipse is found according to some convergence criteria [8,17].

Recently, two methods were introduced for ellipse fitting which combine the two approaches above. The first one [26] is mainly based on edge curvature and convexity to determine groups which potentially form ellipses. A modified 2D HT is used to evaluate these groups and if they fulfill some constraints, an ellipse is fitted using direct least squares [10]. The ellipse candidates are evaluated using three unique saliency measures and an ellipse is selected if its saliency score exceeds the average score for all other ellipses. The second method was proposed by Fornaciari and others [11] and is based on grouping edges that adhere to some geometric constraints. Thus the parameter space for HT is reduced, enabling real-time performance even on smart phones with limited computational power. These two methods were shown to achieve state-of-the-art performance on synthetic data and the Caltech 256 dataset [14]. Both methods will be evaluated on the E-TDCB dataset in Sect. 5.

3   Ellipse Fitting for Visual Cyclists Anaylsis

The Tsinghua-Daimler Cyclists Benchmark [20] provides a comprehensive evaluation for state-of-the-art object detectors on their dataset. The Deformable Parts Model (DPM) object detector [9] was able to achieve a remarkable performance against sophisticated methods such as F-RCNN deep networks [13]. A major advantage of a DPM is its ability to construct a flexible model for the object-of-interest. This model is composed of the prominent parts that form the object, their arrangement, and relationships between them. This makes the DPM approach highly appropriate for visual cyclist analysis as it gives an insight about the structure of the cyclist and facilitates its analysis accordingly. A part of the training process for a DPM detector is to find the optimal locations for different parts and to construct a weight filter for each part. Those filters are convolved with the image during test time and are supposed to produce high output at their corresponding parts.

3.1   Finding the Wheels

We trained a DPM model on the TDCB dataset, restricted to cyclists that are seen from the side view as described in details in Sect. 4. The model has a root filter that locates cyclists as a whole and another eight part filters which locate different parts of the cyclist as illustrated in Fig. 2(a). The state model of the cyclist relies basically on the wheels as described in Sect. 1. Hence, we sample *potential patches* from DPM parts that are located on the wheels (shown in *red* in Fig. 2(a). The Canny edge detector [5], i.e., horizontal and vertical derivatives, is applied to potential patches to get the edge map. Each patch is processed individually on the subsequent steps.

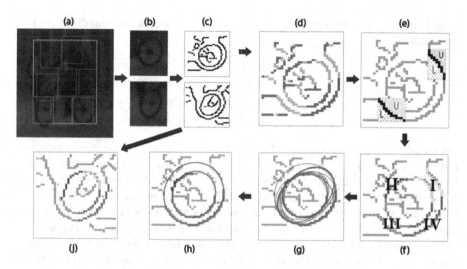

Fig. 2. Method overview. (a) DPM root filter in green, parts filters in white and selected filters in red. (b) Candidate patches sampled at the wheels. (c) Edge maps for candidate patches. (d) Classifying edge pixels according to positive (cyan) and negative (magenta) sign of the orientation tangent. (e) Arcs convexity check. (f) Different arcs quadrants in unique colors. (g) Ellipse hypothesis. (h) Best ellipse hypothesis. (j) A long arc shown in green that is split using inflexion point detection. (Color figure online)

3.2   Arcs Selection

Initially, each edge pixel p in the edge map is classified according to the sign of the orientation tangent of its gradient as follows:

$$\text{sign}(\tan \varphi(p)) = \text{sign}(G_x(p)).\text{sign}(G_y(p)) \tag{1}$$

where $G_x(p)$ and $G_y(p)$ are the horizontal and the vertical derivatives, respectively. Pixels with positive orientations are stored in a set P_{pos} while pixels with negative orientations are stored in P_{neg}. Each set is processed by an edge linking algorithm to define connected arcs. In this work we use the edge linking algorithm by Kovesi [18] and store each arc a_i from $P_{\text{pos}}, P_{\text{neg}}$ in sets $A_{\text{pos}}, A_{\text{neg}}$, respectively. Arcs that have a length less than Θ_{Len} are discarded.

The convexity of each arc is checked to determine if it is up-facing or down-facing. Similarly to [11] and as illustrated in Fig. 2(e), the area below and above the arc determines its convexity:

$$\text{Conv}(a_i) = \begin{cases} 1, & \text{Area}(L^i) > \text{Area}(U^i) \\ -1, & \text{Area}(U^i) > \text{Area}(L^i) \end{cases}, \tag{2}$$

Based on the arc convexity and orientation, the associated quadrant can be determined:

$$Q(a_i) = \begin{cases} I & \text{if} & (a_i \in A_{\text{pos}}) \wedge (\text{Conv}(a_i) = 1) \\ II & \text{if} & (a_i \in A_{\text{neg}}) \wedge (\text{Conv}(a_i) = 1) \\ III & \text{if} & (a_i \in A_{\text{pos}}) \wedge (\text{Conv}(a_i) = -1) \\ IV & \text{if} & (a_i \in A_{\text{neg}}) \wedge (\text{Conv}(a_i) = -1) \end{cases}, \tag{3}$$

note that the signs of G_x and G_y are not sufficient to determine the quadrant as the direction of the gradient is unknown (direction vs. orientation [16]). Figure 2(f) shows different quadrants in unique colors. Occasionally, two arcs are merged due to noise or background clutter as shown in Fig. 2(j). Therefore, we detect inflexion points, where the continuity of the arc is violated. We apply a 3-steps approach for detecting those inflexion points [26]: (a) fit line segments to the arc; (b) calculate their angles with the arc; and (c) check how these angles change between line segments. A sign change for these angles indicates a change in the arc curvature and the arc is split at this point.

3.3 Arcs Grouping

Arcs are grouped into pairs in a anti-clockwise order, e.g., arc $a_i \Leftrightarrow Q(a_i) = I$ is grouped with arc $a_k \Leftrightarrow Q(a_k) = II$. This grouping is constrained to prevent irrelevant arcs from being grouped with arcs from the wheels:

$$Pr(a_i, a_k) = \begin{cases} \top & \text{if } Q(a_i, a_k) = \langle I, II \rangle \wedge \text{abs}(a_i.L_y - a_k.R_y) \leq \Theta_{\text{pair}} \\ \top & \text{if } Q(a_i, a_k) = \langle II, III \rangle \wedge \text{abs}(a_i.L_x - a_k.L_x) \leq \Theta_{\text{pair}} \\ \top & \text{if } Q(a_i, a_k) = \langle III, IV \rangle \wedge \text{abs}(a_i.R_y - a_k.L_y) \leq \Theta_{\text{pair}} \\ \top & \text{if } Q(a_i, a_k) = \langle IV, I \rangle \wedge \text{abs}(a_i.R_x - a_k.R_x) \leq \Theta_{\text{pair}} \\ \bot & \text{else,} \end{cases} \tag{4}$$

where L/R is the leftmost/rightmost point of the arc, abs() is the absolute value, and Θ_{pair} is the pairing threshold. The choice of the value of Θ_{pair} depends on the thickness of the wheel and the size of the bicycle. A good selection for this parameter prevents wrong pairings of arcs that do not belong to the same ellipse. For instance, in Fig. 2(g), a high value for Θ_{pair} will cause arcs belonging to the inner rim to be grouped with arcs from the outer rim. After applying the constraints (4) to all possible pairs, only $\top$-arcs are paired together and added to the set of pairs S_{pairs}. Eventually, all pairs that have a common arc are grouped into triplets and added to a set of triplets S_{triplets}. For example, assume two pairs of arcs $\langle a_c, a_d \rangle$ and $\langle a_d, a_e \rangle$ will be grouped into a triplet $\langle a_c, a_d, a_e \rangle$ and this triplet is added to S_{triplets}.

3.4 Ellipse Fitting, Grouping and Evaluating

For each triplet in S_{triplets}, all arc points in this triplet are used for fitting an ellipse. Direct least squares ellipse fitting [10] is used and the residual error is required to be less than Θ_{LSE} for this ellipse to be considered as an *ellipse hypothesis*, see Fig. 2(g). As a second step, similar ellipses, i.e., ellipses with high overlap with each other, are grouped and their parameters are averaged to get a representing ellipse. Finally, ellipse hypotheses are evaluated by checking their intersection with the edge map. Bicycle wheels are usually occluded and cluttered by the background. Consequently, some parts of the wheels will not form an arc and will be removed as noise which will lead to an imprecise ellipse estimation. Therefore, checking for intersection with all edge pixels results in a better estimation even if some arcs are removed.

4 Dataset

The starting point of this work was to investigate whether the available ellipse detectors work in real-life applications such as visual cyclists analysis, where data is taken from uncontrolled environment, i.e. a camera mounted on the dashboard of a car. We checked the recently published datasets that match the above criteria and we found that the *Tsinghua-Daimler Cyclist Detection Benchmark Dataset (TDCB)* [20] is a perfect match. The images were recorded using a stereo camera setup mounted on a car that drove in the streets of Hong Kong. Bounding box annotations were provided for different classes of objects such as pedestrians, cyclists, motorcycles, and other objects. The dataset has many challenges that only arise in uncontrolled environments. Bicycle wheels are usually cluttered by the background and sometimes occluded by the legs of the cyclist. Motion blur occur occasionally and cause bicycle wheels edges to be smeared. Also if a bicycle is occluded by another bicycle, it becomes tricky to discriminate between their wheels. Some challenging examples are shown in Fig. 1.

We provide wheels annotations on the monocular images for cyclists with bounding box aspect ratio $r < 1.25$ in *training set*, $r < 1.75$ in *test set* where r is defined as $r = \text{bbox}_{height}/\text{bbox}_{width}$. This ratio indicates that the cyclist is seen from the side and the wheels are visible. Other images from the front or the back view are ignored as they do not have visible wheels. The number of cyclists that matched the above ratio is 642 images from training set and 142 image from the test set. This means that the dataset has 1568 manually annotated ellipses for the outer rims of the wheels, while the inner rims were estimated roughly with respect to the cyclist size. Some annotated examples are shown in Fig. 3.

Fig. 3. Examples of wheels annotation.

5   Experiments

As discussed previously, a reliable ellipse fitting is essential for cyclist state estimation. Hence, we evaluate our ellipse detector as well as three state-of-the-art methods on the proposed dataset. The methods are Prasad [26], Basca [2], and Yaed [11]. An overview of each method was provided in Sect. 2, while an overview of the dataset was given in Sect. 4. The source code for Prasad was provided by the author while the source code for Yaed and Basca was found online.

5.1   Evaluation Metrics

For evaluation, following evaluation metrics are used:

$$\text{Precision} = \frac{\text{Number of True Positive Ellipse Hypotheses}}{\text{Total Number of Ellipse Hypotheses}} \tag{5}$$

$$\text{Recall} = \frac{\text{Number of True Positive Ellipse Hypotheses}}{\text{Total Number of Ground-truth Ellipse}} \tag{6}$$

$$\text{F-Score} = \frac{2 \times \text{Precision} \times \text{Recall}}{\text{Precision} + \text{Recall}} \tag{7}$$

where a *true positive hypothesis* is an ellipse which has a certain overlap with any of the ground-truth ellipses. The overlap in case of ellipses is defined as overlap $= 1 - \frac{\text{count}(\text{XOR}(\text{Ellipse, GT}))}{\text{count}(\text{OR}(\text{Ellipse, GT}))}$. High precision indicates that the detector outputs highly confident hypotheses, high recall means that the detector is reliable in finding the ellipses, and f-score incorporates both. A good ellipse detector should combine high precision and a good recall.

5.2   Parameter Selection

Our method has three parameters. The minimum length of arc Θ_{Len} is set to 9, Θ_{pair} can be any value from 5 to 15, and Θ_{LSE} is set to 0.01. In Prasad, we change *minimum_edge_length* to 10 instead of 15 and the rest of parameters are kept unchanged. Basca has many control parameters as it is HT-based. We set *minMajorAxis* to one third of the candidate patch width, *maxMajorAxis* to the largest dimension of the candidate patch, and we only consider at most the best 25 ellipse hypothesis in the evaluation to retain levels of precision. For Yaed, we tried different combinations of its many control parameters and the best f-score was achieved at minimum arc length $Th_{\text{Len}} = 5$, minimum shortest side of arc oriented bounding box $Th_{\text{OBB}} = 1.0$, and the default remaining parameters.

5.3   Quantitative Results

Evaluation metrics are calculated for each method under different overlap ratios from 0 to 1 both on the training and the test set. The test set is more challenging

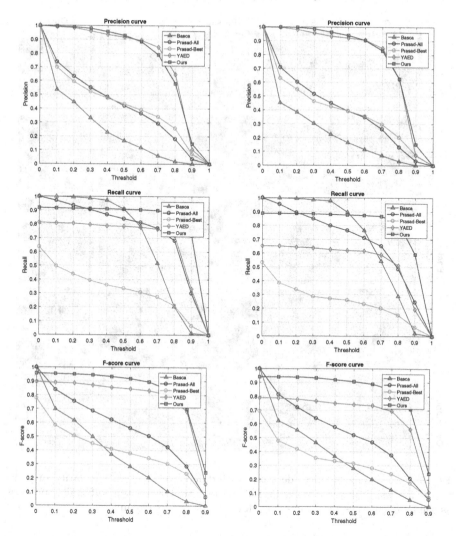

Fig. 4. Evaluation metrics for Basca [2], Prasad [26], Yaed [11], and our proposed method on the training set (Left colum) and test set (Right Column). The test set is more difficult than the training set as the DPM filters sometimes are not perfectly aligned in case of test set, which makes ellipse detection task more challenging.

as the locations of DPM filters are not perfectly aligned as in the training set. Besides, cyclists in the test set have larger yaw angles due to larger value of r which makes ellipse fitting more difficult. Figure 4 summarizes the evaluation metrics for the training and the test set. For Prasad, the final policy for selecting salient ellipses is too strict for this realistic dataset which led to very low recall. Therefore, we evaluated Prasad twice, once on the most salient hypotheses (Prasad-Best) and another on all hypotheses after grouping (Prasad-All).

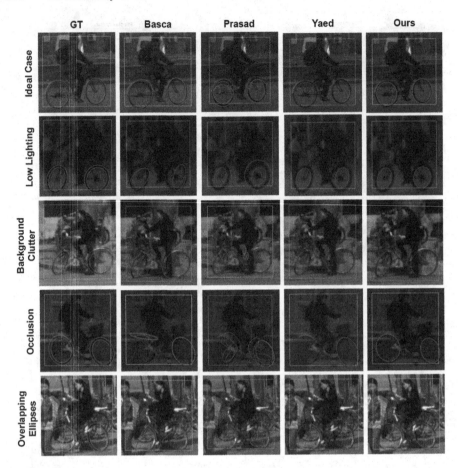

Fig. 5. Performance of ellipse detectors with different challenges. For clarity, only the best hypothesis is drawn. For the Prasad method, we show the top half of all hypothesis as the best hypothesis is always a small ellipse due to its circumference overlap ratio.

As shown, our method and Yaed have a comparably high precision on both sets, while Prasad and Basca have average precision. Our explanation is that their arcs grouping criteria is based on a HT accumulator for finding the potential ellipse center for each arc. In case of concentric ellipses as in bicycle wheels, irrelevant ellipses from outer and inner rims are grouped which leads to false ellipse estimation. On the contrary, the constrained arcs grouping criteria in Yaed and our method alleviates these false groupings. The recall is high for all methods but for different reasons. For Basca, a huge number of hypotheses is produced which leads to a high recall under low thresholds only, while Prasad-All has a high recall due to considering all hypothesis in the evaluation. Yaed has slightly lower recall due to the large number of control parameters that are difficult to determine for balancing precision and recall. Finally, our method was able to retain high recalls even under high thresholds due to its relaxed constraints on dealing with all available edge information. The F-score shows that our method outperforms all other methods, achieving an outstanding balance between accuracy and robustness both on training and test set.

5.4 Qualitative Analysis

Figure 5 shows how each method performs with different challenges. Basca performs well when the wheels are visible in the ideal case, but when some challenges are introduced, a considerable tuning has to be done for the control parameters of HT space. However, it does not perform well with cluttered or occluded wheels as HT-based approaches are sensitive to outliers. Prasad performs similarly to Basca and cannot handle most challenges as its ellipses saliency criteria is either too strict or in favor of small ellipses which has high circumference overlap ratio [26]. YAED noticeably performs better than the latter methods and achieves higher accuracy as it has a reliable arcs selection criteria. However it fails also to detect cluttered and occluded wheels due to its strict ellipses selection criteria and the large number of control parameters that needs to be adjusted to each case. Finally, our method achieves and outstanding performance with all challenges. It succeeds to fit ellipses to occluded and cluttered wheels as it employs a very reliable and relaxed arcs selection criteria that encounters for all available information in the edge map, which is usually treated as noise in other approaches. Also our ellipse selection policy is suitable for real-world scenarios where images are not always ideal. Thereby, our method succeeded in combining both precision and reliability, which is essential for real-world applications.

6 Conclusion and Future Work

Due to the growing emergence of autonomous vehicles and the increasing attention to their safety measures, visual cyclists analysis needs to be investigated more for the sake of cyclists safety. The literature lacks for a realistic real-world dataset for this purpose. Therefore, we introduced a new dataset with wheels annotations for cyclists that contributes to addressing the problem of visual cyclists analysis. We also proposed a robust and reliable method for ellipse fitting on bicycles wheels that is needed for cyclists state estimation. Our method as well as the state-of-the-art methods were evaluated on the new dataset and our method was able to outperform all other methods providing robust and reliable ellipse detection. In the future, more approaches need to be investigated for ellipse fitting. For instance, the use of edge orientation information in ellipse fitting, iterative least square minimization for refined fitting, and further investigations for the most suitable ellipse selection criteria. Also integrating our ellipse fitting method in an existing visual cyclists analysis platform would give some insights on the advantages and drawbacks of our method.

Acknowledgments. This work has been supported by VR (EMC^2, ELLIIT, starting grant [2016-05543]) and Vinnova (Cykla).

References

1. Ardeshiri, T., Larsson, F., Gustafsson, F., Schön, T.B., Felsberg, M.: Bicycle tracking using ellipse extraction. In: 2011 Proceedings of the 14th International Conference on Information Fusion (FUSION), pp. 1–8. IEEE (2011)
2. Basca, C., Talos, M., Brad, R.: Randomized hough transform for ellipse detection with result clustering. In: The International Conference on Computer as a Tool, EUROCON 2005, vol. 2, pp. 1397–1400. IEEE (2005)
3. Belongie, S., Malik, J., Puzicha, J.: Shape matching and object recognition using shape contexts. IEEE Trans. Pattern Anal. Mach. Intell. 24(4), 509–522 (2002)
4. Berg, A.C., Berg, T.L., Malik, J.: Shape matching and object recognition using low distortion correspondences. In: IEEE Computer Society Conference on Computer Vision and Pattern Recognition, CVPR 2005, vol. 1, pp. 26–33 (2005)
5. Canny, J.: A computational approach to edge detection. IEEE Trans. Pattern Anal. Mach. Intell. 6, 679–698 (1986)
6. Chia, A.Y.S., Leung, M.K., Eng, H.-L., Rahardja, S.: Ellipse detection with hough transform in one dimensional parametric space. In: IEEE International Conference on Image Processing, ICIP 2007, vol. 5, p. V-333 (2007)
7. Cooke, T.: A fast automatic ellipse detector. In: 2010 International Conference on Digital Image Computing: Techniques and Applications (DICTA), pp. 575–580. IEEE (2010)
8. Duan, F., Wang, L., Guo, P.: RANSAC based ellipse detection with application to catadioptric camera calibration. In: Wong, K.W., Mendis, B.S.U., Bouzerdoum, A. (eds.) ICONIP 2010. LNCS, vol. 6444, pp. 525–532. Springer, Heidelberg (2010). doi:10.1007/978-3-642-17534-3_65
9. Felzenszwalb, P.F., Girshick, R.B., McAllester, D., Ramanan, D.: Object detection with discriminatively trained part-based models. IEEE Trans. Pattern Anal. Mach. Intell. 32(9), 1627–1645 (2010)
10. Fitzgibbon, A., Pilu, M., Fisher, R.B.: Direct least square fitting of ellipses. IEEE Trans. Pattern Anal. Mach. Intell. 21(5), 476–480 (1999)
11. Fornaciari, M., Prati, A., Cucchiara, R.: A fast and effective ellipse detector for embedded vision applications. Pattern Recogn. 47(11), 3693–3708 (2014)
12. Gal, R., Cohen-Or, D.: Salient geometric features for partial shape matching and similarity. ACM Trans. Graphics (TOG) 25(1), 130–150 (2006)
13. Girshick, R.: Fast R-CNN. In: Proceedings of the IEEE International Conference on Computer Vision, pp. 1440–1448 (2015)
14. Griffin, G., Holub, A., Perona, P.: Caltech-256 object category dataset (2007)
15. Huang, Y.-H., Pan, B.-C., Zheng, S.-L., Pan, J., Tang, Y.: Lip-reading detection and localization based on two stage ellipse fitting. In: International Conference Wavelet Analysis and Pattern Recognition. IEEE (2008)
16. Jähne, B.: Digital Image Processing (2002)
17. Kaewapichai, W., Kaewtrakulpong, P.: Robust ellipse detection by fitting randomly selected edge patches
18. Kovesi, P.: Edge linking and line segment fitting
19. Larsson, F., Felsberg, M., Forssen, P.-E.: Correlating fourier descriptors of local patches for road sign recognition. IET Comput. Vision 5(4), 244–254 (2011)
20. Li, X., Flohr, F., Yang, Y., Xiong, H., Braun, M., Pan, S., Li, K., Gavrila, D.M.: A new benchmark for vision-based cyclist detection. In: 2016 IEEE Intelligent Vehicles Symposium (IV), pp. 1028–1033. IEEE (2016)

21. Ling, H., Jacobs, D.W.: Shape classification using the inner-distance. IEEE Trans. Pattern Anal. Mach. Intell. **29**(2), 286–299 (2007)
22. Liu, H., Ran, B.: Vision-based stop sign detection and recognition system for intelligent vehicles. Transp. Res. Rec. J. Transp. Res. Board **1748**, 161–166 (2001)
23. McLaughlin, R.A.: Randomized hough transform: improved ellipse detection with comparison. Pattern Recogn. Lett. **19**(3), 299–305 (1998)
24. Mori, G., Malik, J.: Recognizing objects in adversarial clutter: breaking a visual captcha. In: 2003 IEEE Computer Society Conference on Computer Vision and Pattern Recognition, Proceedings, vol. 1, p. I. IEEE (2003)
25. Opelt, A., Pinz, A., Zisserman, A.: A boundary-fragment-model for object detection. In: Leonardis, A., Bischof, H., Pinz, A. (eds.) ECCV 2006. LNCS, vol. 3952, pp. 575–588. Springer, Heidelberg (2006). doi:10.1007/11744047_44
26. Prasad, D.K., Leung, M.K., Cho, S.-Y.: Edge curvature and convexity based ellipse detection method. Pattern Recogn. **45**(9), 3204–3221 (2012)
27. Pu, J., Zheng, B., Leader, J.K., Gur, D.: An ellipse-fitting based method for efficient registration of breast masses on two mammographic views. Med. Phys. **35**(2), 487–494 (2008)
28. Radim Halir, J.F.: Numerically stable direct least squares fitting of ellipses (1998)
29. Rocha, J., Pavlidis, T.: A shape analysis model with applications to a character recognition system. IEEE Trans. Pattern Anal. Mach. Intell. **16**(4), 393–404 (1994)
30. Takegami, T., Gotoh, T., Ohyama, G.: An algorithm for model-based stable pupil detection for eye tracking system. Syst. Comput. Japan **35**(13), 21–31 (2004)
31. Teutsch, C., Berndt, D., Trostmann, E., Weber, M.: Real-time detection of elliptic shapes for automated object recognition and object tracking. In: Electronic Imaging 2006, p. 60700J. International Society for Optics and Photonics (2006)
32. Xie, Y., Ji, Q.: A new efficient ellipse detection method. In: 16th International Conference on Pattern Recognition, Proceedings, vol. 2, pp. 957–960 (2002)
33. Xu, L., Oja, E., Kultanen, P.: A new curve detection method: randomized hough transform (RHT). Pattern Recogn. Lett. **11**(5), 331–338 (1990)
34. Zernetsch, S., Kohnen, S., Goldhammer, M., Doll, K., Sick, B.: Trajectory prediction of cyclists using a physical model and an artificial neural network. In: 2016 IEEE Intelligent Vehicles Symposium (IV), pp. 833–838. IEEE (2016)
35. Åström, K.J., Klein, R.E., Lennartsson, A.: Bicycle Dyn. Control **25**(4), 26–47 (2005)

Visual Landmark Based 3D Road Course Estimation with Black Box Variational Inference

Felix Trusheim[1]([✉]), Alexandru Condurache[1], and Alfred Mertins[2]

[1] Robert-Bosch GmbH, Stuttgart, Germany
{Alexandrupaul.Condurache,Felix.Trusheim}@de.bosch.com
[2] ISIP, University of Luebeck, Lübeck, Germany
mertins@isip.uni-luebeck.de

Abstract. In this paper we present an approach which estimates the course of a road over long distances based on static and dynamic scene cues detected by a video camera. The approach is based on a clothoid road model, a probabilistic fusion concept as well as a fast variational inference method. Our experimental results show that the approach outperforms a state-of-the-art road marking-based method in challenging real-world driving situations.

Keywords: Road course estimation · ADAS · Probabilistic environment model · Black box variational inference · Clothoid road-course model · 3D scene reconstruction

1 Introduction

Automated driving requires a robust and precise estimation of the road course. This is a challenging task, which for all conceivable driving situations and environments is still unsolved. State-of-the-art road-course detection systems fuse different complementary sensor signals to receive better estimation results. This are typically signals of radar, lidar and camera systems. Within these setups, camera systems are a valuable source of information. They usually contribute detected road markings to the road-course estimation. Unfortunately, this kind of information is often not sufficiently available in many driving situations like for example on some newly build roads or in dimly lit environments (e.g. at night). Therefore, we propose a camera-based estimation approach, which does not depend on information from road markings but instead on information of different road-course correlated scene cues. Thus, our approach estimates the course of a road within reach of 140 m on the base of static and dynamic scene cues like delineators or other traffic participants. We build our algorithm on a clothoid road model and fuse measured scene cues with the help of a probabilistic model and variational inference. We empirically evaluate our approach in challenging real-world driving situations with reduced light, and prove its performance.

© Springer International Publishing AG 2017
M. Felsberg et al. (Eds.): CAIP 2017, Part I, LNCS 10424, pp. 332–343, 2017.
DOI: 10.1007/978-3-319-64689-3_27

1.1 Related Work

The importance of a robust road-course estimation for the realization of a vehicle-cruise-control system (ADAS) led to a high research interest in this subject early on. The first commercially available systems were based primarily on radar sensors. Such systems detect the road course based on radar-signal-reflecting landmarks such as cars or guardrails. Since then the abilities of such systems progressed constantly. State-of-the-art radar systems depend heavily on increased and detailed environment models, probabilistic estimation approaches [1,7] and advanced radar-hardware designs [2]. Of particular interest is the approach by Hammarstrand et al. [3]. They received a more robust estimation system by integrating a clothoid road model, which explicitly models the course of a road according to real-world design principles in road construction [22]. The first camera-based systems used road markings for a road-course estimation [4,5]. Recent proposals for camera-based systems depend exclusively [11] or additionally [10] on semantic segmentation results computed with the help of a deep convolutional neuronal network. These kind of approaches produce remarkable results in day-light situations. But because these methods mainly exploit surface textures of objects for a scene segmentation, they often lack performance in dimly lit environments (e.g. at night) where surface textures are hardly visible. To overcome limitations of individual sensor systems and methods, fusion-based approaches, that integrate information of multiple methods as well as multiple sources, such as radar, lidar, cameras sensors and digital HDR maps, were proposed [6–8]. Popular representatives of these use an occupancy grid to fuse the information [9]. Another very promising fusion method has been proposed by Geiger et al. [12,13] for a camera-only crossroad structure estimation. Based on graphical probabilistic modeling the approach showed remarkable estimation performances. However, a computational bottleneck of this approach is the used sampling-based inference method. Recent published variational-inference techniques [17–19] promise to solve this problem. Therefore, in this contribution we use a similar probabilistic fusion model combined with an efficient variational inference technique for an estimation of the road course based on camera-detected static and dynamic scene cues. Our versatile probabilistic fusion framework also allows the integration of information provided by other sensors, if these are available.

The remainder of the paper is structured as follows: Sect. 2 describes our proposed estimation approach in detail, then in Sect. 3 we evaluate and discuss our method in challenging driving situations. The Sect. 4 summarizes our approach and offers a brief outlook on future work.

2 Road Course Estimation Using Variational Inference

2.1 Scene Cues

The foundation of our approach are scene cues or landmarks that reflect the course of a road. They are detected by a monoscopic camera system which is mounted behind the rear-view mirror in a car (see Fig. 1).

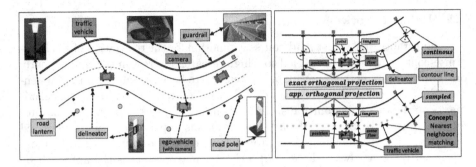

Fig. 1. Schematic road course with regarded evidence types (left) and different data-projection concepts (right)

These scene cues (see Fig. 1) are:

- **Static Objects:** Guardrails, delineators, road poles, road lanterns and road-embedded reflectors.
- **Dynamic Objects:** Bicycle, cars, motorbikes and trucks.

The static and dynamic objects are detected with the help of different classifiers. Each classifer detection is rated by a confidence measure. The detected objects are tracked over time and thus generate tracklet information. The classifier and tracking [16] methods are not the subject of this publication, and therefore will not be discussed in detail.

In addition to the images, our road-course estimation approach makes also use of information from a 6-axis Inertial Measurement Unit (IMU).

2.2   Causality Model

The theoretical foundation of the proposed approach is a generative Bayesian network (BN) model which describes the causal relationship between the road-course defining parameters and the image projections of road-course correlated static and dynamic scene cues. A sketch of the context of our approach together with the proposed probabilistic model are shown in Fig. 2. To properly introduce the complete modeling, we begin with a description of a sub-part of the model. Hence, we start with the causal relationship between the road-course determining parameters and the $3D$ positions of the static landmarks. A look at the context-referencing Fig. 2, suggests that the course of a road can be described by a virtual contour line. This contour line can quantified by a road model $f(C)$, in which the C represents the shape-defining parameters. The $3D$ positions $y_{Stat3DPos}$ of the road-course aligned static landmarks can then be described as objects lateral-shifted to the contour line along the contour-line normal $\tilde{f}(C)$. This is exemplary shown with delineators in Fig. 2. While driving, those $3D$ positions $y_{Stat3DPos}$ get projected on the image sensor according to the $3D$ pose of the user-vehicle camera $y_{UVCam3DPose}$ and thereby generate landmark-corresponding measurements $y_{Stat2DPos}$. The same ideas can be transferred to dynamic objects as well.

Analogous to static landmarks, the $2D$ positions $\boldsymbol{y}_{V2DPos}$ can be modeled as a causal superposition of the road-contour parameter $\boldsymbol{C}$, the vehicle-specific offsets $O_D^{[p]}$ and the camera poses $\boldsymbol{y}_{UVCam3Pose}$. However, unlike static objects, dynamic objects, such as vehicles, change their positions over time and generate a $3D$ scene flow $\boldsymbol{y}_{V3DFlow}$ as well as a corresponding sparse optical flow $\boldsymbol{y}_{V2DFlow}$. Because these generated flows are directly connected with the positions of the vehicles, they can be modeled equally by a causal superposition of the road-contour parameter $\boldsymbol{C}$ and the vehicle lateral offsets O_D. This model, in conjunction with the measured evidence $\boldsymbol{y}_{Stat2DPos}$, $\boldsymbol{y}_{V2DPos}$, $\boldsymbol{y}_{V2DFlow}$ and the IMU data $\boldsymbol{X}$ and $\boldsymbol{\Theta}$, enables the estimation of the road-course parameters $\boldsymbol{C}, \boldsymbol{O}_S$ and $\boldsymbol{O}_D$.

In order to increase the estimation robustness we build the proposed model on additional design principles:

- **Data Buffering:** To obtain a sufficient amount of evidence we accumulate ego-motion compensated data over a time period T.
- **Flat World Assumption:** To simplify the complexity of the contour model $f(\boldsymbol{C})$ during the estimation process we make the assumption that the $3D$ road course is located on flat plane and therefore can be handled as a $2D$ road course.
- **Multi Stage Design:** We use a two stage signal processing pipeline (see Fig. 3) to reconstruct a $3D$ scene from measured $2D$ information and then estimate the course of the road in $3D$.

  - In the **first stage** we reconstruct the $3D$ signals $\boldsymbol{y}_{Stat3DPos}$, $\boldsymbol{y}_{V3DPos}$ and $\boldsymbol{y}_{V3DFlow}$ from the corresponding image measurements $\boldsymbol{y}_{Stat2DPos}$, $\boldsymbol{y}_{V2DPos}$ and $\boldsymbol{y}_{V2DFlow}$. The $3D$ positions of trackable static objects are obtained with the help of standard structure-from-motion (SfM) methods. In detail, we use a combination of an inverse-depth reconstruction [21] and a bundle adjustment calculation [20]. The necessary ego-motion information is obtained from the IMU. For the $3D$ reconstruction of vehicles we use an approach based on prior knowledge about the geometry of the vehicles, as well as the assumption of a distortion-free camera projection model. Hence, we calculate the $3D$ positions and the $3D$ scene flows with the help of the intercept theorem (IcT) [21] and standard tracklet-based differential methods.
  - In the **second stage** of the pipeline we estimate the road course based on the $3D$ evidence. This stage is presented in detail in Sect. 2.3.

2.3   Probabilistic Model

The purpose of our approach is the identification of the contour parameter $\boldsymbol{C}$ as well as the lateral-offset parameters $\boldsymbol{O}_S$ and $\boldsymbol{O}_D$, which explain the measured evidence data $\boldsymbol{Y}$. This directly corresponds to a regression problem. However, due to the chosen clothoid-contour model

$$f(\boldsymbol{C}) = \begin{bmatrix} x(l) \\ y(l) \end{bmatrix} = \begin{bmatrix} x_0 \\ y_0 \end{bmatrix} + \int_0^l \begin{bmatrix} cos(\phi(t)) \\ sin(\phi(t)) \end{bmatrix} \Bigg|_{\phi(t) = \phi_0 + \kappa_0 \cdot t + \frac{\kappa_1}{2} \cdot t^2} dt, \tag{1}$$

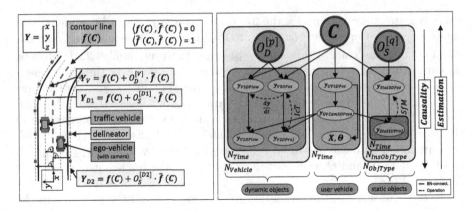

Fig. 2. Schematic road course (left) and proposed probabilistic model (right). The acronyms used in the figures are explained in the text.

this regression problem is not trivial. The integral terms complicates the mathematical handling of the road model within a regression problem. A solution to this problem can be derived by the clothoid-approximation framework of Bertolazzi [23]. This framework allows the definition of a clothoid based on the configuration of its start and end point. Therefore, the contour parameter C is determined by

$$C = [x_{Start}, y_{Start}, \alpha_{Start}, x_{End}, y_{End}, \alpha_{End}].$$

In order to adapt the model to the measured evidence data we need an effective method to project this data orthogonally to road-contour model $f(C)$ (see Fig. 1). However, an optimal orthogonal projection results in a computational heavy regression problem. To avoid that, we apply an approximative orthogonal projection concept similar to the procedure proposed by Geiger et al. [13]. In the first stage, we therefore sample the clothoid road model along its length. Based on the sampling, an evidence point y is then assigned to the closest clothoid-sample point p. Associated with that assignment, the scene flow of an evidence point y (in case of a moving vehicle) is than connected to the corresponding clothoid-tangent at point p (see Fig. 3). Therefore, the regression problem is no longer differentiable and hence can not be solved by an efficient gradient-based optimization method.

Alternatively, we reformulate the regression problem as a probabilistic maximum a-posteriori (MAP) estimation within a graphical model framework and solve that with the help of variational inference.

In this reformulation, the regression parameters (C, O_S, O_D) correspond to hidden random variables and the evidence data Y correspond to observable random variables. The general structure of the joint distribution of this MAP-problem follows the form

$$P(\boldsymbol{H},\boldsymbol{Y})|_{\substack{\boldsymbol{H} = \{\boldsymbol{C},\boldsymbol{O}_S,\boldsymbol{O}_D\} \\ \boldsymbol{Y} = \{\boldsymbol{Y}_S,\boldsymbol{Y}_D\}}} = P(\boldsymbol{C}) \cdot P(\boldsymbol{O}_S) \cdot P(\boldsymbol{O}_D) \tag{2}$$

$$\cdot P(\boldsymbol{Y}_S|\boldsymbol{C},\boldsymbol{O}_S) \cdot P(\boldsymbol{Y}_D|\boldsymbol{C},\boldsymbol{O}_D).$$

Thereby, the evidence data $\boldsymbol{Y}$ consists of static landmarks $\boldsymbol{Y}_S$ and dynamic landmarks $\boldsymbol{Y}_D$. The prior terms of the regression parameters are assumed as Gaussian distributed. Thus, they are defined as

$$P(\boldsymbol{H}_i) = \frac{\exp\left(-\frac{1}{2}(\boldsymbol{H}_i - \boldsymbol{\mu}_{H_i})^T \boldsymbol{\Sigma}_{H_i}^{-1}(\boldsymbol{H}_i - \boldsymbol{\mu}_{H_i})\right)}{\left((2\pi)^{\dim(\boldsymbol{H}_i)} \cdot \det(\boldsymbol{\Sigma}_{H_i})\right)^{\frac{1}{2}}}. \tag{3}$$

Here, the parameters $\boldsymbol{\mu}_{H_i}$ and $\boldsymbol{\Sigma}_{H_i}$ represent manually defined hyperparameters. The likelihood terms of the joint distribution, which reflect the errors between the road model-predicted evidence data $\widehat{\boldsymbol{Y}}$ and the true evidence data $\boldsymbol{Y}$, are modeled similarly in our approach. Their structure is as follows:

$$P(\boldsymbol{Y}_S|\boldsymbol{C},\boldsymbol{O}_S) = \prod_{n=1}^{N_S^{Pos}} P\left(\boldsymbol{Y}_S^{Pos\,[n]}|\boldsymbol{C},\boldsymbol{O}_S\right), \tag{4}$$

with

$$P\left(\boldsymbol{Y}_S^{Pos\,[n]}|\boldsymbol{C},\boldsymbol{O}_S\right) = \frac{\beta_S^{Pos\,\frac{1}{2}} \cdot \exp\left(-\frac{\beta_S^{Pos}}{2} \cdot \frac{errorPos\left(\boldsymbol{C},\boldsymbol{O}_S,\boldsymbol{Y}_S^{Pos\,[n]}\right)^2}{\sigma_{\boldsymbol{Y}_S^{Pos\,[n]}}^2}\right)}{\left(2\pi \cdot \sigma_{\boldsymbol{Y}_S^{Pos\,[n]}}^2\right)^{\frac{1}{2}}}$$

and

$$P(\boldsymbol{Y}_D|\boldsymbol{C},\boldsymbol{O}_D) = \prod_{n=1}^{N_D^{Pos}} P\left(\boldsymbol{Y}_D^{Pos\,[n]}|\boldsymbol{C},\boldsymbol{O}_D\right) \cdot \prod_{n=1}^{N_D^{Flow}} P\left(\boldsymbol{Y}_D^{Flow\,[n]}|\boldsymbol{C},\boldsymbol{O}_D\right) \tag{5}$$

with

$$P\left(\boldsymbol{Y}_D^{Pos\,[n]}|\boldsymbol{C},\boldsymbol{O}_D\right) = \frac{\beta_D^{Pos\,\frac{1}{2}} \cdot \exp\left(-\frac{\beta_D^{Pos}}{2} \cdot \frac{errorPos\left(\boldsymbol{C},\boldsymbol{O}_D,\boldsymbol{Y}_D^{Pos\,[n]}\right)^2}{\sigma_{\boldsymbol{Y}_D^{Pos\,[n]}}^2}\right)}{\left(2\pi \cdot \sigma_{\boldsymbol{Y}_D^{Pos\,[n]}}^2\right)^{\frac{1}{2}}}$$

$$P\left(\boldsymbol{Y}_D^{Flow\,[n]}|\boldsymbol{C},\boldsymbol{O}_D\right) = \frac{\beta_D^{Flow\,\frac{1}{2}} \cdot \exp\left(-\frac{\beta_D^{Flow}}{2} \cdot \frac{errorFlow\left(\boldsymbol{C},\boldsymbol{O}_D,\boldsymbol{Y}_D^{Flow\,[n]}\right)^2}{\sigma_{\boldsymbol{Y}_D^{Flow\,[n]}}^2}\right)}{\left(2\pi \cdot \sigma_{\boldsymbol{Y}_D^{Flow\,[n]}}^2\right)^{\frac{1}{2}}}.$$

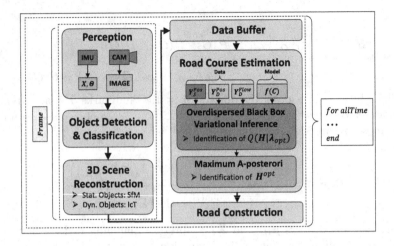

Fig. 3. Schematic representation of the estimation pipeline

Here, the proposed Gaussian character of those distributions shall reflect the typical lateral-fluctuations of landmark positions in the real-world, as well as inaccuracies in our $3D$ reconstruction and small sampling-caused projection errors (see Fig. 1). The value of $errorPos$ corresponds to the euclidean distance between the road model-predicted position and the true position of a evidence data point. Similarly, $errorFlow$ reflects the anti-correlation between the road model-predicted scene flow and the true scene flow of an evidence data point (see Fig. 1). The hyperparameters N_S^{Pos}, N_D^{Pos}, N_D^{Flow} depict the quantities of the different data types within the measured evidence data and the values of $\sigma_{Y_S^{Pos}}$, $\sigma_{Y_D^{Pos}}$, $\sigma_{Y_D^{Flow}}$ are related to the specific confidences of the measured evidence points. In contrast, β_S^{Pos}, β_D^{Pos} and β_D^{Flow} represent design parameters which control the influence of the corresponding evidence data types in the MAP-problem.

2.4 Variational Inference

Solving the MAP-problem requires an analysis of the extreme value of the a-posteriori distribution $P(H|Y)$ which is defined by the joint distribution $P(H,Y)$. However, this analysis is not trivial, because of the structure of the problem and the continuous random variables. Therefore, we propose to approximate the a-posteriori distribution initially and then infer the MAP-solution based on the generated approximation. In detail, we exploit a state-of-the-art variational inference technique, called Overdispersed Black-Box Variational Inference (O-BBVI) [17,19]. This deterministic and fast-converging method lacks the typical high computational cost of popular sampling-based techniques [14,15].

When applied to the MAP-problem, O-BBVI approximates the exact a-posteriori distribution $P(\boldsymbol{H}|\boldsymbol{Y})$ by a factorized distribution with the form

$$Q(\boldsymbol{H}|\boldsymbol{\lambda}) = \prod_{i=1}^{N} q_{H_i}(\boldsymbol{H}_i|\boldsymbol{\lambda}_i),$$

in which each component q_{H_i} is defined as:

$$q_{H_i}(\boldsymbol{H}_i|\boldsymbol{\lambda}_i) = \frac{\exp\left(-\frac{1}{2}\sum_{d=1}^{\dim(\boldsymbol{H}_i)}\left(H_i^{[d]} - \lambda_i^{[d]}\right)^2\right)}{(2\pi)^{\frac{1}{2}\dim(\boldsymbol{H}_i)}}. \tag{6}$$

Based on that, O-BBVI computes this approximation by an optimization over the parameters $\boldsymbol{\lambda}_i$ of the factor terms q_{H_i}. The parameter-decoupling structure of the O-BBVI approximation modifies the solution of the MAP-problem to

$$\boldsymbol{H}^{opt} = \underset{\boldsymbol{H}}{\operatorname{argmax}}\,(P(\boldsymbol{H}|\boldsymbol{Y}))\Bigg|_{P(\boldsymbol{H}|\boldsymbol{Y})\approx Q(\boldsymbol{H}|\boldsymbol{\lambda}_{opt})} \tag{7}$$

$$\approx \underset{\boldsymbol{H}}{\operatorname{argmax}}\,(Q(\boldsymbol{H}|\boldsymbol{\lambda}_{opt}))\Bigg|_{Q(\boldsymbol{H}|\boldsymbol{\lambda}_{opt})=q\left(\boldsymbol{C}|\lambda_{opt}^C\right)\cdot q\left(\boldsymbol{O}_S|\lambda_{opt}^{O_S}\right)\cdot q\left(\boldsymbol{O}_D|\lambda_{opt}^{O_D}\right)} \tag{8}$$

$$\boldsymbol{C}^{opt} = \underset{\boldsymbol{C}}{\operatorname{argmax}}\left(q_C\left(\boldsymbol{C}|\lambda_C^{opt}\right)\right) = \lambda_C^{opt}$$

$$\implies \boldsymbol{O}_S^{opt} = \underset{\boldsymbol{O}_S}{\operatorname{argmax}}\left(q_{O_S}\left(\boldsymbol{O}_S|\lambda_{O_S}^{opt}\right)\right) = \lambda_{O_S}^{opt}$$

$$\boldsymbol{O}_D^{opt} = \underset{\boldsymbol{O}_D}{\operatorname{argmax}}\left(q_{O_D}\left(\boldsymbol{O}_D|\lambda_{O_D}^{opt}\right)\right) = \lambda_{O_D}^{opt}.$$

Finally, this MAP-solution in conjunction with the applied road-contour model $\boldsymbol{f}(\boldsymbol{C})$ returns the full 3D-description of an estimated road course.

3   Experiments and Discussion

In the following section, we will compare our scene cue based fusion approach on three real-world traffic examples with a corresponding state-of-the-art estimation method which is based on road markings [5]. The chosen examples cover various driving situations in which a road marking-based approach shows weaknesses in comparison with the landmark-based approach. The spectrum of the situations varies across different types of roads and different types, numbers and densities of available static and dynamic scene cues. All the situations are scenarios with reduced light.

The course estimations for both methods are post-processed under real-time conditions on the base of recorded images and logged IMU data. However, for a better understanding of the scene structure, we visualize the IMU-recorded trajectory of the ego-vehicle as well as all the $3D$-reconstructed static landmarks upfront. The results are shown in Figs. 4, 5 and 6.

The first scenario in Fig. 4 represents a driving situation on a country road in low-light conditions. The road has roadside-markings and rows of delineators on both sides. The ego-vehicle follows an other vehicle through a curve. The results demonstrate that the range of the landmark-based estimation exceeds the corresponding estimation of road marking-based approach by more than 25 m or 33% because of the detected delineator on the right side. Making use of the delineators on the right side, our approach is even capable of seeing a short distance around the bending of the curve and hence identifies a part of the course which is not detected by the road-marking based system.

The second scenario in Fig. 5 reflects a ride on a highway at night. The scene is only illuminated by the high-beam head lamps of the ego-vehicle. On both sides of the road are reflectors mounted on the guardrails. Road markings are also available. The results illustrate that our approach achieves an estimation of the road course up to 140 m relative to the position of the ego-vehicle. This is accomplished with the help of the detected delineators. As a result, our estimation reaches 70 m or 100% further than the corresponding estimation based on road markings.

The scene in Fig. 6 shows a driving situation in an urban environment at night. The road has no delineators or roadside-markings. The scene is mainly illuminated by a few road-aligned road lanterns on the right and left side. Hence the road lanterns on the right side, our approach generates a virtual road boundary. This information can not be exploited by a road-marking based method. Next to the comparison with the road-marking based method, we also evaluate our approach statically with labeled ground-truth road courses on a database of approximately 14,000 images of various night traffic situations. For this purpose, we labeled the regions in those images where we expect the boundaries of

Fig. 4. Rural road: $3D$ Reconstruction (left), camera image (right)

Fig. 5. Highway: 3D Reconstruction (left), camera image (right)

Fig. 6. Urban road: 3D Reconstruction (left), camera image (right)

the estimated road courses with polygon-shaped tubes. Based on that labeling, we rate road-course estimations which are fully embedded within the labeled polygons as true positives. Estimations which do not fulfill this criterion are rated as false positives. In this evaluation framework our approach achieves a true positive rate (TPR) of 92.18% and a false positive rate (FPR) of 4.68%. With the objective of using our approach on hardware platforms with limited computational resources, we further investigate the influence of the amount of the numerically expensive iterations within the O-BBVI inference method on the quality of the estimated road courses. In detail, we reduce the number of iterations to 75%, 50% and 25% of the amount of iteration which are needed for a complete convergence of the O-BBVI method during the inferences. The reduction to 75% results in a TPR of 81.8% and FPR of 15.4%. A further reduction to 50% causes a TPR of 74.5% and a FPR of 22.9% and a reduction to 25% lowers the TPR to 61.5% and FPR of 36.4%. These results implicate that the O-BBVI based inferences converge fast after a few iterations. Thus, this circumstance allows our approach to easily adapt to available hardware resources without entirely given up on estimation quality.

4   Conclusion and Future Work

In this publication we presented an approach that estimates the course of a road based on images of a monoscopic camera for ranges of 140 m, particularly in difficult situations with reduced light. The method therefore uses static and dynamic scene cues which are correlated to the course of a road. The underlying fusion concept is flexible and hence works with a variety of different landmark types and quantities. This makes the approach highly adaptive to varying evidence in a scene. In order to optimally respond to real-world road designs we proposed to use a clothoid road model. The associated complications with such a road model in a regression problem were addressed with a probabilistic model and a numerically efficient and adaptive variational inference. We demonstrated the performance of our algorithm in challenging low-light driving situations. Thereby, we proved that the approach can achieve larger estimation ranges than a comparable road marking-based method in the same situations. These results reflect that information fusion provides a framework to integrate expert knowledge over the problem setup with data-driven insights into the decision-making process.

In future work we plan to strengthen the presented approach in several areas. At first, we plan to exploit our current set of evidence data more effectively in the fusion process. Therefore, we would like to substitute manual-defined hyperparameters within the probabilistic model (see Eqs. 3, 4 and 5) by data-trained counterparts. In addition to that, we intend to develop a robust strategy to identify lane-changing vehicles. This would allow us to react better to lateral-shifts of vehicles during the estimation process. Beyond these improvements we would like to extend the current set of evidence data by integrating more camera signals, like road markings or semantic segmentation results, as well as signals from other sensors, such as lidar or radar into our probabilistic fusion framework. We expect that this will enhance the robustness and also will allow us to model even more complex driving scenarios, like splitting or reunifying roads. Furthermore, we plan to improve the variational inference procedure in order to achieve even faster and more precise estimates.

References

1. Nieto, J., Hernandez-Gutierrez, A., Nebot, E.: Probabilistic estimation of unmarked roads using radar. J. Phys. Agents, **4** (2010)
2. Sarholz, F., Mehnert, J., Klappstein, J., Radig, B.: Evaluation of different approaches for road course estimation using imaging radar. In: Conference on Intelligent Robots and Systems. IEEE (2011)
3. Hammarstrand, L., Fatemi, M., Angel, F., Svensson, L.: Long-range road geometry estimation using moving vehicles and road-side observations. IEEE Trans. Intell. Transp. Syst. **17**, 2144–2158 (2016). IEEE

4. Goldbeck, J., Huertgen, B.: Lane detection and tracking by video sensors. In: IEEE Intelligent Transportation Systems, pp. 74–99 (1999)
5. Liu, W., Zhang, H., Duan, B.: Vision-based real-time lane marking detection and tracking. In: 11th International Conference on Intelligent Transportation Systems. IEEE (2008)
6. Konrad, M., Szczot, M., Schule, F.: Generic grid mapping for road course estimation. In: Intelligent Vehicles Symposium (IV). IEEE (2011)
7. Konrad, M., Nuss, D., Dietmayer, K.: Localization in digital maps for road course estimation using grid maps. In: Intelligent Vehicles Symposium (IV). IEEE (2012)
8. Schule, F., Schweiger, R., Dietmayer, K.: Augmenting night vision video images with longer distance road course information. In: Intelligent Vehicles Symposium (IV). IEEE (2013)
9. Schule, F., Koch, C., Hartmann, O.: Probabilistic fusion of rural road course estimations. In: 16th International Conference on Intelligent Transportation Systems. IEEE (2013)
10. Limmer, M., Forster, J., Baudach, D., Schule, F., Schweiger, R., Lensch, H.: Robust deep-learning-based road-prediction for augmented reality navigation systems. In: Intelligent Vehicles Symposium (IV). IEEE (2012)
11. Teichmann, M., Weber, M., Zollner, M., Cipolla, R., Urtasun, R.: MultiNet: real-time joint semantic reasoning for autonomous driving. In: arXiv:1612.07695v1
12. Geiger, A., Wojek, C., Urtasun, R.: Joint 3D estimation of objects and scene layout. In: Advances in Neural Information Processing Systems (NIPS) (2011)
13. Geiger, A., Lauer, M., Wojek, C., Stiller, C., Urtasun, R.: 3D traffic scene understanding from movable platforms. IEEE Trans. Pattern Anal. Mach. Intell. **36**, 1012–1025 (2014)
14. Geman, S., Geman, D.: Stochastic relaxation, Gibbs distributions, and the Bayesian restoration of images. IEEE Trans. Pattern Anal. Mach. Intell. **6**, 721–741 (1984)
15. Goodman, J., Weare, J.: Ensemble samplers with affine invariance. Commun. Appl. Mathe. Comput. Sci. **5**, 65–80 (2010)
16. Trusheim, F., Condurache, A., Mertins, A.: Graphical stochastic models for tracking applications with variational message passing inference. In: 6th International Conference on Image Processing Theory, Tools and Applications (IPTA), Oulu, Finland (2016)
17. Ranganath, R., Gerrish, S., Blei, D.: Black box variational inference. In: 17th International Conference on Artificial Intelligence and Statistics (AISTATS), Reykjavik, Iceland (2014)
18. Titsias, M., Lazaro-Gredilla, M.: Local expectation gradients for black box variational inference. In: 28th International Conference on Neural Information Processing Systems (NIPS), Montreal, Canada (2015)
19. Ruiz, F., Titsias, M., Blei, D.: Overdispersed black-box variational inference. In: The Conference on Uncertainty in Artificial Intelligence, New York, USA (2016)
20. Engels, C., Stewenius, H., Nister, D.: Bundle adjustment rules. In: Photogrammetric Computer Vision (2006)
21. Hartley, R., Zisserman, A.: Multiple View Geometry in Computer Vision. Cambridge University Press, New York (2004)
22. Gackstatter, C., Thomas, S., Heinemann, P., Klinker, G.: Stable road lane model based on clothoids. In: Advanced Microsystems for Automotive Applications, pp. 133–143 (2010)
23. Bertolazzi, E., Frego, M.: G^1 Fitting with clothoids. Wiley Online Library (2013)

Multiple Reflection Symmetry Detection via Linear-Directional Kernel Density Estimation

Mohamed Elawady[1(✉)], Olivier Alata[1], Christophe Ducottet[1], Cécile Barat[1], and Philippe Colantoni[2]

[1] Université Jean Monnet, CNRS, UMR 5516,
Laboratoire Hubert Curien, 42000 Saint-Etienne, France
mohamed.elawady@univ-st-etienne.fr
[2] Université Jean Monnet, Centre Interdisciplinaire d'Etudes et de Recherches sur l'Expression Contemporaine no 3068, Saint-Etienne, France

Abstract. Symmetry is an important composition feature by investigating similar sides inside an image plane. It has a crucial effect to recognize man-made or nature objects within the universe. Recent symmetry detection approaches used a smoothing kernel over different voting maps in the polar coordinate system to detect symmetry peaks, which split the regions of symmetry axis candidates in inefficient way. We propose a reliable voting representation based on weighted linear-directional kernel density estimation, to detect multiple symmetries over challenging real-world and synthetic images. Experimental evaluation on two public datasets demonstrates the superior performance of the proposed algorithm to detect global symmetry axes respect to the major image shapes.

Keywords: Multiple symmetry · Symmetry detection · Reflection symmetry · Kernel density estimation · Linear-directional data

1 Introduction

Reflection symmetry is a fundamental principle of visual perception to feel the equally distributed weights within foreground objects inside an image. These weights are inspected respect to textural complexity of their shapes, in such non-identical manner to preserve a well-balanced composition between similar objects and their surrounding background [13,16]. Detection of reflection symmetry has a principal intermediate-level role in recent computer vision applications [1,31,37]. Liu et al. [19] described the global symmetry as top-tier visual features, which are distributed uniformly across the image sides and contributed to define an uppermost similarity behavior. This paper focuses on detecting multiple bilateral symmetry axes inside an image by exploring the geometrical correlation between spatial regions on a global scale.

The baseline algorithm was proposed in 2006 by Loy and Eklundh [21]. They analyzed the bilateral symmetry from image features' constellation by introducing the general scheme: (1) detection of local feature points (i.e. SIFT),

© Springer International Publishing AG 2017
M. Felsberg et al. (Eds.): CAIP 2017, Part I, LNCS 10424, pp. 344–355, 2017.
DOI: 10.1007/978-3-319-64689-3_28

associated with local geometrical properties (location, orientation, scale) and descriptor vectors. (2) pairwise matching and evaluation of a local symmetry magnitude of their descriptors, to generate symmetry candidates. (3) accumulation of their symmetry magnitude in a Hough-like voting space parametrized with orientation and displacement, to identify the dominant reflection axes inside an image. The first survey of symmetry detection algorithms was introduced by the computer vision group of Pennsylvania State University in 2008 [28]. The same group conducted symmetry detection challenges in 2011 [32] and 2013 [18], where the baseline algorithm [21] still outperformed the participated approaches [17,23,26,30]. Other keypoint-based algorithms [5,7] also proposed feature refinement techniques for better results. Edge/contour-based features [3,4,8,9,11,24,36] are modernly used instead of the intensity-based, due to saliency properties in detecting well-defined symmetric structures inside an object. In both approaches, a limited number of feature points are detected in the image, axis candidates are randomly sampled across the voting space. These sparse symmetry candidates further need to be grouped through a smoothing kernel to define relevant mono- or multi- axis hypothesis. Our idea is to formulate the voting problem as a density estimation problem, by computing the probability of detecting symmetry axis at every position and orientation inside the image plane.

Kernel density estimation is one of the most popular techniques in nonparametric statistics. Density estimates are controlled by a smoothing bandwidth and a weighting kernel function. Density estimates with linear kernels have been introduced in 1954 [2], and then have been adapted to deal with directional data since the mid 1980s [15]. Many computer vision applications used kernel density estimation for linear data [10,12,20,25,33,35], and fewer recently used it for directional data [27,34]. Garcia-Portugues et al. [14] derived the general principle of joint kernel density estimator for linear-directional data.

Our contribution is twofold. First we propose a weighted joint density estimator to handle both orientation and displacement information. Second, we introduce a robust linear-directional kernel-based voting representation for reflection symmetry detection. This approach is evaluated for multiple symmetry detection using two public datasets. The remaining sections of this paper are organized as follows. Section 2 describes the proposed algorithm. Sections 3 and 4 present the experimental details and results on two public datasets. Finally, the conclusion is given in Sect. 5.

2   Algorithm Details

Given an image, our algorithm focuses globally to detect all symmetry axes using a dense and regular estimation of linear-directional density, as briefly shown in Fig. 1. First, we extract wavelet-based features with different scales, accompanied with edge and textural characteristics (for better display, only features with high magnitude are displayed over the gray-scale version of the input image). Second, we triangulate each feature pair at each scale with respect to the origin of the

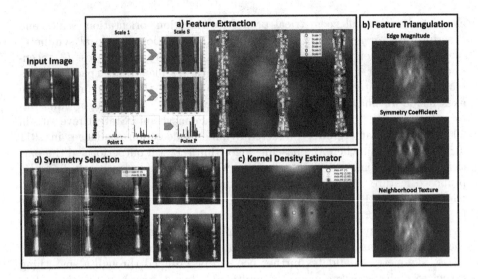

Fig. 1. The proposed framework of reflection symmetry detection, using a weighted linear-directional kernel density estimator (LD-KDE). Best seen on screen (zoom-in for details).

feature space, in order to define symmetrical weights in the polar coordinate system. Third, we formulate a voting representation based on weighted pairs via linear-directional kernel density estimation. Finally, the global symmetry axes are well-chosen by searching for maximum peaks, and spatially defined by the convex hull of the voting features.

2.1   Feature Extraction and Normalization

Upon the application of Morlet wavelet over an image (width W and height H) with multiple scales $\sigma \in \{1, 2, 4, 6, 8\}$ and orientations $\phi \in \{\frac{z\pi}{32}, z = 0 \ldots 31\}$, feature points $\{p^i = [p_x^i, p_y^i]^T, i = 1 \ldots P, P \propto max(W, H)\}$ are sampled, as detailed in [11], along a regular grid with respect to image size ($W \times H$). Each feature point is the center of a neighborhood window $D(p^i)$, which allows to compute its local edge components (maximum wavelet response J^i along side with associated scale σ^i and angle ϕ^i over all orientations) plus neighboring textural histograms h^i of size B:

$$h^i(b) = \sum_{r \in D(p^i)} J^i \delta_{\phi_b - \phi_r}, \; \phi_b \in \{\frac{b\pi}{B}, b = 0 \ldots B - 1, 8 \leq B \leq 32\} \qquad (1)$$

where δ is the Kronecker delta. h^i is $l1$ normalized and circular shifted respect to the maximum magnitude J^i among the neighborhood window $D(p^i)$.

The feature points are normalized with keeping aspect ratio as following:

$$\hat{p}^i = \frac{p^i - c_{W,H}}{max(W, H)} \qquad (2)$$

where $c_{W,H}$ represents the original image center $(\frac{W}{2}, \frac{H}{2})$. So that, the feature space is transferred from the dynamic-sized image plane $[1, W] \times [1, H]$ to a unified version $[-1, 1] \times [-1, 1]$.

2.2 Pairwise Symmetry Triangulation

We first define a set of feature pairs $\{q_n = (\hat{p}^i, \hat{p}^j) \mid n = 1, \ldots, N\}$ such that $i \neq j$ and $\sigma^i = \sigma^j$. Then, we perform a triangulation process (as illustrated in Fig. 2) with respect to the feature origin, producing the symmetry candidate axis as the bisector of the pair segment. The candidate axis is parametrized by angle $\theta_n \in [0, \pi)$, and displacement $\rho_n \in [-\frac{\sqrt{2}}{2}, \frac{\sqrt{2}}{2}])$ and has a symmetry weight ω_n [9,11] defined as follows:

$$\omega_n = \omega(\hat{p}^i, \hat{p}^j) = m(i,j)\, c(i,j)\, d(i,j) \tag{3}$$

$$m(i,j) = J^i J^j \tag{4}$$

$$c(i,j) = |\tau^i S(T_{ij}^{\perp})\tau^j| \tag{5}$$

$$d(i,j) = \sum_{b=1}^{B} min(h^i(b), \tilde{h}^j(b)) \tag{6}$$

where τ^i is a direction associated with angle ϕ_i, $S(T_{ij}^{\perp})$ is the reflection matrix with respect to the perpendicular of line connecting $(\hat{p}^i, \hat{p}^j)$, and $\tilde{h}^j$ is the reverse version of h^j histogram. $l1$ normalization is applied to the symmetry weights $\omega_1, \ldots, \omega_N$. In brief, m is a semi-dense edge magnitude, c is a mirror symmetry coefficient based on the local edge orientation of points of the pair, d is a similarity measure between the local texture around the feature pairs.

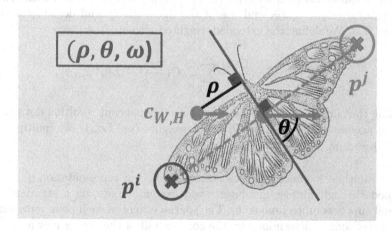

Fig. 2. Symmetric triangulation for a feature pair. Best seen on screen.

2.3   Weighted Linear-Directional Kernel Density Estimation

One dimensional linear random variable ρ represents displacement part of a candidate axis, assuming $\rho_1, \ldots, \rho_N$ samples of ρ with size N. Let μ describes two dimensional directional variable (circular data corresponds to angle θ, representing orientation part of a candidate axis, see Fig. 2), assuming $\mu_1, \ldots, \mu_N$ samples of μ with the same size of ρ. Inspired by [15,29], the linear kernel density estimator $f_l(.)$ is defined as

$$f_l(x; g) = \frac{1}{Ng} \sum_{n=1}^{N} G(\frac{x - \rho_n}{g}), \; x \in \mathbb{R} \tag{7}$$

$$G(u) = \frac{1}{(2\pi)^{\frac{1}{2}}} e^{-\frac{1}{2}|u|^2}, \tag{8}$$

where $G(.)$ is a Gaussian kernel with bandwidth parameter g. The directional kernel density estimator $f_d(.)$ is defined as:

$$f_d(y; k) = C(k) \sum_{n=1}^{N} L(y^T \mu_n; k), \; y \in \Omega_2 \tag{9}$$

$$L(x; k) = e^{kx}, \; C(k) = \frac{1}{2\pi S(0, k)}, \tag{10}$$

$$y = [cos(\theta), sin(\theta)], \; \mu_n = [cos(\theta_n), sin(\theta_n)], \tag{11}$$

where $L(.)$ is a von-Mises Fisher kernel [22] with concentration parameter k, and normalization constant $C(k)$. $S(.)$ is the modified Bessel function of the first kind. y is remarked as directional unit-vector of angle θ, such that $||y|| = 1$. As axis candidate samples $(\rho_1, \mu_1), \ldots, (\rho_N, \mu_N)$ are associated with symmetry weights $\omega_1, \omega_2, \cdots, \omega_N$, and use of the linear-directional density estimator $f_{l,d}(.)$ in [14]. We define the extended weighted version $\hat{f}_{l,d}(.)$ as:

$$\hat{f}_{l,d}(x, y; g, k) = \frac{C(k)}{Ng} \sum_{n=1}^{N} \omega_n G(\frac{x - \rho_n}{g}) L(y^T \mu_n; k) \tag{12}$$

assuming that linear and directional data are independent resulting dot product between accompanying kernels. Previous weights (see Eq. 3) are multiplied by N in order to normalize $\hat{f}_{l,d}$.

The multiple symmetry peaks inside the voting representation $\hat{f}_{l,d}(.)$ are identified through finding non-interleaved extreme spots via a standard non-maximal suppression technique [6]. The spatial extent of each peak representing a symmetry axis is determined by the convex hull of the voting pair associated with the peak [21]. Figures (3a-d) present an example of multiple symmetry detection, using 1D and 2D kernel-based voting maps. Three vertical symmetry axes are shown in the weighted linear kernel density $\hat{f}_l(x; g)$ (Fig. 3b). Two major

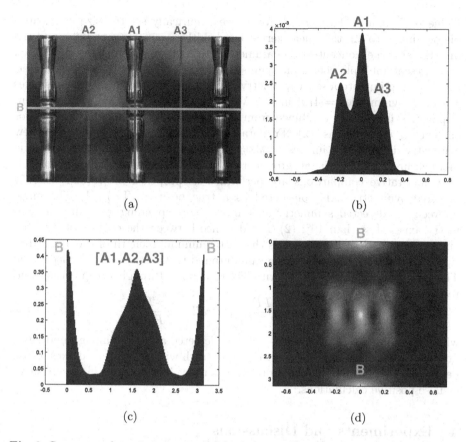

Fig. 3. Symmetry detection process: (a) Input image with global symmetry axis candidates. (b) The output of weighted linear kernel density $\hat{f}_l(x; g)$ over 800 bins. (c) The output of the weighted directional kernel density $\hat{f}_d(y; k)$ over 180 bins. (d) The output of the weighted linear-directional kernel density $\hat{f}_{l,d}(x, y; g, k)$ over 800 × 180 bins. Maximal peaks are associated with global symmetry axes. Best seen on screen.

directional axes appear in the weighted directional kernel density $\hat{f}_d(y; k)$ at angles $\theta = 90°, 180°$ (Fig. 3c). All global symmetry axes are clearly recognized through the combination version of the previous weighted densities $\hat{f}_{l,d}(x, y; g, k)$ (Fig. 3d). To obtain such representation, as θ originally belongs to $[0, \pi)$, each angle value is multiplied by 2 in order to obtain an appropriate periodicity with the directional kernel.

3 Implementation and Evaluation Details

We compare our reflection symmetry detection approach against three different methods (Loy2006 [21], Cicconet2014 [9], and Elawady2016 [11]). We executed their source codes with default parameter values, assigned by the authors for stable performance. For our approach, we empirically set the size of textural histograms B to 32, the linear-kernel bandwidth parameter g to 0.03, and the directional-kernel concentration parameter k to 40. Two public datasets are used to represent multiple reflection symmetry detection results: (1) PSU dataset: Liu's vision group proposed a symmetry groundtruth for Flickr images (# images = 142, # symmetries = 479) in ECCV2010, CVPR2011 and CVPR2013. Non-duplicative images are combined from three previously mentioned versions for challenging comparisons. (2) NY dataset: Cicconet et al. [8] presented a new symmetry database (# images = 63, # symmetries = 188) in 2016[1], providing more accurate and consistent groundtruth for multiple symmetry endpoints.

Quantitative comparisons are performed as proposed in [18,32], where a detected symmetry axis considered as a true positive (TP): (1) The angle between the detected symmetry axis and its corresponding groundtruth symmetry axis is less than 10°; (2) The distance between the centers of detected and same groundtruth axes is less than 20% minimum length of the axes. Multiple detections can match to the same ground-truth axis, but not vice versa. The overall performance of algorithms are defined through the precision and recall rates:

$$precision = \frac{TP}{TP + FP}, \ recall = \frac{TP}{TP + FN}. \tag{13}$$

where false positive (FP) are non-matched detected axes with any groundtruth, false negative (FN) are non-matched groundtruth with any detected axes. Close detections are clustered as one proposal axis to avoid duplicates in true positive or false positive calculations.

4   Experiments and Discussions

Quantitative (Figs. 4, 5) and qualitative (Fig. 6) comparisons are conducted among the proposed method (Our2017), Loy and Eklundh (Loy 2006) [21], Cicconet et al. (Cic2014) [9], and Elawady et al. (Ela2016) [11]. Loy2006 and Ela2016 were reported to have the best performed results for the single symmetry detection in keypoint-based and edge-based methods respectively [11]. Figure 4 presents precision-recall curves for the multiple symmetry datasets (PSUm [18,32], and NYm [8]), to compare the proposed method to three prior algorithms (Loy2006 [21], Cic2014 [9], and Ela2016 [11]). Cic2014 [9] has the lowest performance for precision and recall in both curves. In Fig. 4a, Loy2006 [21] has better precision than Ela2016 [11] over corresponding low recall, while the precision of the proposed method (aka Our2017) outperforms all these methods in most sections of the curve. In Fig. 4b, Ela2016 [11] has a superior precision performance over Loy2006 [21] under the recall rate of 40%, meanwhile the proposed method has the best performance along both precision and recall rates. Additionally, we also compute the F_1 score to define the harmonic mean between precision and recall rates, and used the maximum F_1 score to qualify the overall

---

[1] http://symmetry.cs.nyu.edu/.

performance of different detection algorithms. The values of maximum F_1 score are presented in Fig. 4 to express the precision-recall curve for each method in a single global measure. Figure 5 shows precision and recall rates where the maximum F_1 scores are selected among the corresponding curves in Fig. 4. The proposed method achieved the best performance among all results.

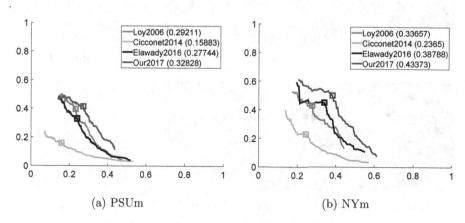

(a) PSUm   (b) NYm

Fig. 4. Precision-Recall curves on (a)PSUm [18,32] and (b)NYm [8] datasets to show the superior performance of our method "Our2017" against the three prior algorithms ("Loy2006" [21], "Cic2014" [9], "Ela2016" [11]). The maximum F1 scores are qualitatively presented as square symbols along the curves, and quantitatively indicated between parentheses inside the top-right legends. Best seen on screen.

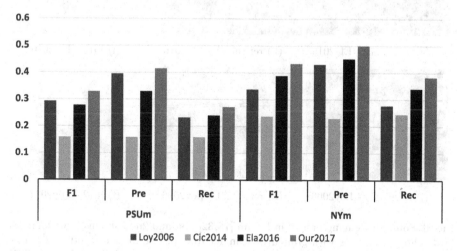

Fig. 5. Comparison of maximum F1 score and its equivalent precision and recall rates among two different datasets (PSUm [18,32], NYm [8]) for our method "Our2017" against the baseline algorithm "Loy2006" [21] and two of the recent algorithms "Cic2014" [9] and "Ela2016" [11].

(a) ref_rm_79 - GT   (b) ref_rm_71 - GT   (c) I012 - GT

(d) ref_rm_79 - Our2017   (e) ref_rm_71 - Our2017   (f) I012 - Our2017

(g) ref_rm_79 - Ela2016   (h) ref_rm_71 - Ela2016   (i) I012 - Ela2016

(j) ref_rm_79 - Loy2006   (k) ref_rm_71 - Loy2006   (l) I012 - Loy2006

Fig. 6. Some challenging images in PSUm [18,32] (1st and 2nd columns) and NYm [8] (3rd column) datasets with groundtruth in 1st row (a, b, c). our method in 2nd row (d, e, f) produces better results among El2016 [11] in 3rd row (g,h,i) and Loy2006 [21] in 4th row (j, k, l). For each algorithm, the top five symmetry results is presented in such order: red, yellow, green, blue, and magenta. Each symmetry axis is shown in a straight line with squared endpoints. Best seen on screen. (Color figure online)

Figure 6 presents some experimental results for multiple symmetry detection from two publicly available datasets (PSUm [18,32], NYm [8]). Groundtruth of the first example (Fig. 6a) shows four symmetries (vertical, horizontal, and diagonals) of a flower within natural background view. The proposed method (Fig. 6d) detects correctly vertical and horizontal axes, while Loy2006 [21] (Fig. 6j) fails to find enough feature points resulting two partial (diagonal and vertical) axes, and Ela2016 [11] (Fig. 6g) breakdown the vertical axis to represent the top five detections. Second and third examples (Figs. 6b and c) display three in-between symmetries of a thinned metal object and five symmetries expressing arms' details of a starfish respectively over texture-less surfaces. These symmetries have been efficiently detected by the proposed method (Figs. 6e and f) over Ela2016 [11] (Figs. 6h and i). However, Loy2006 [21] (Figs. 6k and l) concentrates on local symmetries describing the inner details of the centric objects.

5   Conclusion

This paper proposes a linear-directional kernel-based voting scheme within unified feature representation, in order to support a reliable detection framework for global multiple symmetries. Our approach solves the drawbacks of the previous symmetry detection approaches, by estimating the fixed-sized kernel density with efficient bandwidth parameters, and identifying correctly the symmetrical regions at a global scale. Quantitative and qualitative evaluations present the state-of-the-art performance of our proposed framework among public datasets. This work can be extended to refine the accuracy of the symmetry peaks and the selection of corresponding voting features, using a continuous maxima-seeking technique. The future work is introducing an entropy-based measure, to exploit the global strength of various symmetry axes inside an image.

References

1. Abdolali, F., Zoroofi, R.A., Otake, Y., Sato, Y.: Automatic segmentation of maxillofacial cysts in cone beam CT images. Comput. Biol. Med. **72**, 108–119 (2016)
2. Akaike, H.: An approximation to the density function. Ann. Inst. Stat. Math. **6**(2), 127–132 (1954)
3. Atadjanov, I., Lee, S.: Bilateral symmetry detection based on scale invariant structure feature. In: 2015 IEEE International Conference on Image Processing (ICIP), pp. 3447–3451. IEEE (2015)
4. Atadjanov, I.R., Lee, S.: Reflection symmetry detection via appearance of structure descriptor. In: Leibe, B., Matas, J., Sebe, N., Welling, M. (eds.) ECCV 2016. LNCS, vol. 9907, pp. 3–18. Springer, Cham (2016). doi:10.1007/978-3-319-46487-9_1
5. Cai, D., Li, P., Su, F., Zhao, Z.: An adaptive symmetry detection algorithm based on local features. In: 2014 IEEE Visual Communications and Image Processing Conference, pp. 478–481. IEEE (2014)
6. Canny, J.: A computational approach to edge detection. IEEE Trans. Pattern Anal. Mach. Intell. **6**, 679–698 (1986)

7. Cho, M., Lee, K.M.: Bilateral symmetry detection via symmetry-growing. In: BMVC, pp. 1–11. Citeseer (2009)
8. Cicconet, M., Birodkar, V., Lund, M., Werman, M., Geiger, D.: A convolutional approach to reflection symmetry. Pattern Recogn. Lett. **95**, 44–50 (2017)
9. Cicconet, M., Geiger, D., Gunsalus, K.C., Werman, M.: Mirror symmetry histograms for capturing geometric properties in images. In: 2014 IEEE Conference on Computer Vision and Pattern Recognition (CVPR), pp. 2981–2986. IEEE (2014)
10. Cremers, D., Osher, S.J., Soatto, S.: Kernel density estimation and intrinsic alignment for shape priors in level set segmentation. Int. J. Comput. Vision **69**(3), 335–351 (2006)
11. Elawady, M., Barat, C., Ducottet, C., Colantoni, P.: Global bilateral symmetry detection using multiscale mirror histograms. In: Blanc-Talon, J., Distante, C., Philips, W., Popescu, D., Scheunders, P. (eds.) ACIVS 2016. LNCS, vol. 10016, pp. 14–24. Springer, Cham (2016). doi:10.1007/978-3-319-48680-2_2
12. Elgammal, A., Duraiswami, R., Harwood, D., Davis, L.S.: Background and foreground modeling using nonparametric kernel density estimation for visual surveillance. Proc. IEEE **90**(7), 1151–1163 (2002)
13. Freeman, M., et al.: The Photographer's Eye: Composition and Design for Better Digital Photos. CRC Press, Boca Raton (2007)
14. García-Portugués, E., Crujeiras, R.M., González-Manteiga, W.: Kernel density estimation for directional-linear data. J. Multivar. Anal. **121**, 152–175 (2013)
15. Hall, P., Watson, G., Cabrera, J.: Kernel density estimation with spherical data. Biometrika **74**(4), 751–762 (1987)
16. Hobbs, J.A., Salome, R., Vieth, K.: The Visual Experience. Davis Publications, Worcester (1995)
17. Kondra, S., Petrosino, A., Iodice, S.: Multi-scale kernel operators for reflection and rotation symmetry: further achievements. In: 2013 IEEE Conference on Computer Vision and Pattern Recognition Workshops (CVPRW), pp. 217–222. IEEE (2013)
18. Liu, J., Slota, G., Zheng, G., Wu, Z., Park, M., Lee, S., Rauschert, I., Liu, Y.: Symmetry detection from realworld images competition 2013: summary and results. In: 2013 IEEE Conference on Computer Vision and Pattern Recognition Workshops (CVPRW), pp. 200–205. IEEE (2013)
19. Liu, Y., Hel-Or, H., Kaplan, C.S.: Computational Symmetry in Computer Vision and Computer Graphics. Now publishers Inc., Boston (2010)
20. Liu, Z., Shi, R., Shen, L., Xue, Y., Ngan, K.N., Zhang, Z.: Unsupervised salient object segmentation based on kernel density estimation and two-phase graph cut. IEEE Trans. Multimedia **14**(4), 1275–1289 (2012)
21. Loy, G., Eklundh, J.-O.: Detecting symmetry and symmetric constellations of features. In: Leonardis, A., Bischof, H., Pinz, A. (eds.) ECCV 2006. LNCS, vol. 3952, pp. 508–521. Springer, Heidelberg (2006). doi:10.1007/11744047_39
22. Mardia, K.V., Jupp, P.E.: Directional Statistics, vol. 494. Wiley, New York (2009)
23. Michaelsen, E., Muench, D., Arens, M.: Recognition of symmetry structure by use of gestalt algebra. In: 2013 IEEE Conference on Computer Vision and Pattern Recognition Workshops (CVPRW), pp. 206–210. IEEE (2013)
24. Ming, Y., Li, H., He, X.: Symmetry detection via contour grouping. In: 2013 20th IEEE International Conference on Image Processing (ICIP), pp. 4259–4263. IEEE (2013)
25. Mittal, A., Paragios, N.: Motion-based background subtraction using adaptive kernel density estimation. In: Proceedings of the 2004 IEEE Computer Society Conference on Computer Vision and Pattern Recognition, CVPR 2004, vol. 2, p. II. IEEE (2004)

26. Mo, Q., Draper, B.: Detecting bilateral symmetry with feature mirroring. In: CVPR 2011 Workshop on Symmetry Detection from Real World Images (2011)
27. Pardo, A., Real, E., Krishnaswamy, V., López-Higuera, J.M., Pogue, B.W., Conde, O.M.: Directional kernel density estimation for classification of breast tissue spectra. IEEE Trans. Med. Imaging **36**(1), 64–73 (2017)
28. Park, M., Lee, S., Chen, P.C., Kashyap, S., Butt, A.A., Liu, Y.: Performance evaluation of state-of-the-art discrete symmetry detection algorithms. In: IEEE Conference on Computer Vision and Pattern Recognition, CVPR 2008, pp. 1–8. IEEE (2008)
29. Parzen, E.: On estimation of a probability density function and mode. Ann. Math. Stat. **33**(3), 1065–1076 (1962)
30. Patraucean, V., von Gioi, R.G., Ovsjanikov, M.: Detection of mirror-symmetric image patches. In: 2013 IEEE Conference on Computer Vision and Pattern Recognition Workshops (CVPRW), pp. 211–216. IEEE (2013)
31. Ram, S., Rodriguez, J.J.: Vehicle detection in aerial images using multiscale structure enhancement and symmetry. In: 2016 IEEE International Conference on Image Processing (ICIP), pp. 3817–3821. IEEE (2016)
32. Rauschert, I., Brocklehurst, K., Kashyap, S., Liu, J., Liu, Y.: First symmetry detection competition: summary and results. Technical report, CSE11-012, Department of Computer Science and Engineering, The Pennsylvania State University (2011)
33. Tavakoli, H.R., Rahtu, E., Heikkilä, J.: Fast and efficient saliency detection using sparse sampling and kernel density estimation. In: Heyden, A., Kahl, F. (eds.) SCIA 2011. LNCS, vol. 6688, pp. 666–675. Springer, Heidelberg (2011). doi:10.1007/978-3-642-21227-7_62
34. Vuollo, V., Holmström, L., Aarnivala, H., Harila, V., Heikkinen, T., Pirttiniemi, P., Valkama, A.M.: Analyzing infant head flatness and asymmetry using kernel density estimation of directional surface data from a craniofacial 3d model. Stat. Med. **35**(26), 4891–4904 (2016)
35. Wang, M., Hua, X.S., Mei, T., Hong, R., Qi, G., Song, Y., Dai, L.R.: Semi-supervised kernel density estimation for video annotation. Comput. Vis. Image Underst. **113**(3), 384–396 (2009)
36. Wang, Z., Tang, Z., Zhang, X.: Reflection symmetry detection using locally affine invariant edge correspondence. IEEE Trans. Image Process. **24**(4), 1297–1301 (2015)
37. Yuan, Y., Xiong, Z., Wang, Q.: An incremental framework for video-based traffic sign detection, tracking, and recognition. IEEE Trans. Intell. Transp. Syst. (2016)

Biomedical Image Analysis

CNN-Based Background Subtraction
for Long-Term In-Vial FIM Imaging

Aaron Scherzinger[(✉)], Sören Klemm, Dimitri Berh, and Xiaoyi Jiang

Faculty of Mathematics and Computer Science,
University of Münster, Münster, Germany
scherzinger@wwu.de

Abstract. In recent years, the importance of behavioral studies of model organisms such as *Drosophila melanogaster* has significantly increased in biological research. Recently, a novel monitoring setup for analyzing Drosophila larvae in culture vials was proposed which allows researchers to conduct long-term studies without disturbing the animals' behavioral routine. However, when monitoring larvae in such a setup over several days, dirt accumulates on the vial surface, leading to artifacts in the segmentation process. To overcome this problem and enable researchers to perform experiments involving long-term tracking of the animals, we propose a method for background subtraction which is based on convolutional neural networks (CNNs). Our method produces good results and significantly outperforms other methods. In addition, we show that besides its good performance our compact CNN architecture allows us to apply our method for online-processing on microcomputers in real-time.

1 Introduction

In recent years, the importance of long-term behavioral studies in biological research has significantly increased. Often model organisms such as Drosophila flies [6,11] are monitored to obtain behavioral read-out for evaluating social interactions or observing stimulus-based behavior. Recently, Berh et al. [1] have proposed a novel monitoring setup for long-term analysis of Drosophila larvae in culture vials. This system is based on the FIM imaging technique by Risse et al. [18,19] and consists of five cameras which are connected to Raspberry Pi microcomputers. Each of the microcomputers images a part of the cylindrical culture vial, locally unfolds the vial wall, and sends the processed rectangular view to a central server where the individual images are stitched together to obtain a view of the entire vial wall. The setup and an example of a resulting image are depicted in Fig. 1. Due to the utilization of the principle of frustrated total internal reflection, FIM images generally have a high contrast and foreground segmentation of larvae as a pre-processing step for tracking can thus be reliably performed by a simple thresholding step [1,19]. However, the main challenge arising when imaging a large number of larvae in such a setup for

---

A. Scherzinger and S. Klemm contributed equally to this work.

© Springer International Publishing AG 2017
M. Felsberg et al. (Eds.): CAIP 2017, Part I, LNCS 10424, pp. 359–371, 2017.
DOI: 10.1007/978-3-319-64689-3_29

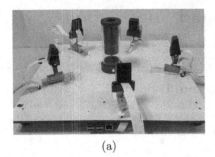
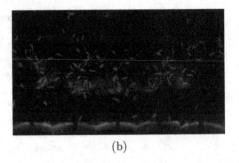

(a)   (b)

Fig. 1. Prototype of the FIM in-vial setup by Berh et al. [1]. (a) A cylindrical culture vial containing Drosophila larvae is imaged using five cameras connected to microcomputers which send their views of the vial to a server. A food bowl and an inner cylinder are inserted into the vial. (b) Example of a stitched view of the vial surface generated by the server from the views of the individual cameras.

several days is the accumulation of dirt on the vial surface. Since a food bowl is inserted into the vial, larvae drag food, dirt, and excretions over the vial, which produces artifacts with similar intensity as larvae in the obtained images. Over time, this dirt accumulates on the vial surface, making the segmentation and thus tracking of the larvae a challenging task (see Fig. 2).

Here, we propose a method for background subtraction in FIM in-vial images by applying a convolutional neural network (CNN). Our network is trained end-to-end using 265 images with manually labeled ground truth data. We evaluate the quality of the results for multiple net configurations and compare our CNN to two other methods, an automatic thresholding approach as well as Gabor filters in combination with a random forest classifier. Moreover, we show that our proposed CNN can be applied in real-time on a Raspberry Pi microcomputer due to its compact structure. This extends the use of the FIM in-vial setup to novel tracking experiments in the context of long-term behavioral studies.

The remainder of this paper is structured as follows. Section 2 provides a brief overview of related work regarding CNN-based segmentation and background subtraction. In Sect. 3, we outline the proposed CNN architecture and describe the loss function we use, which is based on the F1-score. In Sect. 4, we provide an evaluation of the achieved results for different CNN architectures and compare our method to two other approaches. Moreover, we analyze the applicability of our method for real-time processing on a Raspberry Pi microcomputer. In Sect. 5, we draw a conclusion and provide an outlook of some future work.

2   Related Work

Over the last years, deep learning with CNNs has proven a powerful method which has surpassed traditional model-based approaches by a large margin in many image processing and pattern recognition tasks. Despite the fact that the concept of convolutional nets has been known for some time [13], their actual

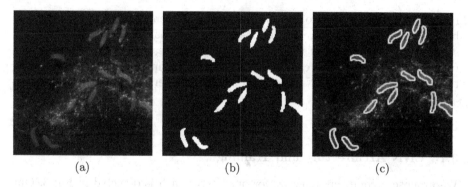

(a)   (b)   (c)

Fig. 2. Exemplary section of an image obtained by long-term monitoring using the FIM in-vial setup (see Fig. 1). The original image (a) depicts the accumulation of dirt on the vial surface, whereas (b) and (c) show the manually labeled ground truth where only larvae are marked as foreground.

breakthrough – starting with the network by Krizhevsky et al. [12] – is due to the availability of large data sets in combination with the processing capabilities of modern GPUs which allow to train large networks in reasonable time.

Convolutional neural networks have also been applied to biomedical segmentation tasks by predicting a class label for each pixel in the images. Ciresan et al. [3] have proposed a deep neural network for segmentation of electron microscopy images by applying a sliding-window approach to classify each pixel in the image. Later on [4], they have extended their former net architecture and applied it to mitosis detection in breast cancer histology images. In contrast to the sliding-window approach, Ronneberger et al. [20] have proposed the *U-net*, a convolutional net architecture specifically designed for biomedical image segmentation which is based on the concept of fully convolutional networks [14]. Their network has won the ISBI 2015 cell tracking challenge, significantly outperforming all other competing approaches. The U-net has recently been extended to three-dimensional image segmentation by Milletari et al. [16] and Çiçek et al. [2].

A large body of work has been published on background subtraction. Usually, this refers to the extraction of moving objects by learning either a static or a dynamic background model from video streams of RGB images. A review of such methods can be found in [22]. The use of different types of neural networks has also been proposed in this context [5,15]. Recently, Xu et al. [24] have proposed the use of deep auto-encoder networks for dynamic background modeling.

In comparison to the aforementioned background subtraction approaches, our problem setting is slightly different. Although the background changes over time due to larvae dragging dirt over the tracking surface [1], it does not contain dynamic movement itself. Moreover, we do not only strive to extract moving objects, since larvae can be inactive for extended periods of time or pupate, in which case they become static objects, but should still be recognized as foreground. Furthermore, learning and maintaining a background model dynamically over an extended period of time is complex and computationally expensive, which hampers its applicability to low-cost, large-scale screening settings in behavioral biological studies. We therefore propose a method that classifies the pixels in a single input image without the context of a video stream. This is especially useful since such an approach can be realized using a compact network which allows the application of the CNN directly on microcomputers such as a Raspberry Pi.

3   Methods

3.1   CNN Architecture and Training

We propose a compact structure for our CNN which is depicted in Fig. 3. Our CNN consists of two consecutive blocks which each comprise a 3×3 convolutional layer, a rectified linear unit (ReLU) activation function, and a max pooling layer. The first convolutional layer consists of 20 filters, i.e., outputs 20 channels, whereas the second one outputs 64 channels. Those two stages are followed by two blocks which comprise a de-convolution layer and a ReLU activation function. De-convolution is performed using 4×4 filter kernels with stride 2 so that in the end the original image size is obtained. Moreover, using 4×4 instead of 2×2 filter kernels in the de-convolution improves the performance of the net (see Sect. 4). In total, the net comprises $32,926$ weights. The de-convolution stages are followed by a softmax layer which normalizes the output to the range $[0, 1]$ so that the resulting output images can be interpreted as probability maps. During training, the probabilities are used as input for the F1-score loss layer. Details about using the F1-score loss function are given in Sect. 3.2. When performing actual classification, the probability map can be thresholded (e.g., using a threshold $t = 0.5$) to obtain a binary segmentation. It should be noted that besides the good results of this CNN architecture, the compact network layout facilitates the use of our proposed method for real-time processing on a microcomputer such as a Raspberry Pi (see Sect. 4). We have tested several alternative network configurations which are comparatively evaluated in Sect. 4.

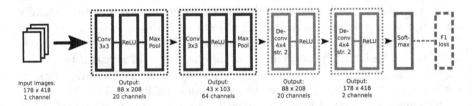

Fig. 3. Our CNN architecture. Each solid box corresponds to a layer in the network. The two convolution and pooling blocks are depicted in red whereas the de-convolution part of the network is depicted in blue. The rightmost layer is the F1-score loss layer, which is depicted using the dashed gray box to indicate that it is only present during the training phase. (Color figure online)

For our implementation we use the Caffe framework [10] which provides most of the layers in our architecture out-of-the-box. The only additional layer not provided by Caffe is the F1-score loss layer which we implemented in Python. For training, we have used Caffe's stochastic gradient descent method.

The CNN is trained end-to-end using 265 images which have been obtained by Berh et al. in a long-term monitoring experiment [1]. We use images of the first three days, more specifically from hours 15 to 63, of this experiment. Images of the hours before that contain only few larvae or do not yet show a lot of dirt accumulation on the tracking surface whereas 63 h are sufficient for obtaining meaningful behavioral read-out. The images have been taken from each of the five cameras in the setup and the lower part containing the boundary of the food bowl has been removed prior to the labeling as it is not relevant for the segmentation and tracking process of the experiment. Each of the images then has an original size of 180×420 pixels, but has been cropped to 178×418 pixels to fit the dimensions required by the CNN when performing the pooling and de-convolution operations. The images have been sorted by their timestamps and were afterwards divided into 4 time slots of which the first 3 slots contain 65 images each and the last slot contains 70 images. Each time slot therefore corresponds to approximately 12 h of monitoring. This allows to compare classification results between different time frames in the experiment (see Sect. 4). For each time slot, the images are randomly partitioned into 5 subsets of 13 images each (14 for the last time slot). From each time slot one of the subsets has then been selected as a set of test images which are not used for training the network. The test set thus comprises 53 images, which corresponds to 20% of the data. Due to the selection of a single subset from each time slot, test images are distributed evenly among the three days of the experiment. This allows an evaluation of the segmentation quality in relation to the monitoring time which corresponds to the amount of dirt on the tracking surface (see Sect. 4). For the remaining 212 images, we perform data augmentation to increase the number of training images by mirroring each image in x-direction, y-direction, as well as both of the axes. Additionally, each of the four obtained images is rotated by $180°$. This procedure increases the number of training images by a factor of 8 so that we obtain $212 \cdot 8 = 1,696$ training images of 180×420 pixels in total. Note that this procedure introduces a slight bias towards the non-mirrored data which is selected twice due to the fact that rotating the image by $180°$ is equivalent to mirroring it in both x- and y-direction.

3.2   Utilizing the F1-Score for Unbalanced Data

Since our images contain notably more background pixels than pixels belonging to foreground objects, i.e., larvae, the occurrences of background and foreground classes are highly imbalanced (see Fig. 2). Unbalanced training data is a common problem in many machine learning tasks and can significantly affect the performance of a classifier [23]. Several strategies to tackle this problem have been proposed, some of which rely on re-sampling of the data to balance the class instances [25] or adjusting the training by using additional weights for the loss of different classes [20].

For binary classification problems (including segmentation or background subtraction) the F1-score, sometimes also referred to as F-measure, is commonly used as an objective quality measure since it is more suitable for unbalanced data than an accuracy-based evaluation [9,21,24]. Pastor-Pellicer et al. [17] have proposed this measure as the loss function for training artificial neural networks with unbalanced data using backpropagation. Milletari et al. [16] have successfully employed the F1-score for training convolutional neural networks in the context of medical image segmentation and show that using the F1-score as the loss function can significantly outperform the use of a weighted logistic loss function in terms of segmentation quality. We therefore adapt the use of the F1-score as the loss function for our convolutional neural network. For convenience, we will give a short summary of the derivation of this loss function.

Given a binary segmentation S and a ground truth labeling G with n pixels, the number of *true positives* (TP) is given by the number of foreground pixels $s_i \in S, 0 \le i < n$, which are also labeled as foreground in G, i.e., for which holds that $s_i = 1 = g_i$, where $g_i \in G$ is the corresponding pixel in G. Consequently, the number of *false positives* (FP) corresponds to the number of pixels $s_i \in S$ for which $s_i = 1 \wedge g_i = 0$, and the number of *false negatives* (FN) is given by the number of pixels $s_i \in S$ for which $s_i = 0 \wedge g_i = 1$. For a probability map O with $o_i \in O, 0 \le i < n, 0 \le o_i \le 1$ (which corresponds to the output of our CNN due to the use of a softmax layer) instead of a binary segmentation S the TP, FP, and FN can be computed as follows:

$$\mathrm{TP} = \sum_{i=0}^{n-1} o_i \cdot g_i \qquad \mathrm{FP} = \sum_{i=0}^{n-1} o_i \cdot (1 - g_i) \qquad \mathrm{FN} = \sum_{i=0}^{n-1} (1 - o_i) \cdot g_i \qquad (1)$$

Based on the TP, FP, and FN the two metrics *Precision* (PR) and *Recall* (RC) are defined as follows:

$$\mathrm{PR} = \frac{\mathrm{TP}}{\mathrm{TP} + \mathrm{FP}} \qquad \mathrm{RC} = \frac{\mathrm{TP}}{\mathrm{TP} + \mathrm{FN}} \qquad (2)$$

The F1-score is then given by the harmonic mean between PR and RC:

$$F_1 = 2 \cdot \frac{\mathrm{PR} \cdot \mathrm{RC}}{\mathrm{PR} + \mathrm{RC}} = 2 \cdot \frac{\sum_{i=0}^{n-1} (o_i \cdot g_i)}{\sum_{i=0}^{n-1} (o_i + g_i)} \qquad (3)$$

In order to apply the F1-score as the loss function for the backpropagation algorithm, it is necessary to compute the gradient given by the partial derivatives with respect to $o_j, 0 \le j < n$, as follows:

$$\frac{\partial F_1}{\partial o_j} = 2 \cdot \frac{g_j \cdot \sum_{i=0}^{n-1} (o_i + g_i) - \sum_{i=0}^{n-1} (o_i \cdot g_i)}{\left[\sum_{i=0}^{n-1} (o_i + g_i) \right]^2} \qquad (4)$$

Pastor-Pellicer et al. [17] have shown that this loss function works well for mini batch training with stochastic gradient descent (SGD), although it should be taken into account that mini batches which only contain background samples in the ground truth do not update the weights since in such cases both the F1-score as well as the partial derivatives will always be 0. Furthermore, it should be noted that in order to use the aforementioned equations in a minimization setting such as SGD, the sign of the partial derivatives has to be inverted.

4 Results and Discussion

4.1 Performance Evaluation

To evaluate the performance of our CNN, we compare the segmentation result to the ground truth data by using the first part of each of the 4 time slots as test data which has not been used for training (see Sect. 3.1). The classification results of the CNN, which are given as probability output maps, are thresholded to obtain a binary image. Segmentation quality is then assessed by computing the pixel-wise F1-score in comparison to the ground truth images. Although we have tested various thresholds on the probability maps ranging from 0.3 to 0.7 in steps of 0.05 for all net architectures (see below), scores were not significantly influenced by the choice of thresholds. In the following, we have therefore only considered the threshold $t = 0.5$ for all CNN architectures.

First, we have compared several net architectures. Besides the proposed network (see Fig. 3) with two 3×3 convolutional layers and 4×4 de-convolutional layers, we have tested different kernel sizes k_c in the convolutional layers and different numbers l_c of convolutional layers (with the corresponding number of de-convolutional layers), all of them with the standard 2×2 de-convolution as opposed to the 4×4 de-convolutional layers in our proposed architecture. All tested nets with their respective input image sizes and number of filter kernels are listed in Table 1. The number of channels is symmetrically reduced in the de-convolutional layers as in our proposed net architecture (see Sect. 3.1).

All networks have been trained end-to-end for 40,000 iterations with a batch size of 20 images. Each net was evaluated on the test data every 500 iterations and a snapshot (i.e., the learned weights at that point) was taken every 1000 iterations. For all nets, the same set of hyper parameters was used for training. The base learning rate was set to 0.001 with a momentum of 0.99 and weight decay set to 0.0005. The learning rate was reduced in a step-wise fashion by multiplication with 0.5 every 7000 iterations. For three-layer nets with larger filter kernels (i.e., *kernel5_layer3* and *kernel7_layer3*) the base learning rate was reduced to 0.0004 in order for the training to converge. Other than that, all parameters stayed the same.

After training, the snapshot with the best F1-score over all of the test data was selected for each network. The results for those selected networks for each time slot as well as over all of the test data are listed in Table 2. It should be noted that, depending on the time slot, other architectures outperform our proposed net slightly (see the difference in F1-score of our proposed net to the

Table 1. Evaluated CNN architectures. For each net, the size k_c of the filters kernels, the number l_c of convolutional layers, the input size, and the number of channels (i.e., filters) in the convolutional layers (*Num. channels*) are denoted.

Label	k_c	l_c	De-conv. size	Input size	Num. channels
kernel3_layer2	3×3	2	2×2	178×418	$20 - 64$
kernel5_layer2	5×5	2	2×2	180×420	$32 - 64$
kernel7_layer2	7×7	2	2×2	178×418	$32 - 64$
kernel3_layer3	3×3	3	2×2	174×414	$20 - 64 - 128$
kernel5_layer3	5×5	3	2×2	180×420	$32 - 64 - 128$
kernel7_layer3	7×7	3	2×2	178×418	$32 - 64 - 128$
Proposed	3×3	2	4×4	178×418	$20 - 64$

best net architecture in the last line of Table 2). Especially in the last time slot which shows the most dirt accumulation on the tracking surface, thus leading to the worst results in all of the nets, deeper nets with more convolutional (and deconvolutional) layers show a better performance. However, the slight difference in performance is negligible since the results are still nearly as good as the best net architecture and one of the main advantages of our very compact architecture is that it can be used in real-time on a microcomputer in long-term tracking experiments (see Sect. 4.3). We therefore propose the architecture in Sect. 3.1 due to its usefulness in the considered application. Overall, our proposed network reaches an F1-score of 0.905 over all of the test data. In the second time slot, the highest score of 0.916 is reached which slightly decreases to 0.906 in the third time slot and drops to 0.878 in the last time slot. Interestingly, whereas (as expected) the performance of all classifiers diminishes with the later time slots due to the increasing accumulation of dirt on the tracking surface, the second time slot displays better results than the first one across all architectures. Despite a careful inspection of the ground truth data to identify potential errors in the labeling, we could not directly determine the origin of this behavior. We assume that dirt accumulation between the two time slots does not significantly increase, while more larvae are visible in the second time slot so that falsely classified pixels do not diminish the results as significant due to more true positive pixels.

4.2 Performance Comparison

In order to compare the performance of our model we use two well-established segmentation algorithms as benchmarks. The first baseline method is Otsu's automatic threshold selection. Despite its simplicity this approach delivers good results on FIM images [1]. However, performance drops significantly after the first day of the experiment due to aggregation of dirt which cannot be segmented by single-pixel intensities. As expected, this leads to a large number of false positives. Overall, the method yields an F1-score of 0.726 for all time slots. The F1-scores of Otsu's method for the individual time slots are listed in Table 3.

Table 2. Results for different CNN architectures. Although the proposed net is slightly outperformed by other architectures, differences (see last line) are negligible considering the applicability to real-time processing (see Sect. 4.3).

Architecture\time slot	1	2	3	4	All
kernel3_layer2	0.896	0.903	0.877	0.831	0.882
kernel5_layer2	**0.920**	0.922	0.909	0.878	0.909
kernel7_layer2	0.789	0.791	0.784	0.755	0.781
kernel3_layer3	0.916	**0.923**	0.913	0.893	**0.913**
kernel5_layer3	0.913	0.920	**0.914**	**0.899**	**0.913**
kernel7_layer3	0.907	0.910	0.909	**0.899**	0.907
Proposed	0.915	0.916	0.906	0.878	0.905
Diff. (proposed to best)	0.005	0.007	0.008	0.011	0.008

Table 3. Baseline segmentation results and proposed method (F1-scores)

Method\time slot	1	2	3	4	All
Otsu	0.717	0.792	0.743	0.679	0.726
Gabor-RF	0.861	0.874	0.855	0.821	0.850
CNN (proposed)	0.915	0.916	0.906	0.878	0.905

An approach based on local texture information is expected to be more robust against dirt on the tracking surface. Hence, our second baseline method (which we refer to as Gabor-RF for the remainder of this paper) combines a random forest classifier which was shown to deliver good results on many different classification tasks [7] with a Gabor filter bank suitable for texture detection [8]. First, all input images are filtered by the filter bank, yielding 72 channels. Second, random forests are trained as pixel-wise classifiers.

While the first baseline relying on automatic thresholding does not require any parameter tuning, Gabor-RF has a number of parameters to be adjusted. To obtain the optimal set of parameters, we performed extensive grid search. Best performance across all time slots according to F1-score was achieved using a Gabor-RF comprising 448 trees with maximum tree depth $h = 24$, and binarization threshold 0.4. Results on individual time slots are listed in Table 3.

Overall, our proposed CNN significantly outperforms both the thresholding method as well as the Gabor-RF on all time slots. Moreover, although the performance drops in the last time slot, the CNN method still performs more consistently over all time slots and has less difficulties dealing with more complex image data. Figure 4 shows some qualitative results for all classifiers.

4.3   Real-Time Processing on Microcomputers

To evaluate the applicability of our CNN for real-time processing in the FIM invial setup, we conducted a runtime study on a microcomputer used for imaging the vial wall. For our tests, we used a Raspberry Pi 3 model B with a 1.2 GHz 64-bit quad-core ARMv8 CPU and the Raspbian Jessie operating system. Caffe was used in CPU mode with OpenBLAS version 0.2.19. We evaluated the classification time for a single image of 178×418 pixels including copying the image to the input of the neural net and retrieving the results. To obtain representative numbers, we measured 5,000 classifications where we observed a mean runtime of 0.3215 s with a standard deviation of 0.0021. This corresponds to 3 frames per second which is more than sufficient for online-processing since the larvae are not recorded with more than approximately 2 frames per second [1]. When increasing the filter kernel size to 5×5 instead of 3×3 filters in the convolutional layers (i.e., using the network denoted as *kernel5_layer2* in Table 1), the mean runtime increases to 0.541 s which is not feasible for real-time processing. The same goes for net architectures with more than two convolutional layers, which increases the runtime even further. Although the performance of the nets with larger filter kernels is slightly superior, we thus propose the net architecture outlined in Sect. 3.1 since differences in performance are only minor and the compact net allows a higher throughput up to real-time processing on a Raspberry Pi. Moreover, due to the use of 4×4 filter kernels in the de-convolution layers, our proposed net retains the dimensions of its input images in the output. In contrast to this, a net with two 5×5 convolutional layers and 2×2 filter kernels in the de-convolution crops the output by 12 pixels in each dimension due to its receptive field while using 7×7 kernels crops the output by as much as 20 pixels in each dimension. This might be problematic when applying the net directly on the individual microcomputers as the subsequent stitching process requires some overlap between the images obtained from different cameras [1].

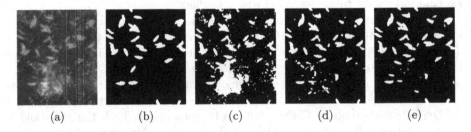

| (a) | (b) | (c) | (d) | (e) |

Fig. 4. Exemplary qualitative results for the proposed CNN and benchmark classifiers. (a) original image, (b) ground truth, (c) Otsu thresholding, (d) Gabor-RF, (e) CNN (proposed). Note that some darker larvae are not labeled in the ground truth as they are assumed by domain experts to be located on the inner cylinder of the setup and should thus not be segmented.

5 Conclusion

We have proposed a method for background subtraction in in-vial FIM images based on convolutional neural networks which provides good results and significantly outperforms other methods such as automatic thresholding or Gabor features in combination with a random forest classifier. To account for unbalanced training data, we use the F1-score as a loss function for training our CNN. We have shown that the use of a relatively small network topology facilitates the application of our method directly on the Raspberry Pi microcomputers which are used in the FIM in-vial setup proposed by Berh et al. [1]. We believe that our method is a significant step towards enabling researchers to perform novel tracking experiments in long-term behavioral studies of Drosophila larvae.

In the future, we are going to extend our method to dynamically learn a background model over time by incorporating the information of the preceding video frames. Moreover, we will extend the classification to a multi-class problem instead of a binary foreground/background classification to specifically label boundaries of larvae and thus provide a means of resolving interactions between animals in the segmentation and tracking process of the FIM setup. Finally, we want to extend our method by including an automatic classification of pupae in the images for allowing a more fine-grained behavioral read-out later on.

Acknowledgments. This work has been partially supported by the Deutsche Forschungsgemeinschaft, DFG EXC 1003 Cells in Motion - Cluster of Excellence, Münster, Germany (PP-2014-05, FF-2016-06).

References

1. Berh, D., Risse, B., Michels, T., Otto, N., Jiang, X., Klämbt, C.: A FIM-based long-term in-vial monitoring system for drosophila larvae. IEEE Trans. Biomed. Eng. (2017). doi:10.1109/TBME.2016.2628203
2. Çiçek, Ö., Abdulkadir, A., Lienkamp, S.S., Brox, T., Ronneberger, O.: 3D U-Net: learning dense volumetric segmentation from sparse annotation. In: Ourselin, S., Joskowicz, L., Sabuncu, M.R., Unal, G., Wells, W. (eds.) MICCAI 2016. LNCS, vol. 9901, pp. 424–432. Springer, Cham (2016). doi:10.1007/978-3-319-46723-8_49
3. Ciresan, D.C., Giusti, A., Gambardella, L.M., Schmidhuber, J.: Deep neural networks segment neuronal membranes in electron microscopy images. In: Proceedings of 26th Annual Conference on Neural Information Processing Systems, pp. 2852–2860 (2012)
4. Cireşan, D.C., Giusti, A., Gambardella, L.M., Schmidhuber, J.: Mitosis detection in breast cancer histology images with deep neural networks. In: Mori, K., Sakuma, I., Sato, Y., Barillot, C., Navab, N. (eds.) MICCAI 2013. LNCS, vol. 8150, pp. 411–418. Springer, Heidelberg (2013). doi:10.1007/978-3-642-40763-5_51
5. Culibrk, D., Marques, O., Socek, D., Kalva, H., Furht, B.: A neural network approach to bayesian background modeling for video object segmentation. In: Proceedings of First International Conference on Computer Vision Theory and Application (VISAPP), pp. 474–479 (2006)

6. Dankert, H., Wang, L., Hoopfer, E.D., Anderson, D.J., Perona, P.: Automated monitoring and analysis of social behavior in drosophila. Nat. Methods 6(4), 297–303 (2009)
7. Delgado, M.F., Cernadas, E., Barro, S., Amorim, D.G.: Do we need hundreds of classifiers to solve real world classification problems? J. Mach. Learn. Res. 15(1), 3133–3181 (2014)
8. Fogel, I., Sagi, D.: Gabor filters as texture discriminator. Biol. Cybern. 61(2), 103–113 (1989)
9. Ge, W., Guo, Z., Dong, Y., Chen, Y.: Dynamic background estimation and complementary learning for pixel-wise foreground/background segmentation. Pattern Recognit. 59, 112–125 (2016)
10. Jia, Y., Shelhamer, E., Donahue, J., Karayev, S., Long, J., Girshick, R.B., Guadarrama, S., Darrell, T.: Caffe: convolutional architecture for fast feature embedding. In: Proceedings of ACM International Conference on Multimedia (MM), pp. 675–678 (2014)
11. Kane, E.A., Gershow, M., Afonso, B., Larderet, I., Klein, M., Carter, A.R., de Bivort, B.L., Sprecher, S.G., Samuel, A.D.T.: Sensorimotor structure of drosophila larva phototaxis. Proc. Natl. Acad. Sci. 110(40), E3868–E3877 (2013)
12. Krizhevsky, A., Sutskever, I., Hinton, G.E.: Imagenet classification with deep convolutional neural networks. In: Proceedings of 26th Annual Conference on Neural Information Processing Systems, pp. 1106–1114 (2012)
13. LeCun, Y., Boser, B., Denker, J.S., Henderson, D., Howard, R.E., Hubbard, W., Jackel, L.D.: Backpropagation applied to handwritten zip code recognition. Neural Comput. 1(4), 541–551 (1989)
14. Long, J., Shelhamer, E., Darrell, T.: Fully convolutional networks for semantic segmentation. In: Proceedings of IEEE Conference on Computer Vision and Pattern Recognition (CVPR), pp. 3431–3440 (2015)
15. Luque, R.M., Domínguez, E., Palomo, E.J., Muñoz, J.: An ART-type network approach for video object detection. In: Proceedings of 18th European Symposium on Artificial Neural Networks (ESANN), pp. 423–428 (2010)
16. Milletari, F., Navab, N., Ahmadi, S.: V-net: fully convolutional neural networks for volumetric medical image segmentation. In: Proceedings of Fourth International Conference on 3D Vision (3DV), pp. 565–571 (2016)
17. Pastor-Pellicer, J., Zamora-Martínez, F., España-Boquera, S., Castro-Bleda, M.J.: F-measure as the error function to train neural networks. In: Rojas, I., Joya, G., Gabestany, J. (eds.) IWANN 2013. LNCS, vol. 7902, pp. 376–384. Springer, Heidelberg (2013). doi:10.1007/978-3-642-38679-4_37
18. Risse, B., Otto, N., Berh, D., Jiang, X., Kiel, M., Klämbt, C.: FIM^{2c}: multicolor, multipurpose imaging system to manipulate and analyze animal behavior. IEEE Trans. Biomed. Eng. 64(3), 610–620 (2017)
19. Risse, B., Thomas, S., Otto, N., Lopmeier, T., Valkov, D., Jiang, X., Klämbt, C.: FIM, a novel FTIR-based imaging method for high throughput locomotion analysis. PloS ONE 8(1), e53963 (2013)
20. Ronneberger, O., Fischer, P., Brox, T.: U-Net: convolutional networks for biomedical image segmentation. In: Navab, N., Hornegger, J., Wells, W.M., Frangi, A.F. (eds.) MICCAI 2015. LNCS, vol. 9351, pp. 234–241. Springer, Cham (2015). doi:10.1007/978-3-319-24574-4_28
21. Scherzinger, A., Kleene, F., Dierkes, C., Kiefer, F., Hinrichs, K.H., Jiang, X.: Automated segmentation of immunostained cell nuclei in 3D ultramicroscopy images. In: Rosenhahn, B., Andres, B. (eds.) GCPR 2016. LNCS, vol. 9796, pp. 105–116. Springer, Cham (2016). doi:10.1007/978-3-319-45886-1_9

22. Sobral, A., Vacavant, A.: A comprehensive review of background subtraction algorithms evaluated with synthetic and real videos. Comput. Vis. Image Underst. **122**, 4–21 (2014)
23. Sun, Y., Wong, A.K.C., Kamel, M.S.: Classification of imbalanced data: a review. Int. J. Pattern Recognit. Artif. Intell. **23**(4), 687–719 (2009)
24. Xu, P., Ye, M., Li, X., Liu, Q., Yang, Y., Ding, J.: Dynamic background learning through deep auto-encoder networks. In: Proceedings of ACM International Conference on Multimedia (MM), pp. 107–116 (2014)
25. Zhou, Z., Liu, X.: Training cost-sensitive neural networks with methods addressing the class imbalance problem. IEEE Trans. Knowl. Data Eng. **18**(1), 63–77 (2006)

Bayesian Diffusion Tensor Estimation with Spatial Priors

Xuan Gu[1,3]([✉]), Per Sidén[2], Bertil Wegmann[2], Anders Eklund[1,2,3],
Mattias Villani[2], and Hans Knutsson[1,3]

[1] Division of Medical Informatics, Department of Biomedical Engineering,
Linköping University, Linköping, Sweden
xuan.gu@liu.se
[2] Division of Statistics and Machine Learning,
Department of Computer and Information Science, Linköping University,
Linköping, Sweden
[3] Center for Medical Image Science and Visualization,
Linköping University, Linköping, Sweden

Abstract. Spatial regularization is a technique that exploits the dependence between nearby regions to locally pool data, with the effect of reducing noise and implicitly smoothing the data. Most of the currently proposed methods are focused on minimizing a cost function, during which the regularization parameter must be tuned in order to find the optimal solution. We propose a fast Markov chain Monte Carlo (MCMC) method for diffusion tensor estimation, for both 2D and 3D priors data. The regularization parameter is jointly with the tensor using MCMC. We compare FA (fractional anisotropy) maps for various b-values using three diffusion tensor estimation methods: least-squares and MCMC with and without spatial priors. Coefficient of variation (CV) is calculated to measure the uncertainty of the FA maps calculated from the MCMC samples, and our results show that the MCMC algorithm with spatial priors provides a denoising effect and reduces the uncertainty of the MCMC samples.

Keywords: Spatial regularization · Diffusion tensor · Spatial priors · Markov chain Monte Carlo · Fractional anisotropy

1 Introduction

Diffusion MRI is a technique used for studying brain connectivity. It has been reported that neighboring voxels show similar diffusion signals, due to their high degree of spatial coherence in terms of underlying fiber microstructure (Aboitiz et al. 1992). The spatial information is ignored by widely used voxel-by-voxel diffusion tensor estimation algorithms, e.g. ordinary least squares (OLS) and weighted least squares (WLS) (Chung et al. 2006), which assume that the data in each voxel is independent. Algorithms with this assumption make the parameter estimation more vulnerable to the image noise. The robustness and reliability

© Springer International Publishing AG 2017
M. Felsberg et al. (Eds.): CAIP 2017, Part I, LNCS 10424, pp. 372–383, 2017.
DOI: 10.1007/978-3-319-64689-3_30

of the estimated diffusion parameters can be improved by incorporating spatial information in the parameter estimation.

Spatial regularization is a technique that encourages similarity between parameters over a neighborhood of voxels. This is typically done by minimizing a cost function which is often the sum of a data error term and a regularization term (Raj et al. 2003; Wang et al. 2011). One of the drawbacks of this method is that the weight of the regularization term must be tuned in order to find the optimal solution. However, determining the appropriate regularization weight is a difficult problem, and there is no general solution. An alternative but similar concept is the use of probabilistic spatial priors in a Bayesian framework to spatially regularise the model fitting procedure (Demiralp and Laidlaw 2011; King et al. 2009; Poupon et al. 2001; Sidén et al. 2017; Walker-Samuel et al. 2011). The regularization strength is then governed by a hyperparameter that scales the prior precision matrix which can be estimated directly during the fitting procedure, rather than being estimated separately. Poupon et al. (2001) used a Markov random field (MRF) framework to obtain a regularized fiber orientation map, in order to improve the diffusion tractography. Martín-Fernández et al. (2003, 2004) proposed both Gaussian and non-Gaussian MRF approaches for the 2D spatial regularization of diffusion tensor fields. Martín-Fernández et al. (2004) extended the 2D Gaussian MRF method to 3D using multivariate Gaussian Markov random field (GMRF) ideas. The maximum a posteriori (MAP) estimator was found by the simulated annealing algorithm in all the three approaches. King et al. (2009) tackled the crossing-fiber problem using an MRF model and Markov chain Monte Carlo (MCMC) sampling for the multifiber ball and stick model outlined in (Behrens et al. 2007), which offered a useful solution for the crossing-fiber problem in diffusion tractography. Walker-Samuel et al. (2011) proposed a MCMC method using GMRF to incorporate spatial information when estimating the apparent diffusion coefficient (ADC).

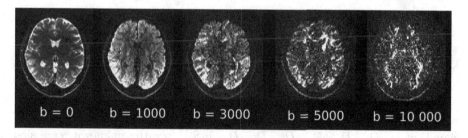

Fig. 1. Diffusion images using b-values of 0, 1000, 3000, 5000, and 10,000 s/mm² show progressively more diffusion weighting, but also reduced SNR.

Figure 1 shows a sequence of diffusion images for 5 different b-values. Increasing the b-value from 0 to 10,000 s/mm² creates greater diffusion weighting of the image, at the expense of a significant reduction of the signal-to-noise-ratio (SNR). The SNR issue becomes important at higher b-value diffusion imaging due to the larger exponential attenuation of the signal, longer echo time and increased susceptibility (Graessner 2011). To use the spatial dependency between the signals in neighbouring voxels for accurate inference, we extend a MCMC algorithm with spatial priors, previously used for fMRI analysis (Sidén et al. 2017), to diffusion MRI. A preconditioned conjugate gradient (PCG) approach is applied to make the MCMC algorithm a practical option for the whole brain analysis. The main contribution of this paper is that we use the PCG approach and MCMC to compute the posterior distribution with spatial priors for the whole brain, which was previously infeasible due to the high computational complexity. We calculate fractional anisotropy (FA) maps from three diffusion tensor estimation methods: least-squares, and MCMC with and without spatial priors for various b-values. The coefficient of variation (CV) is calculated to measure the dispersion of the FA maps from the MCMC samples.

2 Theory

2.1 Diffusion Tensor Estimation

In a diffusion experiment, the diffusion-weighted signal S_i of the ith measurement for one voxel is modeled by

$$S_i = S_0 \exp(-b\mathbf{g}_i^T \mathbf{D}\mathbf{g}_i), \quad \text{for} \quad i = 1, 2, \cdots, T, \tag{1}$$

where S_0 is the signal without diffusion weighting, b is the diffusion weighting factor, $\mathbf{D} = \begin{bmatrix} D_{xx} & D_{xy} & D_{xz} \\ D_{xy} & D_{yy} & D_{yz} \\ D_{xz} & D_{yz} & D_{zz} \end{bmatrix}$ is the diffusion tensor in the form of a 3×3 positive definite matrix, $\mathbf{g}_i$ is a 3×1 unit vector of the gradient direction, and T is the total number of measurements. Using the log transform, the equation above becomes

$$\ln(S_i) = \ln(S_0) - b\mathbf{g}_i^T \mathbf{D}\mathbf{g}_i, \tag{2}$$

which, assuming additive noise on the log scale, can be structured into the well-known multiple linear regression form

$$\mathbf{y} = \mathbf{X}\mathbf{w} + \varepsilon, \tag{3}$$

where $\mathbf{y} = [\ln(S_1), \ln(S_2), \cdots, \ln(S_T)]^T$ represents the logarithm of the measured signal, $\mathbf{w} = [D_{xx}, D_{xy}, D_{yy}, D_{xz}, D_{yz}, D_{zz}, \ln S_0]^T$ are the unknown regression coefficients, $\mathbf{X}$ is a $T \times 7$ design matrix containing the different diffusion gradient directions,

$$\mathbf{X} = -b \begin{bmatrix} g_{1x}^2 & 2g_{1x}g_{1y} & g_{1y}^2 & 2g_{1x}g_{1z} & 2g_{1y}g_{1z} & g_{1z}^2 & \frac{1}{b} \\ g_{2x}^2 & 2g_{2x}g_{2y} & g_{2y}^2 & 2g_{2x}g_{2z} & 2g_{2y}g_{2z} & g_{2z}^2 & \frac{1}{b} \\ \vdots & \vdots & \vdots & \vdots & \vdots & \vdots & \vdots \\ g_{Tx}^2 & 2g_{Tx}g_{Ty} & g_{Ty}^2 & 2g_{Tx}g_{Tz} & 2g_{Ty}g_{Tz} & g_{Tz}^2 & \frac{1}{b} \end{bmatrix}, \tag{4}$$

and $\varepsilon = [\varepsilon_1, \varepsilon_2, \cdots, \varepsilon_T]^T$ are the error terms. We will consider diffusion data containing T volumes with N voxels ordered in a $T \times N$ matrix $\mathbf{Y}$. The linear regression model given in Eq. 3 can be rewritten for simultaneous estimation of all the parameters, according to

$$\mathbf{Y} = \mathbf{X}\mathbf{W} + \mathbf{E}, \tag{5}$$

where $\mathbf{W}$ is a $7 \times N$ matrix of regression coefficients and $\mathbf{E}$ is a $T \times N$ matrix error terms.

2.2   Bayesian Inference

In the Bayesian inference framework, the posterior distribution quantifies the uncertainty of the parameters $\mathbf{W}$ given the data $\mathbf{Y}$, according to Bayes theorem:

$$p(\mathbf{W}|\mathbf{Y}) \propto p(\mathbf{Y}|\mathbf{W})p(\mathbf{W}), \tag{6}$$

where $p(\mathbf{W})$ is the prior information about the parameters, and $p(\mathbf{Y}|\mathbf{W})$ is the likelihood of $\mathbf{Y}$ given $\mathbf{W}$. A closed form expression for the posterior distribution is normally not available, but MCMC can be used to produce samples from the posterior leading to inference that is asymptotically exact.

2.3   MCMC Algorithm

Here we assume that the noise in each voxel is modeled as independent and identically distributed (i.i.d.) Gaussian noise. The likelihood then becomes

$$p(\mathbf{Y}|\mathbf{W}, \lambda) = \prod_{n=1}^{N} \mathcal{N}(\mathbf{Y}_{\cdot,n}; \mathbf{X}\mathbf{W}_{\cdot,n}, \lambda_n^{-1}\mathbf{I}_T), \tag{7}$$

with $\mathbf{Y}_{\cdot,n}$ and $\mathbf{W}_{\cdot,n}$ denoting the nth column of $\mathbf{Y}$ and $\mathbf{W}$, λ_n is the noise precision of voxel n and $\mathbf{I}_T$ is the $T \times T$ identity matrix. The spatial part of the model is incorporated via the following prior on the regression coefficients $\mathbf{W}$

$$p(\mathbf{W}|\boldsymbol{\alpha}) = \prod_{k=1}^{7} p(\mathbf{W}_{k,\cdot}^T|\alpha_k), \quad \mathbf{W}_{k,\cdot}^T|\alpha_k \sim \mathcal{N}(0, \alpha_k^{-1}\mathbf{D}_w^{-1}), \tag{8}$$

where $\mathbf{W}_{k,\cdot}^T$ denotes the transposed kth row of $\mathbf{W}$, $\mathbf{D}_w$ is a $N \times N$ spatial precision matrix and $\boldsymbol{\alpha} = [\alpha_1, \alpha_1, \cdots, \alpha_7]^T$ are hyperparameters that controls the smoothness, which are to be estimated from the data for each regressor in $\mathbf{X}$. We choose the unweighted graph-Laplacian (UGL) (Penny et al. 2005) which has 6's on the diagonal when modeling the 3D brain and 4's when modeling a 2D slice, and $\mathbf{D}_w(i, j) = -1$ if i and j are adjacent. The log likelihood can be expressed as

$$\log p(\mathbf{Y}|\mathbf{W}, \lambda) = \frac{T}{2} \sum_{n=1}^{N} \log(\lambda_n)$$
$$- \frac{1}{2} \sum_{n=1}^{N} \lambda_n \left(\mathbf{Y}_{\cdot,n}^T \mathbf{Y}_{\cdot,n} - 2\mathbf{Y}_{\cdot,n}^T \mathbf{X}\mathbf{W}_{\cdot,n} + \mathbf{W}_{\cdot,n}^T \mathbf{X}^T \mathbf{X}\mathbf{W}_{\cdot,n} \right), \tag{9}$$

where we have ignored everything that is constant with respect to the parameters. Since $\mathbf{Y}$ and $\mathbf{X}$ will not change during the MCMC algorithm, quantities such as $\mathbf{X}^T\mathbf{X}$ can be effectively pre-computed, removing the time dimension from the likelihood which leads to significant speed up. We obtain closed form expressions for all full conditional posteriors, and can therefore perform Gibbs sampling. The full conditional posterior of $\mathbf{W}$ is given by

$$\log p(\mathbf{W}|\mathbf{Y},\boldsymbol{\lambda},\boldsymbol{\alpha}) = -\frac{1}{2}\mathbf{w}_r^T\tilde{\mathbf{B}}\mathbf{w}_r + \mathbf{b}_w^T\mathbf{w}_r, \tag{10}$$

where $\mathbf{b}_w = vec(diag(\boldsymbol{\lambda})\mathbf{Y}^T\mathbf{X})$, $\tilde{\mathbf{B}} = \mathbf{X}^T\mathbf{X} \otimes diag(\boldsymbol{\lambda}) + diag(\boldsymbol{\alpha}) \otimes \mathbf{D}_w$, $\mathbf{w}_r = vec(\mathbf{W}^T)$. The full conditional posterior of $\boldsymbol{\lambda}$ can be written as

$$\log p(\boldsymbol{\lambda}|\mathbf{Y},\mathbf{W},\boldsymbol{\alpha}) = (\tilde{u}_2 - 1)\sum_{n=1}^{N}\log(\lambda_n) - \sum_{n=1}^{N}\frac{\lambda_n}{\tilde{u}_{1n}}, \tag{11}$$

where $\frac{1}{\tilde{u}_{1n}} = \frac{1}{2}\left(\mathbf{Y}_{\cdot,n}^T\mathbf{Y}_{\cdot,n} - 2\mathbf{Y}_{\cdot,n}^T\mathbf{X}\mathbf{W}_{\cdot,n}^T + \mathbf{W}_{\cdot,n}^T\mathbf{X}^T\mathbf{X}\mathbf{W}_{\cdot,n}\right) + \frac{1}{u_1}$, $\tilde{u}_2 = \frac{T}{2} + u_2$. The full conditional posterior of $\boldsymbol{\alpha}$ is given by

$$\log p(\boldsymbol{\alpha}|\mathbf{Y},\mathbf{W},\boldsymbol{\lambda}) = (\tilde{q}_2 - 1)\sum_{k=1}^{7}\log(\alpha_k) - \sum_{k=1}^{7}\frac{\alpha_k}{\tilde{q}_{1k}}, \tag{12}$$

where $\frac{1}{\tilde{q}_{1k}} = \frac{1}{2}\left(\mathbf{Y}_{\cdot,n}^T\mathbf{X}\mathbf{W}_{\cdot,n}^T\right) + \frac{1}{q_1}$, $\tilde{q}_2 = \frac{N}{2} + q_2$. We refer to Sidén et al. (2017) for further details.

2.4 PCG Based Sampling

Cholesky decomposition is usually the bottleneck for algorithms involving high dimensional GMRFs (Rue and Held 2005). In practice it takes too long time to finish the whole 3D brain inference (if $N = 100,000$). Papandreou and Yuille (2010) proposed a sampling method from the posterior for $\mathbf{W}$ that avoids the Cholesky decomposition. The main idea is to minimize $\mathbf{Y} - \mathbf{XB}$ instead of solving $\mathbf{Y} = \mathbf{XB}$. First we construct $\mathbf{B}_{data} = diag(\boldsymbol{\lambda}) \otimes \mathbf{X}^T\mathbf{X}$ and then we calculate $\mathbf{b} = \left(blkdiag\left[\sqrt{\alpha_k}\mathbf{G}_w\right]\right)^T\mathbf{z}_1 + \mathbf{H}_w^T\mathbf{L}_{data}\mathbf{H}_w\mathbf{z}_2 + \mathbf{b}_w$. $\mathbf{z}_1$ and $\mathbf{z}_2$ are random draws from a Gaussian distribution with zero mean and identity covariance. $\mathbf{G}_w$ needs to satisfy $\mathbf{D}_w = \mathbf{G}_w^T\mathbf{G}_w$. $\mathbf{H}_w$ is defined as the permutation matrix such that $vec(\mathbf{W}) = \mathbf{H}_w vec(\mathbf{W}^T)$ (Penny et al. 2007). $\mathbf{L}_{data}$ is the Cholesky factor of $\mathbf{B}_{data}$. Lastly compute $\mathbf{M}$ as the incomplete Cholesky factor of $\tilde{\mathbf{B}}$ and solve $\tilde{\mathbf{B}}\mathbf{w}_r = \mathbf{b}$ approximately using preconditioned conjugate gradient (PCG) with preconditioner $\mathbf{M}$. PCG is a conjugate gradient method with some preconditions set to ensure fast convergence. Please note that $\tilde{\mathbf{B}}$ has to be reordered using a reordering method (Amestoy et al. 1996), and the same reordering has to be used for $\mathbf{b}$. Using the inverse reordering, posterior samples $\mathbf{w}_r$ can be obtained.

3   Data

We use the MGH adult diffusion dataset from the Human Connectome Project (HCP) (Van Essen et al. 2013). Data were collected from 35 healthy adults scanned on a customized Siemens 3T Connectom scanner with 4 different b-values (1000, 3000, 5000 and 10,000 s/mm^2). The data has already been pre-processed for gradient nonlinearity correction, motion correction and eddy current correction (Glasser et al. 2013). The data consists of 40 non-diffusion weighted volumes ($b = 0$), 64 volumes for $b = 1000$ and 3000 s/mm^2, 128 volumes for $b = 5000$ s/mm^2 and 256 volumes for $b = 10,000$ s/mm^2, which yields 552 volumes of $140 \times 140 \times 96$ voxels with an 1.5 mm isotropic voxel size.

Data used in the preparation of this work were obtained from the Human Connectome Project (HCP) database (https://ida.loni.usc.edu/login.jsp). The HCP project (Principal Investigators: Bruce Rosen, M.D., Ph.D., Martinos Center at Massachusetts General Hospital; Arthur W. Toga, Ph.D., University of Southern California, Van J. Weeden, MD, Martinos Center at Massachusetts General Hospital) is supported by the National Institute of Dental and Craniofacial Research (NIDCR), the National Institute of Mental Health (NIMH) and the National Institute of Neurological Disorders and Stroke (NINDS). HCP is the result of efforts of co-investigators from the University of Southern California, Martinos Center for Biomedical Imaging at Massachusetts General Hospital (MGH), Washington University, and the University of Minnesota.

4   Results

In Figs. 2 and 3 we present FA maps obtained from three diffusion tensor estimation methods: a standard least-squares method, and MCMC with and without 2D/3D spatial prior. The formula for FA is

$$\text{FA} = \sqrt{\frac{(\lambda_1 - \lambda_2)^2 + (\lambda_2 - \lambda_3)^2 + (\lambda_3 - \lambda_1)^2}{2(\lambda_1^2 + \lambda_2^2 + \lambda_3^2)}}. \tag{13}$$

The number of MCMC samples we use is 1000, and the number of samples for burn-in is 500. For MCMC with and without spatial priors, we calculated the average FA maps over all MCMC draws. It is known that a higher b-value will result in diffusion data with lower SNR. To investigate the effect of the b-value, we decompose the diffusion data into 5 parts, each of which consists of volumes for a single b-value (0, 1000, 3000, 5000 and 10,000 s/mm^2). In theory it is possible to estimate the diffusion tensor with at least two b-values. For each method, we calculate FA maps for 4 different combinations of b-values ($b = 0/1000, 0/3000, 0/5000, 0/10,000$ s/mm^2), and the entire dataset. Each FA map is normalized by its mean, to adjust the FA maps to a common scale. As showed in the first row of Figs. 2 and 3, the FA maps from the least-squares method are quite noisy. Comparing the least-squares approach to the MCMC approach in Figs. 2 and 3, we can see a denoising effect in MCMC with spatial priors.

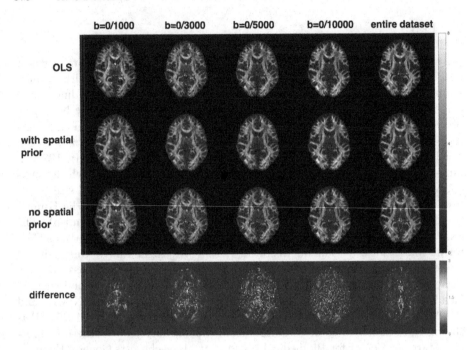

Fig. 2. FA maps from three diffusion tensor estimation methods. **First row**: least-squares method, **second row**: MCMC with 2D spatial priors, **third row**: MCMC without 2D spatial priors, **fourth row**: absolute difference of FA maps from MCMC with and without 2D spatial priors. **Columns from left to right**: $b = 0/1000$, $0/3000$, $0/5000$, $0/10,000$ s/mm^2 and the entire dataset. Each FA map was normalized by its mean, to adjust the FA maps to a common scale. For the FA maps, Values no greater than 0 are mapped to the first color in the colormap, and values no less than 8 are mapped to the last color in the colormap. For the absolute difference of FA maps, the scaling interval is 0 to 3.

Also, one can see a denoising effect for the FA maps from MCMC with spatial priors, compared with those from MCMC without spatial priors, as showed in the second row of Figs. 2 and 3. The MCMC algorithm with a spatial prior provides a better denoising effect for data with higher b-values, since that data has lower SNR. However, the denoising effect of using a spatial prior is clearly weaker for the entire dataset, as we can see from the fifth column in Figs. 2 and 3. The absolute differences between the FA maps from MCMC with and without spatial priors are rather small for the entire dataset. In the fourth row of Figures 2 and 3, one can see that the 3D spatial prior tends to provide a very close absolute difference to its 2D counterpart.

We use the CV (the ratio of the standard deviation to the mean) to measure the uncertainty of the FA maps calculated from the MCMC samples, as showed in Figs. 4 and 5. A lower uncertainty is found for some white matter voxels in the third rows, for both the 2D and 3D priors. For $b = 0/1000$ and $0/3000$ s/mm^2, the

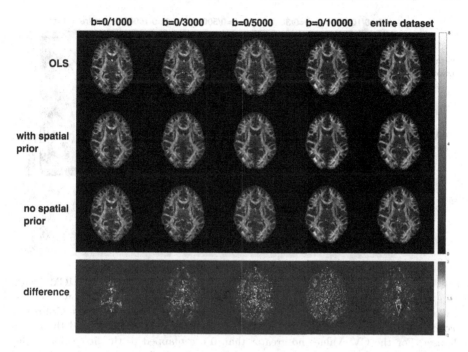

Fig. 3. FA maps from three diffusion tensor estimation methods. **First row**: least-squares method, **second row**: MCMC with 3D spatial priors, **third row**: MCMC without 3D spatial priors, **fourth row**: absolute difference of FA maps from MCMC with and without 3D spatial priors. **Columns from left to right**: $b = 0/1000$, $0/3000$, $0/5000$, $0/10,000$ s/mm^2 and the entire dataset. Each FA map was normalized by its mean, to adjust the FA maps to a common scale. For the FA maps, Values no greater than 0 are mapped to the first color in the colormap, and values no less than 8 are mapped to the last color in the colormap. For the absolute difference of FA maps, the scaling interval is 0 to 3.

decrease in uncertainty was clearly higher than other combinations of b-values. Comparing the third row in Figs. 4 and 5 we see that the 3D prior provides slightly better performance in lowering the uncertainty of the FA maps than the 2D prior for entire dataset.

5 Discussion

Performing Bayesian inference with spatial priors for the whole brain ($N = 294,000$) becomes a practical option with the PCG based MCMC sampling. The results presented in this paper demonstrate that the MCMC algorithm with spatial priors provides improvements by reducing the uncertainty of the MCMC samples. The presented FA maps from three diffusion tensor estimation methods show that the MCMC method provides a denoising effect compared with the least-squares method. The denoising effect comes from the spatial priors. We have seen that MCMC sampling with spatial priors provides a denoising

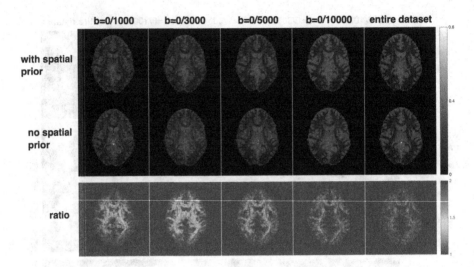

Fig. 4. CV of FA maps calculated from the MCMC samples. **First row**: MCMC with 2D spatial priors, **second row**: MCMC without 2D spatial priors, **third row**: ratios of CV of FA maps calculated from MCMC with and without 2D spatial priors. **Columns from left to right**: $b = 0/1000, 0/3000, 0/5000, 0/10,000$ s/mm^2 and the entire dataset. For the CV, Values no greater than 0 are mapped to the first color in the colormap, and values no less than 0.4 are mapped to the last color in the colormap. For the ratio of the CV, the scaling interval is 1 to 2.

effect for various combinations of b-values. For $b = 0/1000, 0/3000, 0/5000, 0/10000$ s/mm^2, MCMC with spatial priors works much better than MCMC without spatial priors. For the entire dataset, the differences of the performance between MCMC with and without spatial priors become smaller. For diffusion tensor model, MCMC with 3D spatial prior provides very close performance to MCMC with 2D spatial prior. The uncertainty of the FA maps from the posterior distributions generated with MCMC is reduced by the spatial prior, especially for the b-value combinations $b = 0/1000$ and $0/3000$ s/mm^2. We have noticed that most of the voxels with reduced uncertainty are located in white matter, where the FA values are relatively large.

In this paper, we have only considered the isotropic and stationary spatial prior (UGL). One potential improvement of this work is to use an anisotropic and non-stationary spatial prior, e.g. replacing the UGL prior with a weighted graph-Laplacian (WGL) prior. Wegmann et al. (2017) showed that the FA values can be greatly underestimated for methods that take the logarithm of the diffusion measurements. Thus in the formulation of the linear regression model in Eq. 3, using a logarithmic link function (Wegmann et al. 2017) instead of taking the logarithm of the diffusion measurements might be a better option. Also, to make the diffusion tensor positive definite, if it is possible to impose the restriction using a log-Cholesky representation (Koay 2010) of the diffusion tensor. Raj et al. (2011) reported that reconstruction of fiber orientation distribution

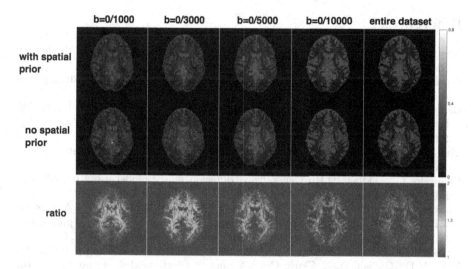

Fig. 5. CV of FA maps calculated from the MCMC samples. **First row**: MCMC with 3D spatial priors, **second row**: MCMC without 3D spatial priors, **third row**: ratios of CV of FA maps calculated from MCMC with and without 3D spatial priors. **Columns from left to right**: $b = 0/1000$, $0/3000$, $0/5000$, $0/10,000$ s/mm^2 and the entire dataset. For the CV, Values no greater than 0 are mapped to the first color in the colormap, and values no less than 0.4 are mapped to the last color in the colormap. For the ratio of the CV, the scaling interval is 1 to 2.

function (ODF) for high angular resolution diffusion imaging (HARDI) using spatial priors provides even stronger advantages, compared with the simpler diffusion tensor model. For the latter case, only extremely noisy data benefit from spatial priors. For future work, we will extend the MCMC method in this paper to HARDI models, such as q-ball imaging and diffusion spectrum imaging, where the ODF can be estimated via a linear estimator.

Acknowledgements. This research was supported by the Information Technology for European Advancement (ITEA) 3 Project BENEFIT (better effectiveness and efficiency by measuring and modelling of interventional therapy) and the Swedish Research Council (grant 2015-05356, "Learning of sets of diffusion MRI sequences for optimal imaging of micro structures" and grant 2013-5229 "Statistical analysis of fMRI data").
Data collection and sharing for this project was provided by the Human Connectome Project (HCP; Principal Investigators: Bruce Rosen, M.D., Ph.D., Arthur W. Toga, Ph.D., Van J. Weeden, MD). HCP funding was provided by the National Institute of Dental and Craniofacial Research (NIDCR), the National Institute of Mental Health (NIMH), and the National Institute of Neuro-logical Disorders and Stroke (NINDS). HCP data are disseminated by the Laboratory of Neuro Imaging at the University of Southern California.

References

Aboitiz, F., Scheibel, A.B., Fisher, R.S., Zaidel, E.: Fiber composition of the human corpus callosum. Brain Res. **598**(1), 143–153 (1992)

Amestoy, P.R., Davis, T.A., Duff, I.S.: An approximate minimum degree ordering algorithm. SIAM J. Matrix Anal. Appl. **17**(4), 886–905 (1996)

Behrens, T.E., Berg, H.J., Jbabdi, S., Rushworth, M., Woolrich, M.: Probabilistic diffusion tractography with multiple fibre orientations: what can we gain? Neuroimage **34**(1), 144–155 (2007)

Chung, S., Lu, Y., Henry, R.G.: Comparison of bootstrap approaches for estimation of uncertainties of DTI parameters. NeuroImage **33**(2), 531–541 (2006)

Demiralp, C., Laidlaw, D.H.: Generalizing diffusion tensor model using probabilistic inference in Markov random fields. In: MICCAI CDMRI Workshop (2011)

Glasser, M.F., Sotiropoulos, S.N., Wilson, J.A., Coalson, T.S., Fischl, B., Andersson, J.L., Xu, J., Jbabdi, S., Webster, M., Polimeni, J.R., et al.: The minimal preprocessing pipelines for the human connectome project. Neuroimage **80**, 105–124 (2013)

Graessner, J.: Diffusion-Weighted Imaging (DWI). MAGNETON Flash, pp. 6–9 (2011)

King, M.D., Gadian, D.G., Clark, C.A.: A random effects modelling approach to the crossing-fibre problem in tractography. NeuroImage **44**(3), 753–768 (2009)

Koay, C.G.: Least squares approaches to diffusion tensor estimation. Diffus. MRI, 272 (2010)

Martín-Fernández, M., Josá-Estépar, R.S., Westin, C.-F., Alberola-López, C.: A novel Gauss-Markov random field approach for regularization of diffusion tensor maps. In: Moreno-Díaz, R., Pichler, F. (eds.) EUROCAST 2003. LNCS, vol. 2809, pp. 506–517. Springer, Heidelberg (2003). doi:10.1007/978-3-540-45210-2_46

Martín-Fernández, M., Westin, C.-F., Alberola-López, C.: 3D bayesian regularization of diffusion tensor MRI using multivariate Gaussian Markov random fields. In: Barillot, C., Haynor, D.R., Hellier, P. (eds.) MICCAI 2004. LNCS, vol. 3216, pp. 351–359. Springer, Heidelberg (2004). doi:10.1007/978-3-540-30135-6_43

Papandreou, G., Yuille, A.L.: Gaussian sampling by local perturbations. In: Advances in Neural Information Processing Systems, pp. 1858–1866 (2010)

Penny, W., Flandin, G., Trujillo-Barreto, N.: Bayesian comparison of spatially regularised general linear models. Hum. Brain Mapp. **28**(4), 275–293 (2007)

Penny, W.D., Trujillo-Barreto, N.J., Friston, K.J.: Bayesian fMRI time series analysis with spatial priors. Neuroimage **24**(2), 350–362 (2005)

Poupon, C., Mangin, J.-F., Clark, C.A., Frouin, V., Régis, J., Le Bihan, D., Bloch, I.: Towards inference of human brain connectivity from MR diffusion tensor data. Med. Image Anal. **5**(1), 1–15 (2001)

Raj, A., Hess, C., Mukherjee, P.: Spatial HARDI: improved visualization of complex white matter architecture with Bayesian spatial regularization. Neuroimage **54**(1), 396–409 (2011)

Rue, H., Held, L.: Gaussian Markov Random Fields: Theory and Applications. CRC Press, Boca Raton (2005)

Sidén, P., Eklund, A., Bolin, D., Villani, M.: Fast Bayesian whole-brain fMRI analysis with spatial 3D priors. Neuroimage **146**, 211–225 (2017)

Van Essen, D. C., Smith, S. M., Barch, D. M., Behrens, T. E., Yacoub, E., Ugurbil, K., Consortium, W.-M. H., et al: The WU-Minn human connectome project: an overview. Neuroimage **80**, 62–79 (2013)

Walker-Samuel, S., Orton, M., Boult, J.K., Robinson, S.P.: Improving apparent diffusion coefficient estimates and elucidating tumor heterogeneity using Bayesian adaptive smoothing. Magnet. Reson. Med. **65**(2), 438–447 (2011)

Wang, Z., Vemuri, B.C., Chen, Y., Mareci, T.: A constrained variational principle for direct estimation and smoothing of the diffusion tensor field from DWI. In: Taylor, C., Noble, J.A. (eds.) IPMI 2003. LNCS, vol. 2732, pp. 660–671. Springer, Heidelberg (2003). doi:10.1007/978-3-540-45087-0_55

Wegmann, B., Eklund, A., Villani, M.: Bayesian heteroscedastic regression for diffusion tensor imaging. In: Modeling, Analysis, and Visualization of Anisotropy. Springer (2017)

Analysis of Multilinear Subspaces
Based on Geodesic Distance

Hayato Itoh[1,2]($\boxtimes$), Atsushi Imiya[3], and Tomoya Sakai[4]

[1] Graduate School of Informatics, Nagoya University, Nagoya, Japan
hitoh@mori.m.is.nagoya-u.ac.jp
[2] Graduate School of Advanced Integration Science, Chiba University, Chiba, Japan
[3] Institute of Management and Information Technologies,
Chiba University, Chiba, Japan
[4] Graduate School of Engineering, Nagasaki University, Nagasaki, Japan

Abstract. Tensor principal component analysis enables the efficient analysis of spatial textures of volumetric images and spatio-temporal changes of volumetric video sequences. To extend the subspace methods for analysis of linear subspaces, we are required to quantitatively evaluate the differences between multilinear subspaces. This discrimination of multilinear subspaces is achieved by computing the geodesic distance between tensor subspaces.

1 Introduction

For computer-assisted diagnosis, inspection and biopsy in precision medicine, abnormality detection based on pattern recognition is a fundamental technique. Organs, cells in organs and microstructures in cells, which are dealt with in biomedical image analysis for these procedures, are spatial textures. From cell to human body, medical data used in biomedical image analysis are multiway data. Furthermore, for longitudinal analysis, the collected of these statistical data are expressed as volumetric video sequences. For the detection of spatio-temporal modifications of these volumetric video sequences, we introduce the geodesic distance between tensor subspaces for longitudinal analysis of data. The proposed geodesic distance is based on tensor-subspace learning methods, which is an extension of principal component analysis to deal with multiway data [1,2].

To measure the difference between subspaces, the canonical angle between two subspaces has been used [3,4]. Furthermore, the Grassmannian distance between two subspaces has been proposed [5–7]. The Grassmannian distance is based on a set of canonical angles between two subspaces. However, the computation of canonical angles between tensor subspaces is computationally expensive from the view point of numerical computation [2] since we need to compute the eigendecomposition of a large projection matrix of size $m^2 n^2$ for images of size $m \times n$. Moreover, the distance between Stiefel manifolds has also been proposed [6,8,9].

© Springer International Publishing AG 2017
M. Felsberg et al. (Eds.): CAIP 2017, Part I, LNCS 10424, pp. 384–396, 2017.
DOI: 10.1007/978-3-319-64689-3_31

To compute the difference between multilinear subspaces, we deal with multilinear subspaces as Stiefel manifolds. We introduce a geodesic distance as the distance between Stiefel manifolds, which are defined by two sets of orthogonal vectors for each modes of two tensor subspaces. In numerical examples, we show the geodesic-distance-based analysis for four-dimensional MRI data [10].

2 Preliminaries

2.1 Tensor Expression for N-way Arrays

We briefly summarise the multilinear projection for N-dimensional arrays from ref. [1]. A Nth-order tensor $\mathcal{X}$ defined in $\mathbb{R}^{I_1 \times I_2 \times \cdots \times I_N}$ is expressed as $\mathcal{X} = (x_{i_1,i_2,\ldots,i_N})$ for $x_{i_1,i_2,\ldots,i_N} \in \mathbb{R}$, using N indices i_n. Each subscript n denotes the n-mode of $\mathcal{X}$. For $\mathcal{X}$, the n-mode vectors, $n = 1, 2, \ldots, N$, are defined as the I_n-dimensional vectors obtained from $\mathcal{X}$ by varying this index i_n while fixing all the other indices. The unfolding of $\mathcal{X}$ along the n-mode vectors of $\mathcal{X}$ is defined as $\mathcal{X}_{(n)} \in \mathbb{R}^{I_n \times (I_1 \times I_2 \times \ldots I_{n-1} \times I_{n+1} \times \cdots \times I_N)}$, where the column vectors of $\mathcal{X}_{(n)}$ are the n-mode vectors of $\mathcal{X}$. Figure 1(a) illustrates unfoldings for a third-order tensor as an example of unfolding of Nth-order tensor. The n-mode product $\mathcal{X} \times_n U$ of a matrix $U \in \mathbb{R}^{J_n \times I_n}$ and a tensor $\mathcal{X}$ is a tensor $\mathcal{G} \in \mathbb{R}^{I_1 \times I_2 \times \cdots \times I_{n-1} \times J_n \times I_{n+1} \times \cdots \times I_N}$, with elements $g_{i_1,i_2,\ldots,i_{n-1},j_n,i_{n+1},\ldots,i_N} = \sum_{i_n=1}^{I_n} x_{i_1,i_2,\ldots,I_N} u_{j_n,i_n}$, by the manner in ref. [11]. A linear projection form of n-mode product is also given by $\mathcal{G}_{(n)} = U\mathcal{X}_{(n)}$. Figure 1(b) shows a linear projection form of a 1-mode projection for a third-order tensor. For the m- and n-mode products by matrices U and V, respectively, we have $\mathcal{X} \times_m U \times_n V = \mathcal{X} \times_n V \times_m U$ since n-mode projections are commutative [11]. We define the inner product of two tensors $\mathcal{X} = (x_{i_1,i_2,\ldots,i_N}), \mathcal{Y} = (y_{i_1,i_2,\ldots,i_N}) \in \mathbb{R}^{I_1 \times I_2 \times \cdots \times I_N}$ by $\langle \mathcal{X}, \mathcal{Y} \rangle = \sum_{i_1} \sum_{i_2} \cdots \sum_{i_N} x_{i_1,i_2,\ldots,i_N} y_{i_1,i_2,\ldots,i_N}$. Using this inner product, we have the Frobenius norm of a tensor $\mathcal{X}$ by $\|\mathcal{X}\|_F = \sqrt{\langle \mathcal{X}, \mathcal{X} \rangle}$. For the Frobenius norm of a tensor, we have $\|\mathcal{X}\|_F = \|\text{vec }\mathcal{X}\|_2$, where vec and $\| \cdot \|_2$ are the vectorisation operator for a tensor and Euclidean norm for a vector, respectively. For the two tensors $\mathcal{X}_1$ and $\mathcal{X}_2$, we define the distance between them by

$$d(\mathcal{X}_1, \mathcal{X}_2) = \|\mathcal{X}_1 - \mathcal{X}_2\|_F. \tag{1}$$

Although this definition is a tensor-based measure, this distance is equivalent to the Euclidean distance between the vectorised tensors $\mathcal{X}_1$ and $\mathcal{X}_2$.

As the tensor $\mathcal{X}$ is in the tensor space $\mathbb{R}^{I_1} \otimes \mathbb{R}^{I_2} \otimes \cdots \otimes \mathbb{R}^{I_N}$, the tensor space can be interpreted as the Kronecker product of N vector spaces $\mathbb{R}^{I_1}, \mathbb{R}^{I_2}, \ldots, \mathbb{R}^{I_N}$. To project $\mathcal{X} \in \mathbb{R}^{I_1} \otimes \mathbb{R}^{I_2} \otimes \cdots \otimes \mathbb{R}^{I_N}$ to another tensor $\mathcal{Y}$ in a lower-dimensional tensor space $\mathbb{R}^{P_1} \otimes \mathbb{R}^{P_2} \otimes \cdots \otimes \mathbb{R}^{P_N}$, where $P_n \leq I_n$ for $n = 1, 2, \ldots, N$, we need N projection matrices $\{U^{(n)} \in \mathbb{R}^{I_n \times P_n}\}_{n=1}^N$. Using the N projection matrices, the tensor-to-tensor projection (TTP) is given by

$$\mathcal{Y} = \mathcal{X} \times_1 U^{(1)\top} \times_2 U^{(2)\top} \cdots \times_N U^{(N)\top}, \tag{2}$$

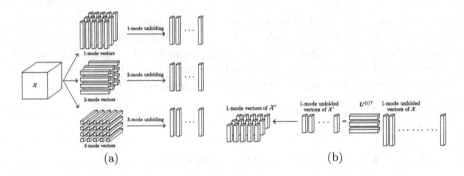

(a)   (b)

Fig. 1. (a) Unfoldings of a third-order tensor showing 1-, 2- and 3-mode unfoldings of the third-order tensor $\mathcal{X} \in \mathbb{R}^{4 \times 5 \times 3}$. (b) 1-mode projection that projects $\mathcal{X} \in \mathbb{R}^{4 \times 5 \times 3}$ to a lower-dimensional tensor $\mathcal{Y} \in \mathbb{R}^{3 \times 5 \times 3}$.

This projection is established in N steps, where at the nth step, each n-mode vector is projected to a P_n-dimensional space by $U^{(n)}$. We call this operation the orthogonal projection of $\mathcal{X}$ to $\mathcal{Y}$.

2.2   Tensor Subspace Derived by Tensor Decomposition

A Nth-order tensor $\mathcal{X} \in \mathbb{R}^{I_1 \times I_2 \times \cdots \times I_N}$, which is the array $\boldsymbol{X} \in \mathbb{R}^{I_1 \times I_2 \times \cdots \times I_N}$, is denoted as a triplet of indices $(i_1, i_2, \ldots, i_N)$. We set the identity matrices $\boldsymbol{I}_j$, $j = 1, 2, \ldots, N$ in $\mathbb{R}^{I_j \times I_j}$. Here we summarise higher-order singular value decomposition (HOSVD) [12] for third-order tensors. For a collection of tensors $\{\mathcal{X}_i\}_{i=1}^M \in \mathbb{R}^{I_1 \times I_2 \times \cdots \times I_N}$ satisfying the zero expectation condition $\mathrm{E}(\mathcal{X}_i) = 0$, we compute

$$\mathcal{Y}_i = \mathcal{X}_i \times_1 \boldsymbol{U}^{(1)\top} \times_2 \boldsymbol{U}^{(2)\top} \cdots \times_N \boldsymbol{U}^{(N)\top}, \tag{3}$$

where $\boldsymbol{U}^{(j)} = [\boldsymbol{u}_1^{(j)}, \ldots, \boldsymbol{u}_{I_j}^{(j)}]$, that minimises the criterion

$$J_- = \mathrm{E}\left(\| \mathcal{X}_i - \mathcal{Y}_i \times_1 \boldsymbol{U}^{(1)} \times_2 \boldsymbol{U}^{(2)} \cdots \times_N \boldsymbol{U}^{(N)} \|_{\mathrm{F}}^2 \right) \tag{4}$$

with respect to the conditions $\boldsymbol{U}^{(j)\top} \boldsymbol{U}^{(j)} = \boldsymbol{I}_j$.

Eigendecomposition problems are derived by computing the extremes of

$$E_j = J_j + tr((\boldsymbol{I}_j - \boldsymbol{U}^{(j)\top} \boldsymbol{U}^{(j)}) \boldsymbol{\Sigma}^{(j)}), \ j = 1, 2, \ldots, N, \tag{5}$$

where we set $J_j = \mathrm{E}\left(\| \boldsymbol{U}^{(j)\top} \mathcal{X}_{i,(j)} \mathcal{X}_{i,(j)}^\top \boldsymbol{U}^{(j)} \|_{\mathrm{F}}^2 \right)$. For $\boldsymbol{C}^{(j)} = \frac{1}{M} \sum_{i=1}^M \mathcal{X}_{i,(j)}$ $\mathcal{X}_{i,(j)}^\top$, $j = 1, 2, \ldots, N$, the optimisation of J_- derives the eigenvalue decomposition

$$\boldsymbol{C}^{(j)} \boldsymbol{U}^{(j)} = \boldsymbol{U}^{(j)} \boldsymbol{\Sigma}^{(j)}, \tag{6}$$

where $\boldsymbol{\Sigma}^{(j)} \in \mathbb{R}^{I_j \times I_j}$, $j = 1, 2, \ldots, N$, are diagonal matrices satisfying the relationships $\lambda_k^{(j)} = \lambda_k^{(j')}$, $k \in \{1, 2, \ldots, K\}$, for $\boldsymbol{\Sigma}^{(j)} = \mathrm{diag}(\lambda_1^{(j)}, \lambda_2^{(j)} \cdots, \lambda_K^{(j)}, 0 \cdots, 0)$. For the optimisation of $\{J_j\}_{j=1}^3$, there is no closed-form solution to this maximisation problem [12]. For practical computation, we use the iterative procedure of multilinear principal component analysis (MPCA) [1].

For $p_k \in \{e_k\}_{k=1}^{K}$, we set orthogonal projection matrices $P^{(j)} = \sum_{k=1}^{k_j} p_k p_k^{\top}$ for $j = 1, 2, \ldots, N$. Using these $\{P^{(j)}\}_{j=1}^{N}$, the low-rank tensor approximation [12] is given by

$$\hat{\mathcal{X}}_i = \mathcal{X} \times_1 (P^{(1)} U^{(1)})^{\top} \times_2 (P^{(2)} U^{(2)})^{\top} \cdots \times_N (P^{(N)} U^{(N)})^{\top}, \qquad (7)$$

where $P^{(j)}$ selects k_j bases of projection matrices $U^{(j)}$. The low-rank approximation using Eq. (7) is used for compression in TPCA.

Setting $\{U_k^{(j)}\}_{j=1}^{N}$ to be orthogonal matrices of a Nth-order tensor projection for the kth category, we have a tensor subspace spanned by $\{U_k^{(j)}\}_{j=1}^{3}$ for the kth category. Therefore, we can define a tensor subspace of a category by

$$\mathcal{C}_k = \{\mathcal{X} \mid \hat{\mathcal{X}} \times_1 U_k^{(1)} \times_2 U_k^{(2)} \cdots \times_N U_k^{(N)} = \mathcal{X}\}. \qquad (8)$$

Since a pattern represented by tensors contains perturbation, we define the kth category by

$$\mathcal{C}_k(\delta) = \{\mathcal{X} \mid \|\hat{\mathcal{X}} \times_1 U_k^{(1)} \times_2 U_k^{(2)} \cdots \times_N U_k^{(N)} - \mathcal{X}\|_{\mathrm{F}} \ll \delta\}, \qquad (9)$$

where a positive constant δ is the bound for a small perturbation of a pattern. Furthermore, using $P^{(j)}$, we define a low-dimensional tensor subspace for the kth category as

$$\mathcal{C}_{\Pi,k}(\delta) = \{\mathcal{X} \mid \|\hat{\mathcal{X}} \times_1 \hat{U}^{(1)} \times_2 \hat{U}^{(2)} \cdots \times_N \hat{U}^{(N)} - \mathcal{X}\|_{\mathrm{F}} \ll \delta + \varepsilon\}, \qquad (10)$$

where $\hat{U}_k^{(1)} = P^{(1)} U_k^{(1)}$, $\hat{U}_k^{(2)} = P^{(2)} U_k^{(2)}, \ldots, \hat{U}_k^{(N)} = P^{(N)} U_k^{(N)}$ and ε is the bound for a small reduction error in a pattern.

3 Distance Between Manifolds

3.1 Grassmann Manifold

The Grassmann manifold (Grassmannian) $\mathcal{G}(m, d)$ is the set of m-dimensional linear subspaces of $\mathbb{R}^d$ [7]. An element of $\mathcal{G}(m, d)$ can be represented by an orthogonal matrix Y of size d by m, where Y comprises the m basis vectors for a set of patterns in $\mathbb{R}^d$. The geodesic distance between two elements on Grassmannian has been defined in terms of principal angles.

Let Y_1 and Y_2 be orthogonal matrices of size $d \times m$. The principal angles $0 \leq \theta_1 \leq \cdots \leq \theta_m \leq \frac{\pi}{2}$ between the two subspaces span (Y_1) and span (Y_2) are defined by

$$\cos \theta_k = \max_{u_k \in \mathrm{span}(Y_1)} \max_{v_k \in \mathrm{span}(Y_1)} u_k^{\top} v_k \quad \text{s.t.} \quad u_k^{\top} u_i = 0, \ v_k^{\top} v_i = 0, \qquad (11)$$

for $i = 1, 2, \ldots, k-1$. These principal angles are related to the geodesic distance by

$$d_G(\boldsymbol{Y}_1, \boldsymbol{Y}_2) = \sqrt{\sum_{i=1}^{k-1} \theta_i^2}. \qquad (12)$$

This geodesic distance represents principal angles between linear subspaces as shown in Fig. 2(a).

3.2   Stiefel Manifold

The Stiefel manifold $S_{m,d}$ is the linear subspace of m orthonormal vectors in $\mathbb{R}^d$, represented by the $d \times m$ matrix $\boldsymbol{Y}$. While the Grassmannian defined by the dimension of the linear subspaces and the dimension of the original space, the Stiefel manifold $S_{m,d}$ is defined by the set of basis vectors. Let two Stiefel manifolds be orthogonal matrices $\boldsymbol{Y}_1$ and $\boldsymbol{Y}_2$. For $p = 1, 2$, the distance between the two Stiefel manifolds is defined by

$$d_S(\boldsymbol{Y}_1, \boldsymbol{Y}_2) = \min_{w_i} \left(\sum_{i=1}^{m} \sum_{j=1}^{m} w_{ij} \theta_{ij}^p \right)^{1/p}, \qquad (13)$$

where $\theta_{ij} \geq 0$ is the angle between the ith basis in span($\boldsymbol{Y}_1$) and the jth basis in span ($\boldsymbol{Y}_2$) and $w_{ij} \geq 0$ is the transportation cost between the two bases. Figure 2(b) shows examples of the distance between Stiefel manifolds. This geodesic distance between two Stiefel manifolds is defined for linear subspaces of a vector space. In the next section, we introduce the distance between tensor subspaces.

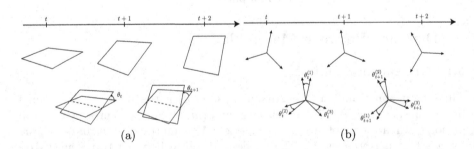

(a)   (b)

Fig. 2. Distances between linear subspaces. For a sequence of images in a time series, we have subspaces of each image. (a) Distance between linear subspaces on the Grassmann manifold computed by using canonical angles. (b) Distance between Stiefel manifolds computed by a pair of bases.

4   Geodesic Distance Between Tensor Subspaces

To simplify the discussion, we first consider second-order tensors, that is, two-dimensional images. For instances $t_1, t_2, \ldots, t_M$ for indices $i = 1, 2, \ldots, M$, we have a set $\{X_i\}_{i=1}^M$ of two-dimensional digital images. For the set $\{X_i\}_{i=1}^M$, we compute orthogonal matrices U and V that minimise

$$J(U, V) = \underset{i}{E} \|X_i - U\Sigma_i V^\top\|_F^2, \tag{14}$$

where Σ_i is a coefficient matrix. If we apply TPCA to each image X_i, we have orthogonal matrices U_i and V_i that minimise

$$J(U_i, V_i) = \|X_i - U_i\Sigma_i V_i^\top\|_F^2, \tag{15}$$

where Σ is a diagonal matrix.

For X_i, $X_{i+1} \in X$, we have pairs of orthogonal matrices $\langle U_i, V_i \rangle$ and $\langle U_{i+1}, V_{i+1} \rangle$, respectively. For these two pairs, we define rotation matrices $R_i^{(1)}$ and $R_i^{(2)}$ by

$$U_{i+1} = R_i^{(1)} U_i, \quad V_{i+1} = R_i^{(2)} V_i, \tag{16}$$

for 1- and 2-mode eigenvectors, respectively. If the orthogonal matrices satisfy $U_i = U_{i+1}$ and $V_i = V_{i+1}$, then the relations $R^{(1)} = R^{(2)} = I$ hold.

For a pair of images $f(x, y)$ and $g(x, y)$, we set

$$p(x, y) = \frac{f(x, y)}{F}, \quad F = \int\int_{\mathbf{R}^2} f(x, y) dx dy, \tag{17}$$

$$q(x, y) = \frac{g(x, y)}{G}, \quad G = \int\int_{\mathbf{R}^2} g(x, y) dx dy. \tag{18}$$

The transportation between $f(x, y)$ and $g(x, y)$ is computed as

$$T(f, g) = \min_c \int\int_{\mathbf{R}^2} \int\int_{\mathbf{R}^2} |d(p, q)| c(x, y; x'y') dx dy dx' dy', \tag{19}$$

where $d(p, q) = p(x, y) - q(x', y')$. For $T(f, g)$, we set $\int\int_{\mathbf{R}^2} c(x, y; x'y') dx' dy' = p(x, y)$ and $\int\int_{\mathbf{R}^2} c(x, y; x'y') dx dy = q(x', y')$. If $f(x, y)$ and $g(x, y)$ are sampled on an $m \times m$ grid, the minimisation of the discrete version of Eq. (19),

$$T(f, g) = \min_{c_{iji'j'}} \sum_{ij=1}^m \sum_{i'j'}^m |p_{ij} - q_{i'j'}| c_{iji'j'}, \tag{20}$$

where $\sum_{i'j'} c_{iji'j'} = p_{ij}$ and $\sum_{ij} c_{iji'j'} = q_{i'j'}$, is achieved using linear programming for $(m \times m)^2$-dimensional vectors.

For two bases u_i and u_j, we define the geodesic distance $d(\cdot, \cdot)$ by

$$d(u_i, u_j) = \begin{cases} \cos^{-1}(u_i^\top u_j), & \text{if } 0 \leq u_i^\top u_j \leq 1 \\ \cos^{-1}(|u_i^\top u_j|), & \text{if } -1 \leq u_i^\top u_j < 0. \end{cases} \tag{21}$$

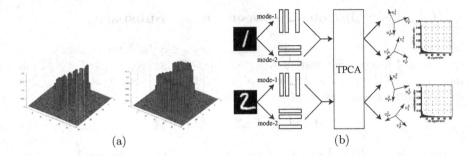

(a)   (b)

Fig. 3. Mathematical properties of expressions of images for computation of the distances between images. (a) Probabilistic distribution of grey values in images. Images are represented by probabilistic distributions normalised so that the L_2-norms are one. The Wasserstein distance is defined by the sum of the transportation costs between two probabilistic distributions. (b) Decomposition of images by TPCA. Images are decomposed to eigenvalues and eigenvectors. The Wasserstein distance is defined by the sum of the transportation costs for the contribution ratios of eigenvalues. In the transportation, the angle between bases is adopted as the cost of transportation.

We set $U^k = [u_1^k, \ldots, u_N^k]$, $V^k = [v_1^k, \ldots, v_N^k]$ and $U^{k+1} = [u_1^{k+1}, \ldots, u_N^{k+1}]$, $V^{k+1} = [v_1^{k+1}, \ldots, v_N^{k+1}]$. Setting $d_{ij}^{(1)} = d(u_i^{k+1}, u_j^k)$ and $d_{ij}^{(2)} = d(v_i^{k+1}, v_j^k)$ and by performing TPCA for X^k, X^{k+1} as preprocessing, we approximate the transportation problem in Eq. (20) as the minimisation of

$$d(X_{k+1}, X_k) = \min_{c_{ij}^{(1)}} \sum_{i,j=1}^m d_{ij}^{(1)} c_{ij}^{(1)} + \min_{c_{ij}^{(2)}} \sum_{i,j=1}^m d_{ij}^{(2)} c_{ij}^{(2)}. \tag{22}$$

For this minimisation problem, we give constraint conditions

$$\sum_j c_{ij}^{(1)} = \sum_j c_{ij}^{(2)} = \lambda_i^{k+1} / \sum_{i=1}^m \lambda_i^{k+1}, \quad \sum_i c_{ij}^{(1)} = \sum_i c_{ij}^{(2)} = \lambda_j^k / \sum_{j=1}^m \lambda_j^k. \tag{23}$$

where λ_j^k and λ_i^{k+1} are eigenvalues for X^k and X^{k+1}, respectively. Therefore, the problem is transformed to one of linear programming for $m \times m$-dimensional vectors. Figure 3 summarises mathematical properties of image expressions. Figure 4 illustrates the Wasserstein distance based on mode-1 and -2 bases for a second-order tensor.

Next, we define the Wasserstein distance among Nth-order tensors. For two Nth-order tensors $\mathcal{X}_1$, $\mathcal{X}_2 \in \mathbb{R}^{I_1 \times I_2 \times \cdots \times I_N}$, using HOSVD or TPCA, we have the decompositions

$$\mathcal{X}_1 = \mathcal{Y}_1 \times_1 U_1^{(1)} \times_2 U_1^{(2)} \cdots \times_N U_1^{(N)}, \mathcal{X}_2 = \mathcal{Y}_2 \times_1 U_2^{(1)} \times_2 U_2^{(2)} \cdots \times_N U_2^{(N)}. \tag{24}$$

As the result of these HOSVDs, we have sets of orthogonal matrices. In these decompositions, each base $u_{k,l}^{(n)}$ corresponds to eigenvalue $\lambda^{(n)k,l}$ for $k = 1, 2,$

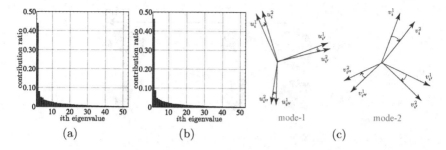

Fig. 4. Wasserstein distance between tensor subspaces for second-order tensors. The Wasserstein distance is the solution of the transportation problem between the contribution ratios of eigenvalues. (a) and (b) show the contribution ratios of eigenvalues obtained by singular value decomposition for two different images. The angle between bases is computed as the transportation cost between eigenvalues as shown in (c). The Wasserstein distance is obtained by minimisation of the total transportation cost among the eigenvalues for two tensor subspaces.

$l = 1, 2, \ldots, I_n$ and $n = 1, 2, \ldots, N$. Using these bases of orthogonal matrices, we define the Wasserstein distance between two Nth-order tensors by

$$d(\mathcal{X}_1, \mathcal{X}_2) = \min_{c_{ij}^{(1)}} \sum_{i,j=1}^{I_1} d_{ij}^{(1)} c_{ij}^{(1)} + \min_{c_{ij}^{(2)}} \sum_{i,j=1}^{I_2} d_{ij}^{(2)} c_{ij}^{(2)} + \cdots + \min_{c_{ij}^{(N)}} \sum_{i,j=1}^{I_N} d_{ij}^{(N)} c_{ij}^{(N)}, \quad (25)$$

where we set the constraints

$$\sum_j c_{ij}^{(n)} = \lambda_i^{(n)} / \sum_{i=1}^{I_n} \lambda_i^{(n)}, \quad \sum_i c_{ij}^{(n)} = \lambda_j^{(n)} / \sum_{j=1}^{I_n} \lambda_j^{(n)}. \quad (26)$$

We can use this Wasserstein distance to compute the distance between two tensor subspaces of Nth-order tensors spanned by $\{U_1^{(j)}\}_{j=1}^{N_1}$ and $\{U_2^{(j)}\}_{j=1}^{N_2}$.

5 Numerical Examples

In this section, we show three examples of analysis of multilinear subspaces. Throughout this section, we use the cardiac dataset [10], which contains 17 volumetric video sequences of beating hearts. Figure 5 shows frames of sequences. For each video sequence, we compute tensor subspaces by the MPCA. For the computation of geodesic distance, we use the Wasserstein distance between tensor subspaces that obtained by the MPCA.

For the first example, we compute the Wasserstein distance between third-order tensor subspaces. By expressing a volumetric video sequence as a set of 20 third-order tensors, we compute 20 tensor subspaces for 20 volumetric frames. We then compute geodesic distance between subspaces of the first and jth frames for 20 volumetric frames. For comparison, we also compute the Euclidean distance,

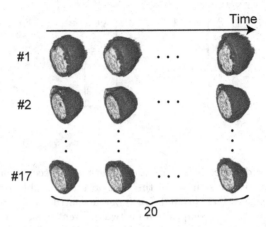

Fig. 5. Illustration of volume sequences of human ventricles. These sequences of volumetric data are extracted from cardiac MRI dataset with landmarks of endocardium of left ventricles [10]. We have 17 sequences of volumetric data of left ventricle for 17 patients. Each sequence of volumetric data represents one cardiac beat by 20 frames of 81 × 81 × 63 voxels.

which given by Frobenius norm for tensors, between the first frame and ith frame of a sequence of a beating heart. Figure 6(a) shows the comparison among distances.

In Fig. 6(a), the graph of the Wasserstein distance represents changes in the time series more clearly than the graph of the Euclidean distance, although those two graphs are similar. The graph of the Wasserstein distance for mode-2 is similar to the graph of the Wasserstein distance. This result implies that tensor subspaces of mode-2 mainly represent the difference between frames. On the other hand, the graph of the Wasserstein distance for mode-3 represents small changes. These results imply that the shrink and expansion of a beating heart are mainly happen in the direction of mode-1 and -2 to send blood to direction of mode-3.

For the second example, by expressing 17 volumetric video sequence as 17 sets of 20 third-order tensors, we compute 17 third-order tensor subspaces for each sequence by the MPCA. For these 17 third-order tensor subspaces, we compute Wasserstein distances between tensor subspaces of the first and jth sequences for $j = 2, 3, \ldots, 17$. Figure 6(b) shows the results for tensor subspaces for sets of third-order tensors. Furthermore, by expressing 17 volumetric video sequence as 17 fourth-order tensors, we compute 17 fourth-order tensor subspaces for each sequence by the MPCA. For these 17 fourth-order tensor subspaces, we compute Wasserstein distances between tensor subspaces of the first and jth sequences for $j = 2, 3, \ldots, 17$. Figure 6(c) shows the results for tensor subspaces for sets of fourth-order tensors.

By expressing volumetric video sequences as fourth-order tensors, we can compute distances that represent changes in a time series, that is distance for

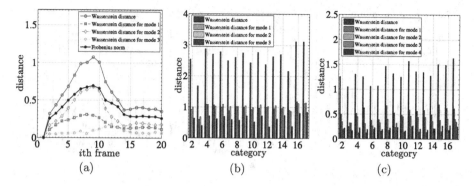

Fig. 6. Wasserstein distance between tensor subspaces for third-order tensors. (a) Wasserstein distances and relative distance between first and ith frames for $i = 1, 2, \ldots, 20$. In (a), the horizontal and vertical axes represent the index number of the frame and the distance, respectively. The plotted relative distance for the Frobenius norm of the first frame is defined by $\|X_1 - X_i\|_F / \|X_1\|_F$, where X_1 and X_i are the first and ith frames, respectively, and $\| \cdot \|_F$ is the Frobenius norm. (b) Wasserstein distances between the first and ith categories obtained using the third-order tensor expression for $i = 1, 2, \ldots, 17$. We compute the distances by using the eigenvectors of mode-1, -2, -3 and all modes. (c) Wasserstein distances between the first and ith categories obtained using the fourth-order tensor expression for $i = 1, 2, \ldots, 17$. We compute the distances by using the eigenvectors of mode-1, -2, -3, -4 and all modes. In (b) and (c), the horizontal and vertical axes represent the index number of the category and the Wasserstein distances, respectively.

mode-4. In Fig. 6(c), the Wasserstein distance for mode-4 represents the distances for a tensor subspace of changes in a time series in addition to mode-1, 2 and 3, which represent spatial changes of slice images. Comparing Figs. 6(b) and (c) shows that the Wasserstein distance between fourth-order tensor subspaces measures the difference between volumetric sequences more clearly than the Wasserstein distance between third-order tensor subspaces.

For the third example, we compute Wasserstein distances between fourth-order tensor subspaces. For each mode, we adopt one, five and fifteen eigenvectors that corresponds to one, five and fifteen largest eigenvalues. Then, we refer the residual principal components except major components as minor principal components. Figure 7(a)–(c) summarise the Wasserstein distances between three-mode tensor subspaces in the case of using one, five and fifteen major principal components, respectively. Figure 7(d)–(f) summarise the Wasserstein distances between fourth-order tensor subspaces in the case of using minor principal components.

In Figs. 7(a)–(c), the Wasserstein distance measures the differences among categories. On the other hand, in Figs. 7(d)–(f), the Wasserstein distance is unable to measure the difference among categories for minor principal components. These results imply that the differences among the categories are mainly concentrate on only 15 major principal components.

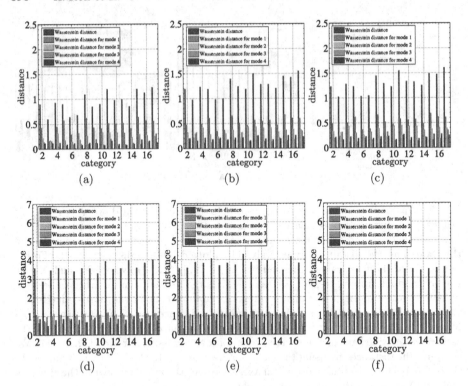

Fig. 7. Wasserstein distances between the subspaces of the first and ith categories for 17 categories. The horizontal and vertical axes represent the number of the category and the Wasserstein distance, respectively. The top and bottom rows show the Wasserstein distances obtained using major and minor principal components, respectively. In (a)–(c), 1, 5 and 15 major principal components, respectively, are used for the computation. In (d)–(f), minor components, that is, the principal components except 1, 5, and 15 major principal components, are used for the computation, respectively.

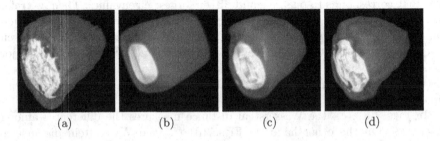

Fig. 8. Rendered frame of a volumetric volume sequence of a beating heart. From the left to right, the volumetric heart reconstructed by using first, five and fifteen major principal components and original volumetric heart are shown.

Figure 8 shows the sequences reconstructed from only major principal components. Figure 8(a) shows that the first major principal components of each mode represent the mean of each mode. The major components from the second to the fifth principal components represent the changes in the directions of the width, height, depth and time as shown in Fig. 8(b). Figure 8(c) illustrates that The next ten principal components represent smooth changes, which are given by second derivatives, in space and time.

6   Conclusions

We introduced the geodesic distance to tensor subspaces. In numerical examples, the proposed distance measures the difference between tensor subspaces. The results clarified that the analysis of volumetric video sequences requires fourth-order tensor subspaces for data expression. Furthermore, we showed that only fifteen principal components mainly express the differences among fourth-order tensor subspaces of volumetric video sequences.

This research was supported by "Multidisciplinary Computational Anatomy and Its Application to Highly Intelligent Diagnosis and Therapy" project funded by a Grant-in-Aid for Scientific Research on Innovative Areas from MEXT, Japan, and "Object oriented data-analysis for understanding and recognition of higher-dimensional multimodal data" by grant for Scientific Research from JSPS, Japan.

References

1. Lu, H., Plataniotis, K.N., Venetsanopoulos, A.N.: MPCA: multilinear principal component analysis of tensor objects. IEEE Trans. Neural Netw. **19**(1), 18–39 (2008)
2. Itoh, H., Imiya, A., Sakai, T.: Approximation of N-way principal component analysis for organ data. In: Chen, C.-S., Lu, J., Ma, K.-K. (eds.) ACCV 2016. LNCS, vol. 10118, pp. 16–31. Springer, Cham (2017). doi:10.1007/978-3-319-54526-4_2
3. Cock, K.D., Moor, B.D.: Subspace angles between ARMA models. Syst. Contr. Lett. **46**, 265–270 (2002)
4. Knyazev, A.V., Argentati, M.E.: Principal angles between subspaces in an a-based scalar product: algorithms and perturbation estimates. SIAM J. Sci. Comput **23**(6), 2009–2041 (2002)
5. Wong, Y.-C.: Differential geometry of Grassmann manifolds. Proc. Natl. Acad. Sci. **57**, 589–594 (1967)
6. Absil, P.-A., Mahony, R., Sepulchre, R.: Riemannian geometry of Grassmann manifolds with a view on algorithmic computation. Acta Applicandae Mathematicae **80**(2), 199–220 (2004)
7. Hamm, J., Lee, D.D.: Grassmann discriminant analysis: a unifying view on subspace-based learning. In Proceedings of International Conference on Machine Learning, pp. 376–383 (2008)
8. Edelman, A., Arias, T.A., Smith, S.T.: The geometry of algorithms with orthogonality constraints. SIAM J. Matrix Anal. Appl. **20**(2), 303–353 (1998)

9. Turaga, P., Veeraraghavan, A., Chellappa, R.: Statistical analysis on Stiefel and Grassmann manifolds with applications in computer vision. In: IEEE Conference on Computer Vision and Pattern Recognition, pp. 1–8, June 2008

10. Andreopoulos, A., Tsotsos, J.K.: Efficient and generalizable statistical models of shape and appearance for analysis of cardiac MRI. Med. Image Anal. **12**, 335–357 (2008)

11. Cichoki, A., Zdunek, R., Phan, A.H., Amari, S.: Nonnegative Matrix and Tensor Factorizations. Wiley, Chichester (2009)

12. Lathauwer, L.D., Moor, B.D., Vandewalle, J.: On the best rank-1 and rank-(r_1, r_2, r_n) approximation of higher-order tensors. SIAM J. Matrix Anal. Appl. **21**(4), 1324–1342 (2000)

Streaming Algorithm for Euler Characteristic Curves of Multidimensional Images

Teresa Heiss[1,2(✉)] and Hubert Wagner[1]

[1] IST Austria, Am Campus 1, 3400 Klosterneuburg, Austria
{teresa.heiss,hubert.wagner}@ist.ac.at
[2] Vienna University of Technology, Karlsplatz 13, 1040 Vienna, Austria

Abstract. We present an efficient algorithm to compute Euler characteristic curves of gray scale images of arbitrary dimension. In various applications the Euler characteristic curve is used as a descriptor of an image.

Our algorithm is the first streaming algorithm for Euler characteristic curves. The usage of streaming removes the necessity to store the entire image in RAM. Experiments show that our implementation handles terabyte scale images on commodity hardware. Due to lock-free parallelism, it scales well with the number of processor cores.

Additionally, we put the concept of the Euler characteristic curve in the wider context of computational topology. In particular, we explain the connection with persistence diagrams.

1 Introduction

The Euler characteristic curve is a powerful tool in image processing [7]. It has been used in a variety of fields including astrophysics[1] [2,13], medical image analysis [11,17], and image processing in general [14,18]. Its wide applicability stems from simplicity and efficient computability.

However, with the advances in image acquisition technology, there is need to handle very large images. For example the state-of-the-art micro-CT scanner Skyscan 1272 creates images of size $14450 \times 14450 \times 2600$ with 14-bit precision. Therefore, a single scan yields more than half a trillion voxels. It is also possible to combine multiple scans of the same object which further multiplies the size of data. As loading the resulting multi-terabyte image into RAM of a commodity computer is infeasible, a streaming approach is needed.

We present the first streaming algorithm for computing Euler characteristic curves.[2] Our algorithm divides a multidimensional image into chunks that fit into RAM, calculates the Euler characteristic curves for each chunk separately and merges them in the end. Since these chunks can be made arbitrarily small, commodity hardware can be used to compute Euler characteristic curves of arbitrarily large images. The fact that the chunks are not dependent on each other makes lock-free parallelism possible.

---

[1] The astrophysics community refers to the Euler characteristic curve as the genus.
[2] We review related work at the end of the paper.

© Springer International Publishing AG 2017
M. Felsberg et al. (Eds.): CAIP 2017, Part I, LNCS 10424, pp. 397–409, 2017.
DOI: 10.1007/978-3-319-64689-3_32

For defining the Euler characteristic curve we first need to explain what the Euler characteristic is. There are two ways to define the Euler characteristic and the Euler-Poincaré formula states that they are both equivalent. For discrete two-dimensional surfaces, like a triangulation of a sphere or a torus, the definitions are: first, the number of vertices minus the number of edges plus the number of faces; second, the number of connected components minus the number of tunnels plus the number of voids. Originally the Euler characteristic was defined for the surface of a convex polyhedron where it always equals two. To see this, consider that such a surface consists of one connected component, no tunnels and one void.

The equivalence between these two definitions seems to be the reason for the usefulness of the Euler characteristic: It captures global topological structures—like holes—although it can be computed locally—by adding up vertices, edges and faces.

The Euler characteristic curve of an image is the vector of Euler characteristics of consecutive thresholded images. We illustrate this for the example image of a bone[3] in Fig. 1a with values ranging from 0 (black) to 255 (white). The Euler characteristic curve of this image maps each $t \in \{0, 1, ..., 255\}$ to the Euler characteristic of the set of pixels with gray value smaller or equal to t. Figure 1 illustrates this process. This concept can be extended to more general settings, e.g., images with floating point gray values (see Sect. 3).

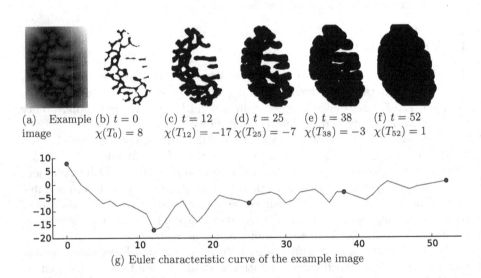

(a) Example (b) $t = 0$ (c) $t = 12$ (d) $t = 25$ (e) $t = 38$ (f) $t = 52$
image $\chi(T_0) = 8$ $\chi(T_{12}) = -17$ $\chi(T_{25}) = -7$ $\chi(T_{38}) = -3$ $\chi(T_{52}) = 1$

(g) Euler characteristic curve of the example image

Fig. 1. Definition of the Euler characteristic curve illustrated with an image showing a 2D slice of the distance transform of a segmented bone. Subfigures (b)–(f) show thresholded images T_t for different thresholds t. For $t = 38$, there is 1 component and 4 holes, hence $\chi = -3$. Starting from $t = 52$, there are no holes, so χ stabilizes at 1.

---

[3] We thank Reinhold Erben and Stephan Handschuh from Vetmeduni Vienna for providing micro-CT scans of rat vertebrae.

2   Theoretical Background

We give basic definitions needed in Sect. 3 using the language borrowed from computational topology [10]. With this we can provide precise definitions in arbitrary dimension and explain the connection between the Euler characteristic curve and other topological descriptors.

Cubical Cell. A k-dimensional **cubical cell** (short: **cell**) c of embedding dimension d is defined as the Cartesian product of intervals and singletons:

$$c := I_1 \times I_2 \times \cdots \times I_d$$

where exactly k of the sets $(I_i)_{i \in \{1,2,...,d\}}$ are intervals of the form $I_i = [a_i, a_i + 1]$ with integers $a_i \in \mathbb{Z}$ and the remaining $d - k$ sets are singletons $I_i = \{b_i\}$ with integers $b_i \in \mathbb{Z}$. A zero-dimensional cell is called a **vertex**, a one-dimensional cell an **edge**, a two-dimensional cell a **square**, a three-dimensional cell a **cube**.

Face. A cell c_1 of embedding dimension d is called a **face** of a cell c_2 of embedding dimension d if c_1 is a subset of c_2.

Cubical Complex. A p-dimensional **cubical complex** (short: **complex**) of embedding dimension d is a finite set of cubical cells of embedding dimension d such that

1. The faces of each cell are also elements of the complex
2. The intersection of any two cells is also an element of the complex[4]

where p is the highest dimension of all cells in the complex. The complexes that appear in our algorithm always fulfill $p = d$. Figure 2 shows an example of a cubical complex.

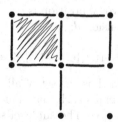

Fig. 2. Example of a two-dimensional cubical complex of embedding dimension two with one square, eight edges and eight vertices.

---

[4] The second condition is implied by the first condition since we allow only consecutive integers as interval endpoints in the definition of cells.

Filtration of Cubical Complexes. A sequence of complexes $K_1, K_2, \ldots, K_m$ is called a **filtration** if the complexes are monotonically increasing: $K_1 \subseteq K_2 \subseteq \cdots \subseteq K_m$.

Sublevel Set Filtration. Let K be a cubical complex. A cell that is not a face of any other cell than itself is called a **maximal cell**. Let $f : M \to \mathbb{R}$ be a function, where M is the set of maximal cells. This function f can be extended to a function $\tilde{f} : K \to \mathbb{R}$ defined on all cells:

$$\tilde{f} \colon K \to \mathbb{R}$$

$$c \mapsto \tilde{f}(c) := \begin{cases} f(c) & \text{if } c \in M \\ \min_{\substack{m \in M \\ c \subseteq m}} f(m) & \text{otherwise.} \end{cases}$$

For each $t \in \mathbb{R}$ the **sublevel set** $\tilde{f}^{-1}((-\infty, t])$ of this extended function is the set of cells that are a face of at least one maximal cell with f-value smaller or equal to t. As K consists of only a finite number of cells, $\tilde{f}$ can only have a finite number of different function values $\{t_1, t_2, \ldots, t_m\}$. The sublevel sets $\tilde{f}^{-1}((-\infty, t_1]), \ldots, \tilde{f}^{-1}((-\infty, t_m])$ form a filtration of cubical complexes—the **sublevel set filtration** induced by the function f.

To see this, notice that the definition of $\tilde{f}$ implies that for each $t \in \mathbb{R}$ the sublevel set $\tilde{f}^{-1}((-\infty, t])$ is a cubical complex: all faces of a cell c belong to the same sublevel set as c. Furthermore the sublevel sets are monotonically increasing $\tilde{f}^{-1}((-\infty, t_1]) \subseteq \tilde{f}^{-1}((-\infty, t_2]) \subseteq \cdots \subseteq \tilde{f}^{-1}((-\infty, t_m])$. Therefore, the sublevel sets form a filtration.

Consecutive Thresholded Images as Sublevel Set Filtrations. A d-dimensional gray scale image with $n_1 \times n_2 \times \cdots \times n_d$ voxels[5] can be interpreted as a d-dimensional cubical complex K of embedding dimension d with a function f on its maximal cells: for each voxel index $(i_1, \ldots, i_d)$, $i_1 \in \{1, \ldots, n_1\}, \ldots, i_d \in \{1, \ldots, n_d\}$ the corresponding voxel position is represented by the d-dimensional cell $c_{i_1, \ldots, i_d}$ of embedding dimension d:

$$c_{i_1, \ldots, i_d} := [i_1 - 1; i_1] \times [i_2 - 1; i_2] \times \cdots \times [i_d - 1; i_d].$$

The cubical complex K is defined as the set of all these cells $c_{i_1, \ldots, i_d}$ along with all their faces. The function f maps each maximal cell $c_{i_1, \ldots, i_d}$ to the gray value of the voxel with index $(i_1, \ldots, i_d)$. The sublevel set filtration[6] induced by the function f is formed by consecutive thresholdings of the image.[7]

---

[5] Throughout this paper we use "voxel" as multidimensional generalization of "pixel".

[6] Another interpretation of voxel data is via the dual complex (voxels become vertices) using the lower star filtration. The way we use appears more natural in image processing context. The two approaches yield similar but not necessarily identical Euler characteristic curves.

[7] Defining cells as products of *closed* intervals implies $(3^d - 1)$-connectivity for the voxels of the thresholded images. This corresponds to 8-connectivity for 2D images.

Euler Characteristic. The **Euler characteristic** χ of a p-dimensional complex K of embedding dimension d is defined as

$$\chi(K) := \sum_{k=0}^{p} (-1)^k n_k$$

where n_k is the number of k-dimensional cells in K.

The Euler-Poincaré formula states

$$\chi(K) = \sum_{k=0}^{p} (-1)^k n_k = \sum_{k=0}^{d-1} (-1)^k \beta_k$$

where β_k is the kth Betti number (the number of k-dimensional holes). For a formal definition of cubical homology and the involved Betti numbers, see [10]. In three-dimensional space, β_0 is the number of connected components, β_1 is the number of tunnels and β_2 is the number of voids.

Euler Characteristic Curve. The **Euler characteristic curve** e of a filtration of cubical complexes $K_1 \subseteq K_2 \subseteq \cdots \subseteq K_m$ is the vector

$$e = (\chi(K_1), \chi(K_2), \ldots, \chi(K_m)).$$

This vector can also be interpreted as a function

$$e \colon \{1, 2, \ldots, m\} \to \mathbb{Z}$$
$$t \mapsto \chi(K_t),$$

which is used to visualize the Euler characteristic curve as in Fig. 1g.

Euler Characteristic Curve of an Image. We already saw that for an arbitrary gray scale image the sequence of consecutive thresholded images is a filtration—the sublevel set filtration. The Euler characteristic curve of this filtration is the **Euler characteristic curve of an image**.

Connection to Other Topological Descriptors. We want to put the above considerations in the wider context of computational topology. Two popular topological descriptors of a filtration are Betti curves and persistence diagrams [6], which both capture information about holes at different thresholds.

For each hole in the image the persistence diagram tracks the first and last threshold at which the hole occurs. The kth Betti curve, which counts the k-dimensional holes at each threshold, is easily computable from the persistence diagram. Furthermore, the alternating sum of the Betti curves yields the Euler characteristic curve. Therefore, the Euler characteristic curve summarizes a persistence diagram, but it can be computed locally.

The usefulness of the Euler characteristic curve suggests that the two richer descriptors may also be useful in image processing. However, large images are out of reach of the currently available persistence diagram software. For now, the Euler characteristic curve remains the only feasible option.

3   Algorithm

The input for our algorithm is a gray scale image of arbitrary dimension d with n voxels. The output is the Euler characteristic curve of this image, as defined in Sect. 2.

Range of Values. In Sect. 2 the function f maps to $\mathbb{R}$. However, in practice the gray values of an image are in a predefined range, usually $\{0, 1, \ldots, 255\}$ or $\{0, 1, \ldots, 65535\}$. If the range contains negative numbers, it can be shifted so that it starts from zero. For this reason we focus on ranges of the form $\{0, 1, \ldots, m-1\}$ with a positive integer m. A version of our algorithm that can handle floating point values will be discussed at the end of this section.

Tracking the Changes. It is suboptimal to compute the Euler characteristic for each threshold separately, as already noted in [18]. To avoid redundant computations we track the changes between consecutive thresholds. More precisely, we determine how each voxel contributes to the change in Euler characteristic. Therefore we first compute a vector of changes in Euler characteristic (VCEC) $(a_0, a_1, \ldots, a_{m-1})$ whose entries $a_t := \chi\left(\tilde{f}^{-1}\left((-\infty, t]\right)\right) - \chi\left(\tilde{f}^{-1}\left((-\infty, t-1]\right)\right)$ are the difference between the Euler characteristics of two consecutive thresholded images.[8] The Euler characteristic curve is then:

$$\left(a_0, a_0 + a_1, \ldots, \sum_{t=0}^{m-1} a_t\right) =$$
$$= \left(\chi\left(\tilde{f}^{-1}\left((-\infty, 0]\right)\right), \chi\left(\tilde{f}^{-1}\left((-\infty, 1]\right)\right), \ldots, \chi\left(\tilde{f}^{-1}\left((-\infty, m-1]\right)\right)\right).$$

When changing from one thresholded image $\tilde{f}^{-1}\left((-\infty, t-1]\right)$ to the next $\tilde{f}^{-1}\left((-\infty, t]\right)$, all voxels with gray value t are included, along with all their faces that have not already been included at a previous threshold. We say that these new faces are introduced by these new voxels. More precisely, a face c of a voxel v is **introduced** by v if all other voxels w that have c as a face fulfill one of the following two conditions:

1. $f(w) > f(v)$
2. $f(w) = f(v)$ and $w \succeq v$,

where $\succeq$ is any total order of the voxel positions, e.g., the lexicographical order[9]. The output of our algorithm is independent of the chosen total order. However, it is necessary to require $w \succeq v$ in condition 2 to ensure that each face is introduced by a unique voxel.

Now we can decompose the cubical complex of the input image into blocks such that each block contributes to exactly one change of threshold. A block

---

[8] where $\chi\left(\tilde{f}^{-1}\left((-\infty, -1]\right)\right) = \chi(\emptyset) = 0$.

[9] In lexicographical order a voxel at position $(i_1, \ldots, i_d)$ succeeds a voxel at position $(j_1, \ldots, j_d)$ if $i_k > j_k$ for the first k where i_k and j_k differ.

 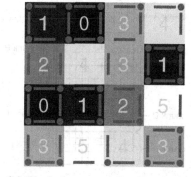

(a) The cubical complex consists not only of the voxels but also of their faces (red).

(b) Block decomposition shows which voxels introduce which faces, e.g., the upper left voxel introduces 3 edges and 2 vertices.

Fig. 3. This is an illustration for the block decomposition. (Color figure online)

consists of a voxel together with all the faces it introduces. Which faces are introduced by a certain voxel is determined only by the gray values of the voxel's $3^d - 1$ neighbors. We exploit this locality in the design of our parallel streaming algorithm. Figure 3 shows the decomposition of a two-dimensional example image into blocks.

Storage. For the computations we store only the gray values of the voxels. The geometric information and the adjacency relations between cells are implicit in the voxel grid and calculated locally whenever needed. Similarly the function $\tilde{f}$ and the block decomposition it induces are never explicitly stored. Apart from storing the result vector, the memory overhead is essentially zero.

Streaming. If the entire image does not fit into RAM, we divide it into chunks that fit into RAM. In our implementation we use a simple strategy: an image of size $n_1 \times n_2 \times \cdots \times n_d$ is divided into c chunks of size $\frac{n_1}{c} \times n_2 \times \cdots \times n_d$ (see Fig. 4, left). As these correspond to contiguous memory regions, streaming the chunks from a single input file is easy. We then separately compute the VCEC for each chunk either sequentially or in parallel.

Parallel Computations. For parallelism we use a thread pool. Each worker thread is assigned memory for a single chunk and one initially empty VCEC vector. One task is to read a chunk from disk, update the worker thread's VCEC vector by the VCEC of this chunk and discard the chunk. At any given time at most w chunks reside in RAM, where w is the number of worker threads. Because different worker threads work with disjoint memory regions we achieve lock-free parallelism. The underlying data structure for the collection of VCECs is a vector of vectors, called `euler_changes`.

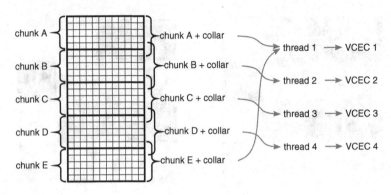

Fig. 4. Whenever a worker thread is free it loads one chunk along with a one voxel thick collar into its space in RAM and computes the VCEC of this chunk.

Processing One Chunk. Along with the chunk we read a one voxel thick collar surrounding it. This way we have access to all neighbors of the voxels in the chunk (see Fig. 4). With this information we compute the VCEC of this chunk as specified in Algorithm 1.

Algorithm 1. Computing the VCEC

Input: One chunk of an image along with a one voxel thick collar surrounding it and current_thread, the index of the worker thread processing this chunk.

Output: An updated version of the vector euler_changes[current_thread] which is the VCEC of all chunks this thread has processed.

1: **for all** voxels v in the chunk **do**
2:     t = gray value of v
3:     change = 0
4:     **for all** faces c introduced by v **do**
5:         **if** dim(c) is even **then**
6:             change++
7:         **else**
8:             change−−
9:     euler_changes[current_thread][t] += change
10: remove chunk and collar from RAM

Post-Processing. In the end, when all chunks have been processed, a single thread sums up the VCECs yielding the VCEC of the whole image, which is a vector $(a_0, a_1, \ldots, a_{m-1})$. The Euler characteristic curve is then computed as $\left(a_0, a_0 + a_1, \ldots, \sum_{t=0}^{m-1} a_t\right)$.

Analysis. To analyze the complexity we remind that d is the dimension of the image, n is the number of voxels, m is the number of gray values[10] and w is the

---

[10] The input size is $\log_2(m)n$.

number of worker threads. We introduce a new variable s, the number of voxels per chunk including the collar.

Assuming perfect parallelization, the worst case running time of our algorithm is $\mathcal{O}(\frac{3^d n}{w} + mw)$ because for each voxel we visit all its neighbors and sequentially post-process the VCECs. We analyze the practical scaling behavior in Sect. 4. As the dimension d is usually small (mostly 2 or 3), the exponential term is usually not a problem in practice.

For each worker thread, we need s integers of $\log_2(m)$ bits to store the gray values of a chunk. Additionally, the euler_changes data structure consists of wm 64-bit integers. Therefore, the total storage is $\log_2(m)ws + 64wm + \mathcal{O}(1)$ bits in RAM. By decreasing the chunk size s, the dominant part, $\log_2(m)ws$, can be made arbitrarily small. Because of this, our algorithm works for arbitrarily large images on commodity hardware.

Other Ranges of Gray Values. When the range of gray values is not of the form $\{0, 1, \ldots, m - 1\}$—for example for floating point values—one option is to use a hash map to store the euler_changes. If the number of different input values approaches n, the output size dominates the overall storage and the advantage of a streaming approach disappears. In this situation, it is preferable to transform (i.e., round, scale, shift) the input values to obtain a range of the form $\{0, 1, \ldots, m-1\}$. Running our standard algorithm on the transformed input yields the same result as transforming the domain of the Euler characteristic curve computed for the original data.

4 Experiments

We implemented the above algorithmic scheme in C++14. We made experiments on two different machines: a laptop with Intel core i5-5200U CPU with two physical cores clocked at 2.2 GHz with 8 GB of RAM and a workstation with Intel Xeon E5645 CPU with 12 physical cores clocked at 2.4 GHz and 72 GB of RAM. Table 1 shows the running time and memory usage for different 3D input images ran on the laptop. We use images from a standard data set[11], see [4,21]. The names' suffixes distinguish between 8- and 16-bit precision images. The last column shows that the running time is linear in n and does not depend on the content of the image.

Due to the above, we show the scaling behavior using a single image. The computations were performed on the workstation for a $512 \times 499 \times 512$ image with 16-bit precision. We used from 1 to 12 threads taking the mean running time and standard deviation across 20 runs. Figure 5 shows the speed-up gained using w threads instead of one. In particular using 10 threads is 7.1 times faster than using a single thread.

---

[11] Most of the images are available at www.byclb.com/TR/Muhendislik/Dataset.aspx.

Table 1. Running time and memory usage.

Name	Size	Million voxels	Memory[MB]	Time[s]	Time[s]/million voxels
prone16	512 × 512 × 463	121.4	70.4	24.7	0.20
xmastree16	512 × 499 × 512	130.8	72.9	26.7	0.20
vertebra16	512 × 512 × 512	134.2	74	28.4	0.21
random8	512 × 512 × 512	134.2	51	28.9	0.22
random8	1024 × 1024 × 1024	1073.7	93.1	236.7	0.22
random8	2048 × 2048 × 2048	8589.9	261.6	1767.1	0.21

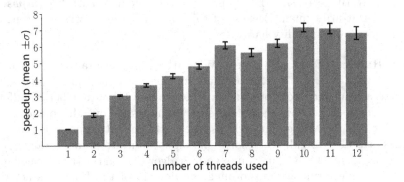

Fig. 5. Scaling with number of threads

Table 2 shows that the Euler characteristic curve of terabyte scale images can be computed on a single computer with limited memory. Memory usage could be further decreased by changing the chunking scheme. However, the experiments demonstrate that—for the foreseeable future—our implementation is a reasonable trade-off between performance and simplicity.

5   Related Work and Discussion

We review the work related to computing the Euler characteristic (curve) of images. We embed this in the context of computing other topological descriptors of images, particularly persistence diagrams.

Algorithms related to the Euler characteristic received a lot of attention in image processing [5,16,19,22], starting from the seminal work by Gray [8]. Many modern implementations aim at real-time processing of small 2D images [18]. Our goal is different, namely handling large multidimensional images.

Computing other topological descriptors of images is a more recent advancement [4,9,10,12,15,20,21], which has however not entered mainstream image processing. Specialized methods for computing persistence diagrams handle 3D images up to 500^3 voxels [4,9] on what we consider commodity hardware. The

Table 2. Running time and memory usage for computations on large 3D images, performed on the 12-core workstation. The voxel values are generated independently from a uniform random distribution in the range $\{0, 1, \ldots, 250\}$. As we already showed, using other images of the same size will exhibit almost identical performance.

Size	Threads	Time	Memory
$4096 \times 4096 \times 4096$	12	1.8 h	1.93 GB
$14\,500 \times 14\,500 \times 2\,600$	24	9 h	4.5 GB
$14\,500 \times 14\,500 \times 2\,600$	12	13 h	2.27 GB
$10\,000 \times 10\,000 \times 10\,000$	8	32 h	3.98 GB

main limitation is the storage of the entire image in memory. There exist distributed implementations [1], which alleviate the storage problem per machine, but are not specialized to image data, resulting in large overall memory overhead. The largest reported computed instances are in the range of 1000^3 on 32 server nodes.

Overall it is clear that a specialized, streaming approach is necessary for handling large images. We offer a robust implementation for the Euler characteristic curve, with possible future extensions to other topological descriptors.

We expect that these more complex topological descriptors will be computable for this terabyte scale data in the future but currently we are limited to using Euler characteristic curves. Let us discuss the properties of our algorithmic scheme and mention limitations of our current implementation.

The advantages of our algorithm are:

- It can handle arbitrarily large images on commodity hardware.
- It can handle images of arbitrary dimension.
- Linear running time.
- Predictable running time and memory usage.
- Due to lock-free parallelism, running time scales well with increasing number of threads.
- Our algorithm can be easily adapted to a massively-distributed setting using a map-reduce framework [3].

Some limitations of the current implementation:

- It uses $(3^d - 1)$-connectivity. For other types (e.g., 6-connectivity for 2D images), modifications on the algorithm can be made.
- For simplicity we use slices of the image as chunks. For very large images even a one voxel thick slice may not fit into memory.
- For technical reasons we surround each chunk with a second collar of voxels with value ∞. Effectively a five voxel thick slice has to fit into memory, which may become a problem for very large images.
- To include the value ∞ we may need a larger data type. For example if the input contains all values from 0 to 255 we use a 16-bit data type to store the original values along with an extra value for infinity.

Despite the limitations, our implementation is robust and can handle even the largest data produced by state-of-the-art image acquisition technology. We plan to release this software as open source.

References

1. Bauer, U., Kerber, M., Reininghaus, J.: Distributed computation of persistent homology. In: 2014 Proceedings of the Sixteenth Workshop on Algorithm Engineering and Experiments (ALENEX), pp. 31–38. SIAM (2014)
2. Colley, W.N., Richard Gott III, J.: Genus topology of the cosmic microwave background from WMAP. Mon. Not. R. Astron. Soc. **344**(3), 686–695 (2003)
3. Dean, J., Ghemawat, S.: Mapreduce: simplified data processing on large clusters. Commun. ACM **51**(1), 107–113 (2008)
4. Delgado-Friedrichs, O., Robins, V., Sheppard, A.: Skeletonization and partitioning of digital images using discrete morse theory. IEEE Trans. Pattern Anal. Mach. Intell. **37**(3), 654–666 (2015)
5. Dyer, C.R.: Computing the euler number of an image from its quadtree. Comput. Graph. Image Process. **13**(3), 270–276 (1980)
6. Edelsbrunner, H., Harer, J.: Computational Topology: An Introduction. American Mathematical Society (2010)
7. Gonzalez, R., Wintz, P.: Digital Image Processing. Addison-Wesley Publishing Co. Inc., Reading (1977)
8. Gray, S.B.: Local properties of binary images in two dimensions. IEEE Trans. Comput. **100**(5), 551–561 (1971)
9. Günther, D., Reininghaus, J., Wagner, H., Hotz, I.: Efficient computation of 3D Morse–Smale complexes and persistent homology using discrete Morse theory. Vis. Comput. **28**(10), 959–969 (2012). Springer
10. Kaczynski, T., Mischaikow, K., Mrozek, M.: Computational Homology. Springer, New York (2004)
11. Odgaard, A., Gundersen, H.: Quantification of connectivity in cancellous bone, with special emphasis on 3-d reconstructions. Bone **14**(2), 173–182 (1993)
12. Pikaz, A., Averbuch, A.: An efficient topological characterization of gray-levels textures, using a multiresolution representation. Graph. Models Image Process. **59**(1), 1–17 (1997)
13. Rhoads, J.E., Gott, J.R., Postman, M.: The genus curve of the abell clusters. Astrophys. J. **421**, 1–8 (1994)
14. Richardson, E., Werman, M.: Efficient classification using the euler characteristic. Pattern Recognit. Lett. **49**, 99–106 (2014)
15. Robins, V., Wood, P.J., Sheppard, A.P.: Theory and algorithms for constructing discrete morse complexes from grayscale digital images. IEEE Trans. Pattern Anal. Mach. Intell. **33**(8), 1646–1658 (2011)
16. Saha, P.K., Chaudhuri, B.B.: A new approach to computing the euler characteristic. Pattern Recognit. **28**(12), 1955–1963 (1995)
17. Ségonne, F., Pacheco, J., Fischl, B.: Geometrically accurate topology-correction of cortical surfaces using nonseparating loops. IEEE Trans. Med. Imaging **26**(4), 518–529 (2007)
18. Snidaro, L., Foresti, G.: Real-time thresholding with euler numbers. Pattern Recognit. Lett. **24**(9–10), 1533–1544 (2003)

19. Sossa-Azuela, J.H., et al.: On the computation of the euler number of a binary object. Pattern Recognit. **29**(3), 471–476 (1996)
20. Verri, A., Uras, C., Frosini, P., Ferri, M.: On the use of size functions for shape analysis. Biol. Cybern. **70**(2), 99–107 (1993)
21. Wagner, H., Chen, C., Vuçini, E.: Efficient computation of persistent homology for cubical data. In: Workshop on Topology-based Methods in Data Analysis and Visualization (2011)
22. Ziou, D., Allili, M.: Generating cubical complexes from image data and computation of the euler number. Pattern Recognit. **35**(12), 2833–2839 (2002)

Author Index

Printed in the United States
By Bookmasters

Printed in the United States
By Bookmasters